Solids in Sewers

Scientific and Technical Report No.14

Solids in Sewers

Characteristics, effects and control of sewer solids and associated pollutants

Edited by R.M. Ashley, J.-L. Bertrand-Krajewski, T. Hvitved-Jacobsen and M. Verbanck

JOINT COMMITTEE ON URBAN DRAINAGE

Sewer Systems and Processes Working Group

IWA Publishing

Published by IWA Publishing, Alliance House, 12 Caxton Street, London SW1H 0QS, UK
Telephone: +44 (0) 20 7654 5500; Fax: +44 (0) 20 7654 5555; Email: publications@iwap.co.uk
Web: www.iwapublishing.com

First published 2004

Edited and typeset by Steve Turrington, London, UK.
Index prepared by Indexing Specialists, Hove, UK.
Printed by TJ International (Ltd), Padstow, Cornwall, UK.

British Library Cataloguing in Publication Data
A CIP catalogue record for this book is available from the British Library

Library of Congress Cataloging- in-Publication Data
A catalog record for this book is available from the Library of Congress

ISBN 1 900222 91 4

Contents

Foreword

The recognition of problems associated with the presence and behaviour of solids in sewer systems can be traced back to the first sewer systems built over two millenia ago by the ancient Romans. However, it is only within the last two decades that major research programmes have been carried out to investigate and develop an understanding of the origins, nature, behaviour and impact of solids entering, within and being discharged from sewer systems. The prime objective of this Scientific and Technical Report is to present state-of-the-art knowledge about the origins, nature and effects of sewer solids for use by engineers, scientists, planners, designers and operators of sewerage systems. This knowledge is focused on the two key issues related to solids in sewer systems. These are the more efficient design and maintenance of sewer systems and the reduction of the risk of environmental damage arising, in particular, from combined sewer overflows and other urban wastewater system discharges resulting from rainfall on to impermeable areas.

The management of sewer solids is a key component in developing an holistic approach to the design and operation of wastewater systems. This Report, produced by active researchers and practitioners, forms a prime source of information to support the development of effective and efficient solutions to control and mitigate the effects of sewer solids on wastewater systems and, consequently, their impact in the wider environment. In addition, the Report identifies deficiencies in current knowledge and planning tools that must be addressed before robust computational models can be developed to account fully for all issues related to solids in sewer systems.

As a synthesis of current knowledge the Report is recommended to all professionals involved in the design and operation of sewerage systems and others with an interest in the environmental impact of sewer solids. It is to be hoped that the Report can be used as a vehicle to promote the dissemination of the results of recent research to sewerage system managers and operators.

Bob Crabtree
Pollution Management Business Manager, WRc plc, UK, July 2003

Contributors

Richard M. Ashley
Pennine Water Group
School of Engineering, Design and Technology
University of Bradford, BD7 1DP
United Kingdom

Jean-Daniel Baladès
CETE du Sud-Ouest
24 rue Carton, BP 58
33019 Bordeaux Cedex
France

Jean-Luc Bertrand-Krajewski
URGC - Hydrologie Urbaine
INSA de LYON
34 avenue des Arts
69621 Villeurbanne Cedex
France

Hansjoerg Brombach
Umwelt und Fluidtechnik
UFT
Umwelt- und Fluid-Technik
Dr. H. Brombach GmbH
Steinstrasse 7
97980 Bad Mergentheim
Germany

David Butler
Urban Water Research Group
Department of Civil and Environmental Engineering
Imperial College London
South Kensington Campus
London SW7 2AZ
United Kingdom

Ghassan Chebbo
CEREVE/ENPC
6-8 avenue Blaise Pascal
Cité Descartes – Champs sur Marne
77 455 Marne La Vallée Cedex 2
France

Cecile Delattre
c/o URGC – Hydrologie Urbaine
INSA de LYON
34 avenue des Arts
69621 Villeurbanne Cedex
France

Peter Hedges
Department of Civil Engineering
Aston University
Birmingham
United Kingdom

Reimer Herrmann
Lehrstuhl für Hydrologie
Universität Bayreuth
95440 Bayreuth
Germany

Thorkild Hvitved-Jacobsen
Section of Environmental Engineering
Department of Life Sciences
Aalborg University
Sohngaardsholmsvej 57
DK-9000 Aalborg
Denmark

Benoît Le Guennec
Institut de Mécanique des Fluides de Toulouse
Avenue du Professeur Camille Soula
31400 Toulouse
France

Emmanuelle Lucas-Aiguier
c/o CEREVE
ENPC
6-8 avenue Blaise Pascal
Champs sur Marne
77455 Marne la Vallée cedex 2
France

Richard May
HR Wallingford
Wallingford
Oxfordshire OX10 8BA
United Kingdom

Roberto Mayerle
Institute of Geosciences
University of Kiel
Otto-Hahn-Platz 3
24118 Kiel
Germany

Steven Michelbach
UFT
Umwelt- und Fluid-Technik
Dr. H. Brombach GmbH
Steinstrasse 7
97980 Bad Mergentheim
Germany

Chandra Nalluri
c/o University of Newcastle-upon-Tyne
Department of Civil Engineering
Newcastle-Upon-Tyne
NE1 7RU
United Kingdom

Per Halkjaer Nielsen
Section of Environmental Engineering
Department of Life Sciences
Aalborg University
Sohngaardsholmsvej 57
DK-9000 Aalborg
Denmark

Vladimir Novotny
Marquette University
Dept of Civil and Environmental Engineering
1515 West Wisconsin Avenue
Milwaukee W1 53233
USA

Gustavo Perrusquía
Department of Water Environment Transport
School of Civil Engineering
Chalmers University of Technology
SE-412 96 Göteborg
Sweden

William Pisano
Montgomery Watson Harza
100 Boylston Street, 6th floor
Boston, MA 02116
USA

Robert Pitt
Department of Civil and Environmental Engineering
University of Alabama
Tuscaloosa, AL 35487
USA

Christopher Pratt
School of Built Environment
University of Coventry
Coventry
United Kingdom

Kamma Raunkjaer
City Engineer
Municipality of Frederikshavn
Denmark

Helmut Reiff
Ibu
Bahnhofstrasse 6
D-76137 Karlsruhe
Germany

Erik Ristenpart
IFS Ingenieurgesellschaft für Stadthydrologie GmbH
Stiftstrasse 12
D-30159 Hannover
Germany

Adrian Saul
Pennine Water Group
Department of Civil and Structural Engineering
University of Sheffield
Mappin Street
Sheffield
United Kingdom

Virginia Stovin
Pennine Water Group
Department of Civil and Structural Engineering
University of Sheffield
Mappin Street
Sheffield
United Kingdom

Simon Tait
Pennine Water Group
Department of Civil and Structural Engineering
University of Sheffield
Mappin Street
Sheffield
United Kingdom

Naomi Tyack
IMC Consulting Engineers
PO Box 18
Huthwaite
Sutton-inAshfield
Nottingham NG17 2NS
United Kingdom

Matthias Uhl
Ingenieurgesellschaft für Stadthydrologie
Vahrenwalderstrasse 7
30165 Hannover
Germany

Michel Verbanck
Université Libre de Bruxelles
Laboratory of Water Pollution Control
Campus de la Plaine, CP 208
1050 Brussels
Belgium

Jes Vollertsen
Section of Environmental Engineering
Department of Life Sciences
Aalborg University
Sohngaardsholmsvej 57
DK-9000 Aalborg
Denmark

Executive summary

BENEFITS

The management of sewer solids is intrinsic to an holistic integrated approach to waste water system control. A concerted international assessment of knowledge has resulted in this report which may be used as a prime source of information for minimising and controlling the effects of sewer solids on wastewater systems.

OBJECTIVE

The objective of this report is to present information concerning the current state of knowledge regarding the origins, occurrence, nature and effects of sewer solids for use by engineers, scientists and others for the planning, design and operation of sewerage systems. The report provides a *state-of-the-art* in terms of knowledge and addresses both sewer maintenance requirements and environmental protection.

REASONS

Tougher environmental standards, coupled with public expectations, have led to increasingly stringent water quality standards. In response to this, it has been necessary to develop new methodologies and computer-based analytical techniques to model and understand the performance of all aspects of wastewater systems. Fundamental to these techniques is the understanding of the way in which sewer solids contribute to the poor performance of wastewater systems and consequential environmental damage. Awareness of the extent,

© 2004 IWA Publishing. *Solids in Sewers.* Edited by R.M. Ashley, J.-L. Bertrand-Krajewski, T. Hvitved-Jacobsen and M. Verbanck.. ISBN: 1 900222 91 4.

nature and effects of solids in sewers has grown only over the past two decades in most countries of the world and knowledge in this area is still limited.

A summary and distillation of knowledge about sewer solids is required in order to provide the foundation on which to develop effective and efficient solutions to the problems caused by these solids. The increasing adoption by developing countries of many of the techniques utilised, often from long-established tradition, by developed countries, also requires an examination of current knowledge and ideas for understanding and controlling the effects of sewer solids.

CONCLUSIONS

Knowledge about the origins, nature, movement, hydraulic and polluting effects of solids in sewers has enabled strategies and rules to be developed for the management of sewerage systems to minimise the deleterious effects of these solids and associated pollutants. New ideas and on-going research is identifying optimum methods for the design and operation of sewer networks within the context of an approach which is based on the integrated management of all of the components of wastewater systems. Further knowledge is needed ultimately to produce robust computational models which will account fully for all aspects of solids in sewer systems.

Methods are described for the control and management of solids entering sewer systems, for the design and operation of sewers to minimise the effects of sewer solids, or allow for *designed-maintenance*, for selection, design and operation of sewer overflow devices with and without storage, and for settling basin design and operation. The problems and possibilities for disposal of solids removed from sewer inlets, sewers, tanks and basins are also considered. New and innovative systems and techniques for sewer solids control not yet in general use are also summarised and possibilities for the future indicated.

RECOMMENDATIONS

This report, together with the key cited source material, should be used by consultants, municipal bodies and sewerage undertakers for detailed studies of the design and operation of sewerage systems. It may also be used by administrators and water quality planners to understand the nature of the problem, and by others, such as academics and postgraduate students, who wish to increase their knowledge of this subject.

RESUME OF CONTENTS

The report introduces initially in outline the nature of the problem in terms of the fundamental concepts of sewerage design and operation set within a framework of functional requirements. The scope of the report and interface to the other components of wastewater systems is defined. The contemporary approach to the holistic assessment and management of such systems is seen as a first step toward sustainable wastewater management within the overall framework of the report.

The origins, nature and movement of sewer solids is described in Section 2 in terms of surfaces, domestic, industrial and other types of input. Following this description of the characteristics of sewer solids, Section 3 describes the effects of sewer solids in terms of the problems caused to the hydraulic conveyance of sewers, the build-up and release of solids and solids associated pollutants in sewers and tanks, and the problems caused by gas generation and release. The implications for maintenance needs and other controls lead into

Section 4 where sediment management options are described. These include source control measures, within the context of domestic and cultural habits and education, industrial effluent controls, new ideas for *limited deposition* or other approaches to 'self-cleansing' sewer design, tank design to either encourage (settling basin) or discourage sedimentation, innovative and traditional maintenance systems, and indications where there may be opportunities for Real Time Control to minimise and control sewer solids. New ideas, not yet fully developed, such as bed-load traps and automatic sewer flushing, are also introduced.

Throughout the report the various approaches used to model aspects of solids in sewers are described.

1

Introduction

An awareness of the problems caused by solids in sewer systems has been apparent since the first *cloacae* (sewer) system was built in Rome in the 6th century BC. However, it was only with the advent of industrial society and the increasing concentration of humanity into cities, that the problems became acute (Halliday, 1999). The 18th century saw a number of developments that gave a compelling impetus to the construction in cities of 'designed' underground tunnels (sewers) for the collection of sewage (Markham, 1994); for example, utilisation of the WC (generally acknowledged as invented in 1589, although there is earlier evidence for some form of water closet) and linkage of cholera transmission directly to inadequate sanitation facilities in London by Dr John Snow in the 1850s. In the year 1854 alone, some 1000 miles of new sewer pipes were laid in towns in Britain. At this time a number of innovations were introduced to attempt to ensure that the solids collected in the drains and sewers of London were 'conveyed most cheaply and innoxiously to any distance out of towns' (Chadwick, 1842). These included the invention of egg-shaped sewers and the use of in-sewer flushing systems (Bertrand-Krajewski, 2002). Some engineers, however, refused to acknowledge that smaller sewers, by giving the best means of transporting the solids, were preferable to larger, and maintained that larger entry sizes were essential for maintenance. Others propounded the use of flat-bottomed sewers and eschewed the use of 'circular pipes', as it was thought that the shape of the cross section was not particularly important for sediment transport (Binnie, 1981). Such disagreements are still not entirely resolved even today, as discussed in Sections 2.3 and 4.2 of this report. It was also realised at this time that the nature and operation of sewerage systems was inseparably linked to the supply of water, as intermittency in supply resulted in more solids problems; hence the view that water and wastewater systems had to be considered holistically was recognised as early as the 19th century. Figure 1.1 illustrates a brick culvert more than 150 years old still in use in Dundee until the 1990s. Figure 1.2 illustrates, more typically, that many of the early sewers and those constructed more recently, differ little.

Figure 1.1. Sediments in a brick culvert in Dundee – constructed in 1800s (photograph taken in 1990).

Figure 1.2. Typical concrete sewer in Marseille with sediment deposits (photograph courtesy of Dominic Laplace).

Since these early developments, advances in sewerage and methods for managing the solids have been more modest. While the fabric has been made stronger, new materials such as vitrified clay and concrete have been utilised and the use of flexible joints has become widespread, system performance understanding has advanced mostly in terms of the hydrology and hydraulics of the inputs, and the hydraulics of the flows in the system.

Knowledge about behaviour of the solids and the best means of managing them remained largely as it was at the end of the 19th century until very recently. It is true that the problems caused by in-sewer solids are often unacknowledged, even today, and differentiated investment policies which favour capital rather than operational expenditure discriminate against a realistic look at these problems, leading to many controlling authorities and utilities preferring not to be aware of in-sewer solids problems.

The use of computer-based hydrologic and hydraulic deterministic simulation models for the above- and below-ground drainage processes is now ubiquitous, and the first generation of models linking flows to solids and pollutants being transported has emerged. Nonetheless many questions remain, and surprisingly it was only within the past 20 years that concerted research was initiated in a number of European countries to investigate the origins, movement, nature and effects of the solids entering, moving and discharged from sewerage systems. These initiatives, together with studies in the USA and Japan, were primarily undertaken to determine and control the discharges from combined sewer overflows, and were driven by aspirations to develop predictive computer models.

Sewer solids comprise a wide variety of very small (sub-micron) to large (centimetre) particles, which may be classified in terms of their physical, chemical and biological characteristics; they originate from a variety of sources as shown diagrammatically in Figure 1.3. It is important to appreciate that very small microorganisms, which are of major significance in wastewater systems, are a part of this overall group of 'sewer solids', as illustrated in Figure 1.4 (adapted from Sections 2.2.1 and 2.2.2 of this report).

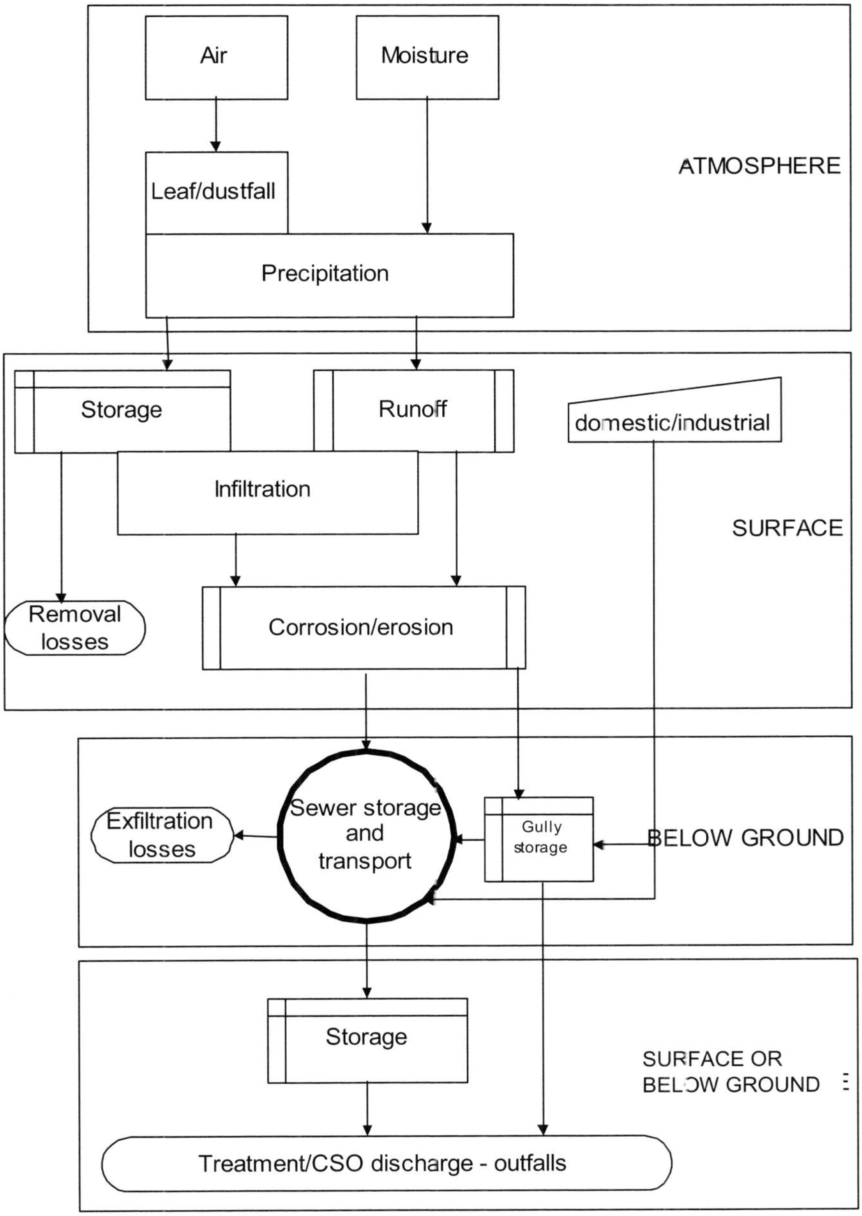

Figure 1.3. Pathways for sewer solids in urban environment.

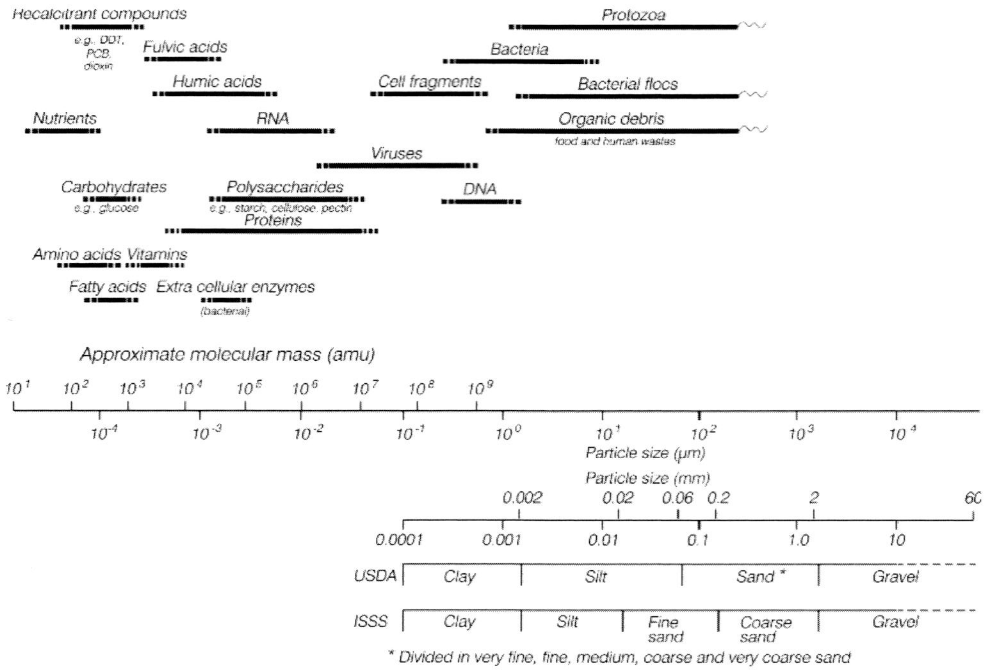

Figure 1.4. Solids of significance in wastewater systems. Particle size characteristics for organic and inorganic matter applied to sewer solids. The organic matter relates to typical organic constituents in settled municipal wastewater (adapted from Levine *et al.*, 1985).

Early studies in the 1970–1980s (Lindholm and Balmer, 1978; Dauber and Novak, 1982; Krejci *et al.*, 1987) showed that sediments in sewers were easily resuspended during wet weather flows in combined sewers, contributing significantly to overflow spill loads of solids and pollutants. Later studies continue to confirm the importance of sediments both for efficient sewer operation (Crabtree, 1989) and receiving watercourse impact (Rees and White, 1993).

From the continuing work a number of important results have emerged. In the UK, for example, it was concluded in 1986 that the sediment deposits in the nation's sewers were costing some £60 million per annum to manage (CIRIA, 1986). Sewer solids have been shown to be reservoirs of pollutants, often readily available for re-release by even fairly small increases in flow (Gromaire *et al.*, 2001).

Many current studies are concerned with minimising the effects from these 'eroding events', or 'flushes', on receiving water bodies; new ideas have emerged for better sewer design and operation, as well as for the design and operation of storage systems both in-sewer and at the ends of pipes (e.g. Pisano *et al.*, 2003).

Notwithstanding these new initiatives, there are many areas of uncertainty related to sewer solids; as well as influencing thinking about system operation, these uncertainties affect the way in which concepts of 'sustainability' may be introduced. In industrialised countries the possibility of a 'sewerless' society appears to be very remote, despite its apparent potential to be *more* sustainable in many cases.

Areas where research which includes aspects of sewer solids is underway or is needed are outlined in Table 1.1.

Table 1.1. Sewer systems and research topic needs that relate to solids

Topic	Sub-topic	Number of countries in which studies have been undertaken: Few	More than 10	More than 20	Further studies needed
Surface inputs	Nature			X	De-icing effects
	Modelling			X	Deterministic compared with conceptual and other types of models
	Surface control			X	Gully inlets and SUDS/BMPs effectiveness
Sanitary inputs	Gross solids	X			More on nature and amounts needed
	Sewage quality			X	
	Modelling		X		Current models in juvenile stages Risk assessment for control
Other inputs	Construction sites and other		X		More data required
General	Human factors		X		Individual behaviour effects Public, professional and institutional acceptability of new ideas
Sewers	Types/nature of system		X		The need for sewers at all
			X		Alternative types of system
				X	Combined vs separate, 3 pipe systems
In-sewer processes	Sediment transport			X	Still not well understood
	Modelling			X	Current models use derivatives of river and similar derived models – need to be further developed
	Bioprocesses		X		Both aerobic and anaerobic processes need more data from real systems
	Modelling		X		New applications of bioprocesses and models equivalent to those traditionally associated with treatment plants Better integration of aerobic and anaerobic process models
In sewer management	Sediment control			X	New techniques, whole life costs, sustainability and optimisation
	Sewers as 'treatment plants'	X			Cost-effectiveness, techniques, rules, viability and acceptability
Discharge systems	Tanks and chambers performance (inc. CSOs)			X	Optimum layouts. Pollutant efficiency performance
	Interaction sewers – treatment plants			X	Examples of effects. Benefits and disbenefits, storage, washout and drain-down problems for WwTW compliance
Sustainability issues	Long-term viability		X		Understanding, definitions and rules
	New paradigms		X		Integrated, holistic urban water management: multi-disciplinarity. Cross disciplines and infrastructure systems Human acceptability etc

This Scientific and Technical Report presents a review of knowledge relating to solids and sewer systems. It is written to be accessible to a broad audience, not only those responsible for wastewater systems. Chapter 2 deals with sewer solids and associated pollutants: the origins (Section 2.1), the characteristics and transformative processes related to sewer solids (Section 2.2) and the movement, transport, deposition and erosion of the solids (Section 2.3). Chapter 3 considers the effects of sewer solids. The effects are introduced in Section 3.1, with the conveyance and potential blockage problems considered in Section 3.2. Section 3.3 considers the formation of sulphides as related to sewer solids and the effects these may cause. The potential contribution of solids to flushes and wet weather pollution are considered in Section 3.4; Section 3.5 considers the effects of sewer solids on the performance of ancillary structures and wastewater treatment facilities. Other effects such as those caused by fats and greases are considered in Section 3.6. Chapter 4 deals with the management of sewer systems to minimise the occurrence and impacts caused by solids. Section 4.1 deals with controls at source and input options, Section 4.2 considers sewer design, operation and maintenance and Section 4.3, the control of solids in tanks, CSO and in settlement devices. Chapter 5 concludes, briefly reviewing new ideas and future directions in relation to solids in sewers.

REFERENCES

Bertrand-Krajewski, J-L. (2002) Sewer sediment management: some historical aspects of egg-shape sewers and flushing tanks. *Proc. Conference on Sewer Processes and Networks*, Paris, April.

Binnie, G.M. (1981) *Early Victorian Water Engineers*. Thomas Telford, London.

Chadwick, E. (1842) *The sanitary condition of the labouring population of Gt. Britain*. (Republished 1964, Edinburgh University Press).

CIRIA (1986) *Sediment movement in combined sewerage and storm-water drainage systems. Phase 1. Project report*. CIRIA research project, No. 366.

Crabtree, R.W. (1989) Sediment in sewers. *Journal of the Institution of Water and Environmental Management*, 3(6), 569–578.

Dauber, L. and Novak, B. (1982) *Quellen und Mengen der Schmutzstoffe in Regenabflüssen einer städtischen Mischkanalisation*. Separatum Nr. 927 der Eidgenössischen Anstalt für Wasserversorgung, Abwasserreinigung und Gewässerschutz (EAWAG), Zürich.

Gromaire, M.-C., Garnaud, S., Saad, M. and Chebbo, G. (2001) Contribution of different sources to the pollution of wet weather flows in combined sewers. *Water Research*, 35(2), 521–533.

Halliday, S. (1999) *The great stink of London: Sir Joseph Bazalgette and the cleansing of the Victorian Metropolis*. Sutton Publishing, Glos., UK.

Krejci, V., Dauber, L., Novak, B. and Gujer, W. (1987) Contribution of different sources to pollutant loads in combined sewers. *Proc. 4th Int. Conf. Urban Storm Drainage*, Lausanne, 34–39.

Levine, A.D., Tchobanoglous, G. and Asano, T. (1985) Characterization of size distribution of contaminants in wastewater: treatment and reuse implications. *J. WPCF*, 57, 805–816.

Lindholm, O. and Balmer, P. (1978) Pollution in storm runoff and combined sewer overflows. *Proc. 1st Int. Conf. Urban Storm Drainage*, Southampton, 575–585.

Markham, A. (1994) *A Brief History of Pollution*. Earthscan, London. (Reprinted 1995).

Pisano, W.C., O'Riordan, O.C., Avotte, F.J. and Barsanti, J.R. (2003) Automated sewer and drainage flushing systems in Cambridge, Massachusetts. *Journal of Hydraulic Engineering*, 129(4), 260-266.

Rees, A. and White, K.N. (1993) Impact of combined sewer overflows on the quality of an urban watercourse. *Regulated Rivers: Research and Management*, 8, 83–94.

Ristenpart, E., Ashley, R.M. and Uhl, M. (1995) Organic near-bed fluid and particulate transport in combined sewers. *Wat. Sci. Tech.*, 31(7), 61–68.

2

Sewer solids – characteristics and processes

2.1 ORIGINS

2.1.1 Introduction

Solids entering sewer systems originate from a variety of sources. Five principal sources are typically defined:

- the atmosphere, which contains fine dust and aerosols;
- the surfaces of the catchment, where solids accumulate during dry weather periods and are washed off during storm events: roofs, streets, parkings, highways, etc.;
- domestic sewage, which contains the largest proportion of organic solids;
- the environment and processes inside the drainage/sewer system: natural water body interactions, infiltration/exfiltration, decay and degradation of solids;
- industrial and commercial effluents and solids from construction sites, which typically may contribute very significantly to the solids loads entering sewers.

In the following sections, the characteristics and processes relative to each source of solids will be briefly considered.

2.1.2 The atmosphere

The atmosphere contains fine dust particles and aerosols which contribute to raindrop formation. These particles originate from different sources which can sometimes be remote from the catchment where they will precipitate: heating, automobile traffic, waste incineration, industry, construction sites, erosion of natural soils, etc. During storm events, these particles are transported within raindrops. As they reach the ground at the very beginning of the event, the pollutant load is usually independent of the duration and of the intensity of the rainfall (Randall *et al.*, 1978; Göttle, 1978a, 1978b).

The contribution of these particles to the total mass of solids during storm events is usually less than 10 %, and the suspended solids concentration in rainfall water ranges from 1 to 10 mg/l, with mean values of about 3–4 mg/l (Göttle, 1978a, 1978b; Novotny *et al.*, 1985; Deutsch and Grange, 1986; Artières, 1987, Uchimura *et al.*, 1996). Göttle (1978a, 1978b) and Ellis (1986) proposed a simple relationship giving the mass of solids transported by raindrops:

$$Me = 1-(1-e_d \cdot D_g^n) \qquad (2.1.1)$$

where Me is mass of solids transported by rainfall (expressed as a percentage of available mass in the atmosphere)
e_d is efficiency of the collision between solids of diameter d and raindrops of diameter D_g (e_d is correlated with rainfall intensity and duration)
n is number of raindrops per unit surface.

If the mass of atmospheric solids transported by raindrops is not very significant in comparison with other sources, the contribution to the pollutant load can be high or even very high for some pollutants associated with the finest particles, such as PAHs (polycyclic aromatic hydrocarbons) and PCBs (polychlorobyphenyls) (Bertrand-Krajewski, 1993).

Many measurements of pollutant concentrations in rainwater have been undertaken. General ranges for COD (chemical oxygen demand), heavy metals and hydrocarbon mean concentrations are summarised in Table 2.1.1. The broad ranges are due to the significant variability of local conditions. A significant proportion of the pollutants is associated with the particulate fractions (Section 2.1.3.2.8).

Table 2.1.1. Mean COD, metals and hydrocarbon concentrations observed in rainwater (after Bertrand-Krajewski, 1993)

Pollutant	Unit	Minimum	Maximum
COD	mg/l	20	160
Cd	µg/l	0.15	17
Cr	µg/l	0.16	14
Cu	µg/l	0.50	78
Fe	µg/l	0.10	300
Mn	µg/l	1	15
Ni	µg/l	1.5	50
Pb	µg/l	1.6	110
Zn	µg/l	5	800
Hydrocarbons	mg/l	0.02	0.07

2.1.3 Catchment surfaces

Solids washed off from the different surfaces of a catchment are the main contribution to the pollutant load of runoff water. For practical reasons, four primary sources of 'surface solids' are usually distinguished:

- roofs
- streets, highways and car parks
- gullies
- de-icing.

In addition, permeable and other 'natural' surfaces also contribute, especially during heavy rainfall events.

A recent study by Armitage and Roseboom (2000) reviewed a number of reports about urban litter that finds its way into watercourses. It was concluded (p. 183) that 'there was an infinite variety of the types and quantities of litter washed off a catchment'. In a South African study, in a catchment of 299 ha (85 % commercial/industrial), of some 1,467 m^3 (139 365 kg) of litter that found its way onto streets per year, some 18 % (25 000 kg) was eventually found in the storm drainage system. The litter was reported to be transported in the open channel drainage system with distinct foul flushes following long dry periods. The first flush of the wet season was comprised of dense solids: bricks, stones, tyres, down to fine sand, together with suspended debris which was mostly plastic bags. Animal carcasses and a lot of floating material were also present.

2.1.3.1 Roofs

The solids washed off from roofs during storm events have a number of different origins:

- solids from atmosphere deposited during dry weather periods;
- solids from chemical and physical degradation of the roof and gutter material itself;
- solids from bird droppings, branches and leaves.

Solids from the atmosphere result mainly from human activities: heating, automobile traffic, waste incineration, industries, etc., as described in Section 2.1.2. These solids are very fine and their median diameter is lower than 30 µm (Novotny *et al.*, 1985). Depending on meteorological and local conditions, they can be transported by winds over long distances, and their deposition rates are highly variable. Some assessment of dry weather period deposition rates have been carried out, and results are presented in Table 2.1.2. As explained above, the wide ranges are due to the variability of local conditions.

Table 2.1.2. Ranges of dry fallout of solids

Dry precipitation (g/m^2/year)	Location	Source
8–40	various sites	Novotny *et al.*, 1985; Ellis, 1986
30–40	rural areas	Göttle, 1978a, b
40–75	urban areas	Göttle, 1978a, b
66–360	urban city centre	Göttle, 1978a, b
150–650	industrial areas	Göttle, 1978a, b

Illustrative solids concentrations in roof runoff are presented in Table 2.1.3. The ranges are broad as the concentrations depend on the site, rainfall intensity, and nature, shape and state of the roof.

Table 2.1.3. Ranges of suspended solids concentration in roof runoff (mean value in brackets)

Suspended solids (mg/l)	Source
12–216	Ellis, 1986
64–204	Yaziz *et al.*, 1989
4–68	Förster, 1990
23	Shinoda, 1990
0–46 (12)	Sakakibara, 1996

Ellis (1986) concluded that the solids in roof water constitute about 15 to 30 % of the total mass of solids in runoff water during storm events. Roof water also contains other pollutants, which are frequently mainly associated with the particulate phase (Förster, 1996). Some ranges of concentrations for heavy metals and organic compounds are given in Table 2.1.4.

Table 2.1.4. Range of mean concentration of pollutants in roof water (after Bertrand-Krajewski, 1993)

Pollutant	Range of mean concentration (μg/l)
Cd	0.50–1.20
Cr	5.0
Cu	5–230
Fe	0.05–0.15
Mn	0–40
Pb	10–100
Zn	20–1000
PAHs	0.50
PCBs	0.30

In addition to the direct deposition of particles onto roofs, vapour phase pollutants can also be adsorbed and absorbed by falling particles. For thermodynamical reasons, these pollutant gases in the atmosphere do not remain in this phase but interact with water molecules and dissolve in small water particles, resulting in high gas concentrations. These water particles also interact with aerosols, which are defined as a combination of solid and concentrated aqueous phases produced from various sources, particularly combustion and photochemical pollution (Hough, 1988).

Complex models combining fundamental aspects of the physics, chemistry and thermodynamics of gases and small atmospheric water particles have been proposed to describe the atmospheric fallout of aerosols and gases (see, for example, Hough, 1988; Chamberlain, 1986). However, these sophisticated models have not been widely used in the context of urban hydrology where 'lumped' or aggregated approaches are preferred.

2.1.3.2 Streets

2.1.3.2.1 Introduction

Streets, highways and parking areas constitute the most common impervious surfaces in urban catchments, and are the main source of solids transported, via runoff, into the sewer system. The solids originate from:

- erosion of road material;
- wear of vehicles tyres;
- dry deposition of fine solids from the atmosphere;
- solid wastes and litter (cans, broken glass, bottles, pull tabs, papers, building materials, plastic, garbage, etc.) which can also be classified as 'gross solids';
- animal droppings and vegetation;
- de-icing materials;
- construction sites.

In addition, in coastal areas, wind blown sand may arise from all types of surface.

A study by Shaheen (1975) in the Washington, D.C. Metropolitan area quantified the traffic contributions. This study estimated that approximately 0.7 g/axle.km of roadway of solids can be directly attributed to traffic. In another US Environmental Protection Agency (USEPA) study, direct traffic emissions were reported to be 0.2 g/vehicle.km from vehicle exhaust and 0.125 g/vehicle.km from tyre wear (EPA, 1977). Heavy traffic also degrades road surfaces and increases solids loads.

In residential areas, fallen leaves and vegetation residues, including grass clippings, dominate street refuse composition during the autumn. A mature tree can produce from 15 to 25 kg of organic leaf residue (dry weight) that contains significant amounts of nutrients (Heaney and Huber, 1973). A typical value of foliage and leaf fallout for a forested area in Minnesota with some 420 trees/ha is about 3800 kg/ha.year or approximately 9 kg/tree. Of the yearly values, about 65 % of the fallout occurs during the autumn. The fallen leaves are about 90 % organic and contain about 0.04 to 0.28 % of phosphorus. Figure 2.1.1. shows the important elements of solids transport and accumulation on street surfaces.

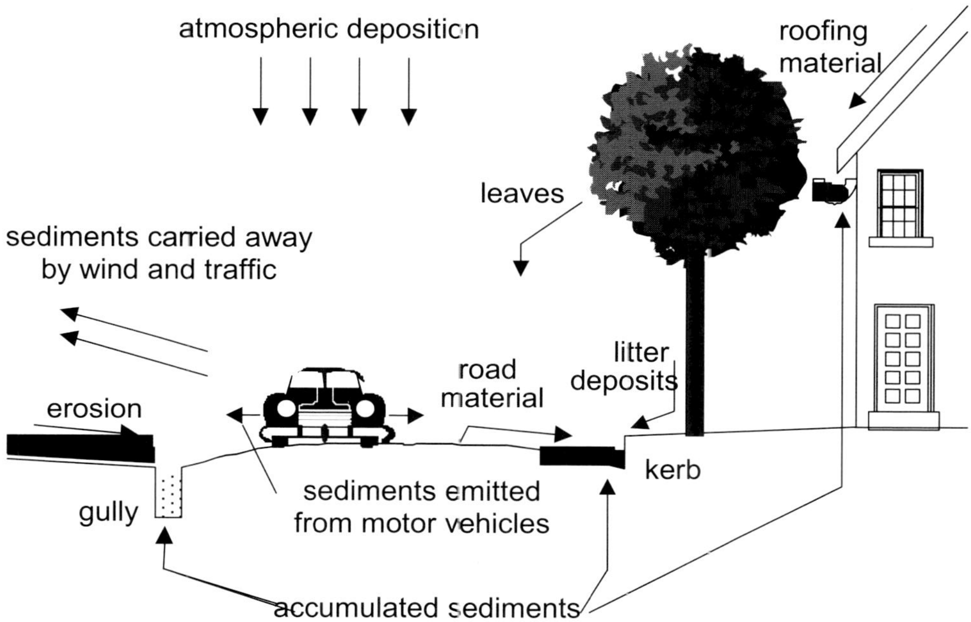

Figure 2.1.1. Solids transport and accumulation on urban surfaces (adapted from Butler and Clark, 1995).

2.1.3.2.2 Solids location

The solids are not uniformly distributed on street surfaces. For a typical street, three regions may be distinguished: the road, the gutter and the footpath. More than 80–85 % of the solids are accumulated in the first metre from the footpath (Sartor *et al.*, 1974). This result has been confirmed by many researchers (e.g. Artières, 1987; Novotny *et al.*, 1985; Leduc and Ouldali, 1989). According to Artières (1987):

- 50 % of the total mass is located in the gutter;
- 90 % of particles less than 80 μm in size are in the gutter, with only a few fine particles on the road and on the footpath.

This distribution is due to the turbulence and air currents caused by traffic and street sweeping techniques.

2.1.3.2.3 Accumulation

A number of studies have been undertaken since the end of the 1960s to understand and describe solids accumulation on street surfaces. The first study to provide extensive information on solids accumulation on urban streets was conducted by the American Public Works Association (APWA) in eight American cities. The 1969 APWA study (APWA, 1969) was followed, notably by Sartor and Boyd (1972) reporting a cross section of data from ten US cities and Shaheen (1975) reporting data for highways in the Washington, D.C. area. A drawback of these earlier studies is that it is difficult to draw conclusions about the relationship between street surface accumulation and stormwater concentrations since the two were seldom measured simultaneously. The two objectives of these initial studies were: to assess accumulation ratios (expressed as kg of solids/ha/year or as kg of solids/km/day); and to propose empirical relationships for modelling purposes.

(a) Accumulation rates

Accumulation of solids depends on many factors (urbanisation, traffic, way of life, street texture and nature, street cleaning practices, etc.); hence rates are highly variable. Consequently a classification is utilised based on urbanisation density. However, this density is very roughly estimated and very simple categories are used: residential, urban, city centre, commercial, industrial, etc. Typical values are given in Table 2.1.5.

Table 2.1.5. Solids accumulation rates on street surfaces for different types of urbanisation in different countries

Urbanisation	Accumulation rate	Unit	Source
Residential areas	10–210	kg/km of curb/day	Ellis, 1986
	35	kg/km of curb/day	Qing, 1984
	339	kg/km of curb/day	Sartor and Boyd, 1972
	560	kg/km of curb/day	Leduc and Ouldali, 1989
	95–147	kg/ha/year	Novotny *et al.*, 1985
	620–3200	kg/ha/year	Ellis, 1986
	276	kg/ha/year	Leduc and Ouldali, 1989
	515	kg/ha/year	Waller and Hart, 1986
	470–1700	kg/ha/year	Hémain, 1985
	500–1650	kg/ha/year	Philippe and Ranchet, 1987
	1685	kg/ha/year	Verbanck (1995)
Commercial areas	13–180	kg/km of curb/day	Ellis, 1986
	26	kg/km of curb/day	Qing, 1984
	82	kg/km of curb/day	Sartor and Boyd, 1972
	360–1220	kg/ha/year	Novotny *et al.*, 1985
	50–840	kg/ha/year	Ellis, 1986
Industrial areas	80–288	kg/km of curb/day	Ellis, 1986
	790	kg/km of curb/day	Sartor and Boyd, 1972
	72	kg/km of curb/day	Qing, 1984
	400–1700	kg/ha/year	Ellis, 1986
	537	kg/ha/year	Novotny *et al.*, 1985

Other surface accumulation data may be gleaned, somewhat less directly, from references on loading functions such as McElroy *et al.* (1976), Huber *et al.* (1981) and Ball *et al.* (1996). The traffic intensity is also an important parameter, with accumulation rates increasing with the traffic expressed in vehicle/day (Payne and Osborne, 1996).

Detailed studies have shown that the accumulated mass of solids on streets is the result of deposition and removal rates, plus 'permanent' storage (Pitt, 1979; also Section 4.1). The permanent storage component is a function of street texture and condition, and is the quantity of street solids that cannot be removed naturally (by rainfall) or by street cleaning equipment. It is literally *trapped* in the texture, or cracks, of the street. The street solids loading at any time is this initial permanent loading plus the accumulated mass corresponding to the exposure period. Measurements in two residential areas in San Jose, California, USA, have shown that accumulation rates for both sites were similar, but that permanent storage values were very different depending on the street texture (Pitt, 1979). In comparison with previous studies, the calculated accumulation rates in San Jose were then about 10 times lower than the proposed Sartor and Boyd (1972) accumulation rates. The difference was mainly due to the permanent storage. Some experimental estimations of accumulation rates including initial values or permanent storage are given in Table 2.1.6, and illustrated in Figure 2.1.2.

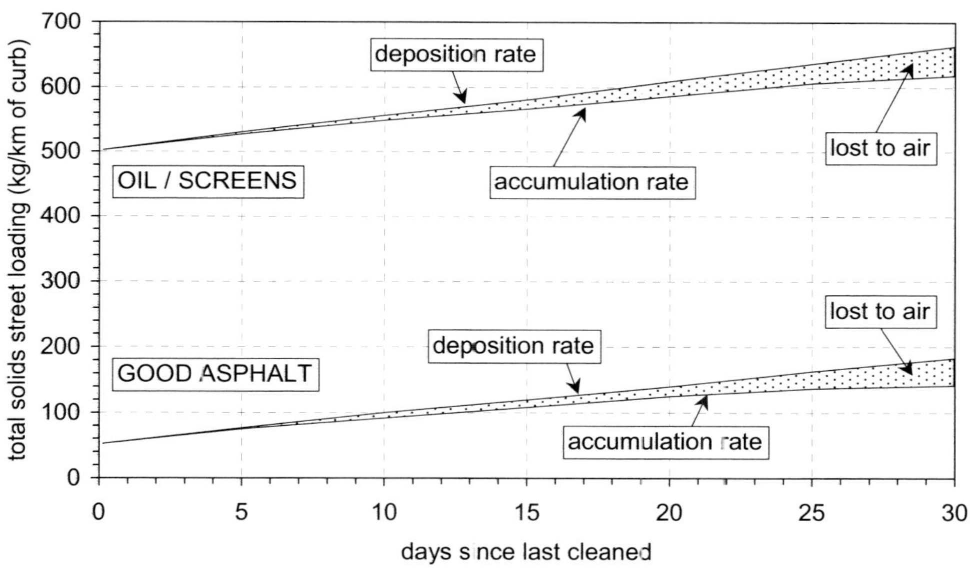

Oil/screens correspond to roughly repaired street surfaces.

Figure 2.1.2. Measured rates of road surface solids loading as a function of cleaning frequency (after Pitt, 1979).

(b) Accumulation modelling

Many factors are involved in the solids accumulation process: the antecedent dry weather period, the street sweeping, the automobile traffic, the vegetation, the urbanisation. The physical phenomena are too complex to be easily described. Therefore more simple descriptions have been used for modelling purposes. Three empirical approaches are commonly proposed, with linear, asymptotic and power relationships between the time and the accumulated mass of solids, as shown in Figure 2.1.3.

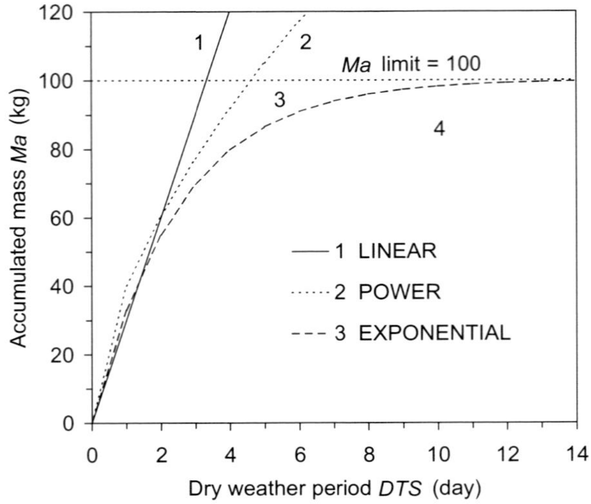

Figure 2.1.3. Comparison of different solids build-up equations (after Huber and Dickinson, 1992).

The linear accumulation process is the simplest one and has been validated by, for example, Servat (1984) using long term on-site measurements:

$$Ma = A.ACCU.DTS \tag{2.1.2}$$

where *Ma* is mass of solids accumulated over the catchment (kg)
A is area of the catchment (ha)
ACCU is accumulation rate of solids (kg/ha/day)
DTS is dry weather period (day).

After calibration on four different French residential catchments, *ACCU* was found to range between 1.7 and 3.6 kg/ha/day. This simple model was chosen after testing other formulae: asymptotic, power, parabolic, etc. The results are accurate for an annual balance, with a precision of about 5 %, but for some storm events the discrepancies were found to be greater, increasing to 30 % or more (Servat, 1984).

The asymptotic (exponential) relationship has been derived from the experiments of Sartor *et al.* (1974) and is widely used in several models (Bertrand-Krajewski *et al.*, 1993). It assumes that an equilibrium is reached between the supply of solids and their removal by wind effects, traffic, street sweeping, natural degradation of solids, etc.

The relationship is written (Alley and Smith, 1981):

$$\frac{dMa(t)}{dt} = A.ACCU - DISP.Ma(t) \tag{2.1.3}$$

where *Ma(t)* is accumulated mass of solids over the catchment at time *t* (kg)
DISP is removal coefficient (per day)
t is time since the previous rainfall (day).

Table 2.1.6. Street solids loadings and accumulation rates (see also Section 4.1)

Site	Condition	Initial loading (g/m of curb)	Accumulation rate (g/m of curb /day)	Maximum observed loading (g/m of curb)	Days to maximum observed loading (days)	Source
Smooth and intermediate textured streets						
Reno, NV	good	80	1	85	5	Pitt and Sutherland, 1982
Reno, NV	good, with smooth gutters	250	7	400	30	Pitt and Sutherland, 1982
San Jose, CA	good	35	4	> 140	> 50	Pitt, 1979
US nation-wide (residential streets)	good	110	6	140	5	Sartor and Boyd, 1972 (corrected*)
US nation-wide (commercial streets)	good	85	4	140	5	Sartor and Boyd, 1972 (corrected*)
Reno, NV	moderate to poor	200	2	200	5	Pitt and Sutherland, 1982
Reno, NV (new residential area – construction)		710	17	910	15	Pitt and Sutherland, 1982
Reno, NV	poor, with lipped gutters	370	15	630	35	Pitt and Sutherland, 1982
San Jose, CA	fair to poor	80	4	230	70	Pitt, 1979
Castro Valley, CA	moderate	85	10	290	70	Pitt and Shawley, 1982
Ottawa, Ontario	moderate	40	20	na	na	Pitt, 1983
Toronto, Ontario (residential)	moderate	40	32	100	> 10	Pitt and McLean, 1986
Toronto, Ontario (industrial)	moderate	60	40	351	> 10	Pitt and McLean, 1986
Bellevue, WA (dry period)	moderate	140	6	> 230	20	Pitt, 1984
Bellevue, WA (heavy traffic)		60	1	110	30	Pitt, 1984
Bellevue, WA (other residential sites)		70	3	140	30	Pitt, 1984
Average		150	9	> 270	> 25	
Range		35–170	1–40	85–910	5–70	
Rough and very rough textured streets						
San Jose, CA (oil and screens overlay)		510	6	> 710	> 50	Pitt, 1979
Ottawa, Ontario	very rough	310	20	na	na	Pitt, 1983
Reno, NV -		630	10	860	35	Pitt and Sutherland, 1982
Reno, NV (windy)		540	34	> 1400	> 40	Pitt and Sutherland, 1982
San Jose, CA	poor	220	6	430	30	Pitt, 1979
Ottawa, Ontario	rough	200	20	na	na	Pitt, 1983
US nation-wide - industrial streets	poor	190	10	370	10	Sartor and Boyd, 1972 (corrected*)
Average		370	15	> 750	> 30	
Range		190–630	6–34	370–1400	10–50	

* Initial values of accumulation rates from Sartor and Boyd (1972) have been corrected to account for permanent storage, according to Pitt (1979)

The limit of the asymptotic accumulation is then *A.(ACCU/DISP)*. Depending on the catchment, this limit is reached after 10 to 20 days. *ACCU* and *DISP* need to be calibrated for each catchment. *DISP* is empirical and has been related to the wind velocity, curb height, traffic intensity and speed, and to the character of the areas surrounding the street or the highway of the contributing area (Novotny *et al.*, 1985; Novotny and Olem, 1994). Based on observed data from experimental watersheds located in Milwaukee metropolitan area, Novotny *et al.* (1985) reported that the removal coefficient in medium density residential areas with moderate traffic was fairly constant, attaining values around 0.2 to 0.4 per day, i.e. that approximately 20 to 40 % of the solids accumulated near the curb on the street surface is removed daily by wind and traffic. After integration and taking into account the residual mass of solids *MR* remaining over the catchment after the previous rainfall, equation 2.1.3 becomes:

$$Ma(DTS) = \frac{ACCU}{DISP} \cdot A \cdot \left(1 - \exp(-DISP \cdot DTS)\right) + MR \cdot \left(\exp(-DISP \cdot DTS)\right) \qquad (2.1.4)$$

The choice of a relationship for accumulation modelling is always site specific and should be made only after local measurements. As an example, Ball *et al.* (1996) have used a power relationship to fit experimental measurements of solids accumulation for a road in a semi-residential area. Ellis (1986) explained that highway studies in London have suggested that an equilibrium level is reached within 4–5 days when vehicle induced resuspension is dominant, but that where resuspension is negligible, such as in residential side streets, the mass accumulates linearly with time. It should also be noted that the choice of a relationship is more important for continuous modelling than for single event modelling. Unfortunately, there are few data to support the empirical relationships, and the entire process might more easily be considered a calibration exercise, in which the build-up parameters for whichever functional form is chosen will be calibrated as part of the comparison of predicted and measured runoff quality (Huber, 1986).

2.1.3.2.4 Washoff by rainfall

During storm events, many of the accumulated particles on streets are washed off. A number of parameters are involved in this phenomenon: rainfall intensity, rainfall height, rainfall duration, runoff peaks and volume, topography, particle characteristics, type and condition of the street surface. The most important among these parameters is the rainfall intensity, and especially the maximum rainfall intensity (Shivalingaiah and James, 1984). Particles are washed off by raindrop impact, and transported into sewers by surface runoff (e.g. Young and Wiersma, 1973; Huber, 1986; Aalderink *et al.*, 1990).

Observations of solids washoff during controlled tests using actual street material and debris are affected by street solids distribution and armouring. The earliest controlled street washoff experiments were conducted by Sartor and Boyd (1972) during the summer of 1970 in Bakerfield, California, USA. Their data were used in many stormwater models, such as SWMM, STORM or HSPF (Bertrand-Krajewski *et al.,* 1993), to estimate the percentage of available street solids that would be washed off during rainfall of different intensities. The tests used artificial rainfall and were conducted on concrete and old and new asphalt. This led to the well-known equation given below (2.1.10).

More recently, based on a variety of field experiments, Pitt (1987) concluded that:

- the masses of washed off solids, expressed in g/m^2, increase with rainfall height and reach an asymptotic limit that depends on site conditions;
- during washoff, only a small fraction (typically about 15 % according to Pitt (1984)) of the accumulated street solids are eroded and transported by runoff: the ratio

between the measured total load of street solids and the 'available' load for washoff ranges from 1/3 to less than 1/10;

- the finest solids (d < 100 μm) are normally easily washed off even for small rainfall events, whereas coarse solids (d > 500 μm) are only washed off by strong events;
- solids washoff should be divided into two main categories, one during high intensity rainfall with dirty streets (possibly subdivided according to street texture) and another category for all other conditions.

Unfortunately, little research has been carried out on the detailed physical processes of washoff over urban catchments, and the models proposed for soil erosion in agricultural or natural sites are not easily adaptable for urban hydrology.

Studies in Belgrade and Lund, monitoring washoff of solids and dissolved pollutants from asphalt surfaces (Deletic *et al.*, 1997), have been used to show the applicability of a deterministic model for solids washoff, accounting for non-uniform sediment build-up over the contributing surface. One solution method utilises an entrainment relationship which accounts separately for the rainfall and runoff processes, and was shown to produce reasonable results even when the rainfall intensity was low (< 6 mm/h).

By calculation of the kinetic energy E_k (J/m^2/s) of raindrops:

$$E_k = 5.86\, x^{-3}\, i^{1.156} \tag{2.1.5}$$

where x is a spatial coordinate (m)
i is the rainfall intensity (mm/h)

and accounting for the penetration depth h_p (m) of raindrops:

$$h_p = 0.69\, x\, i^{0.182} \tag{2.1.6}$$

the solids entrainment relationship is given as:

$$\frac{\partial M_r}{\partial t} = -M_r\, b_1\, E_k \left(1 - e^{-h_p/h}\right) + C \tag{2.1.7}$$

with

$$C = 0 \;\text{ for }\; \tau_b \le \tau_{cr} \tag{2.1.8}$$

$$C = -M_r b_2 \left(\tau_b - \tau_{cr}\right)^{1.5} \;\text{ for }\; \tau_b > \tau_{cr} \tag{2.1.9}$$

where M_r is available material on the road surface (g/m^2)
b_1, b_2 are calibration coefficients (= 0.013)
h is depth of overland flow (m)
τ_b is bed shear stress (N/m^2)
τ_{cr} is critical shear stress (N/m^2), given from the Shields diagram (see Section 2.3).

The model assumes that all material entrained passes ultimately to the drainage outlet point, being conveyed either across the road surface or in gutter flow. Dispersion is accounted for in both of the types of flow path, although it is concluded that it may be considered negligible for the transport across the road surface itself. More recently the approach has been extended to estimate the transport of particles through grass filter strips (Deletic, 2001). Modelling using a classical suspended sediment transport approach was found to be effective.

Physically based models for washoff are usually too complex to use, and they have typically been displaced by conceptual or global models. Such models usually do not separate erosion by raindrops and transport by runoff. Sometimes, these models utilise a simple 'washoff coefficient'. Some examples are briefly presented here.

In the SWMM (Storm Water Management Model) (Huber and Dickinson, 1992), the washoff process is presumed to be in direct ratio to the available mass of accumulated particles and to the rainfall intensity. This assumption is written (Jewell and Adrian, 1978; Alley, 1981):

$$\frac{dMa(t)}{dt} = -Ke \cdot I(t) \cdot Ma(t) \qquad (2.1.\ 10)$$

where *Ma* is accumulated mass on impervious surfaces at time *t* (kg)
I(t) is rainfall intensity at time *t* (mm/h)
Ke is washoff coefficient (per mm).

The initially proposed standard value of *Ke* = 0.18 per mm. However it has been shown that *Ke* needs to be calibrated for each catchment (Alley, 1981) as it depends on many parameters: street texture and condition (Sartor and Boyd, 1972), rainfall characteristics (Alley, 1981), particle size distribution, etc. For different rainfall and street conditions in Toronto, Canada, Pitt (1987) found *Ke* ranging from 0.077 to 1.007 per mm.

If *Me* is the mass of particles entering into the drainage system during a time step, equation 2.1.10 becomes:

$$Me(t+\Delta t) = Ma(t) \cdot \left(1 - exp^{(-Ke \cdot I(t+\Delta t) \cdot \Delta t)}\right) \qquad (2.1.11)$$

An adaptation to very low or very heavy rainfall is possible by introducing an availability factor *Kd* (Bujon, 1988):

$$Me(t+\Delta t) = Kd \cdot Ma(t) \cdot \left(1 - exp^{(-Ke \cdot I(t+\Delta t) \cdot \Delta t)}\right) \qquad (2.1.12)$$

with $Kd = 0.057 + 0.04.I(t)^{1.1}$ for suspended particles
$Kd = 0.028 + 0.003.I(t)^{1.8}$ for deposited particles.

Several authors have refined and/or adapted these equations for a range of software in order to get better calibration and simulation of experimental data. The main modifications are:

- introduction of an exponent ω for rainfall intensity, ranging from 0.8 to 2 and with mean values about 1.4 to 1.8, in order to provide better calculations of peak values;
- introduction of washoff equations specifically adapted to rainfall intensity classes, e.g. Hydropol (Briat, 1995);
- introduction of two terms in the expression of *Ke* with different exponent values in order to improve the calibration results (Zug *et al.*, 1998).

However, all these refinements are very site specific and not easily transposed or generalised.

More recently, the MOUSETRAP model (SAFEGE, 1996) was developed with the following washoff relationship :

$$V = Kw \left(\frac{i}{i_d}\right)^n L\,W(1-\varepsilon)A_s \qquad (2.1.13)$$

where *V* is volume of washoff sediments (m^3/h)
Kw is washoff coefficient (m/h)
i is rainfall intensity (mm/h)
i_d is reference rainfall intensity (mm/h), default value = 25 m/h
n is numerical exponent (-), default value = 2
L is length of the catchment (m)

W is width of the catchment (m)
ε is sediment porosity (-)
A_s is fraction of the surface that is covered with sediment (-).

The variables i_d or n can be modified and adjusted by the user to calibrate the model. Calibration tests in France have shown that the exponent n rather than the reference rainfall intensity i_d should be modified in order to fit the experimental results. In the case of Boulogne-Billancourt in France, n has been set to 1.3 (SAFEGE, 1996).

Using substantial field data and statistical analysis, Servat (1984) proposed the following expression :

$$Me = Ks.Ma^a.I_{max5}^b.Vr^c \tag{2.1.14}$$

where Me is washed off mass (kg)
Ma is accumulated mass (kg)
Ks is washoff coefficient
I_{max5} is maximum rainfall intensity during a time step of 5 minutes (mm/h)
Vr is runoff volume (m^3)
a, b, c are numerical coefficients.

The model predicts total mass to within some +/-5 % for washoff over a long period (several months), however, for individual events, the prediction is only within some 10-30 %. Servat (1984) used a maximum intensity for a 5 minute time interval, although one minute would be expected to produce better results, as the washoff phenomenon depends on instantaneous peaks in rainfall intensity.

As in the estimation of solids accumulation, the washoff empirical relationships should be considered as tools for model calibration, to be adapted and calibrated for each catchment. Alternative expressions can be found in Huber (1986) and Bertrand-Krajewski *et al.* (1993).

2.1.3.2.5 Size distribution

Size distribution of the solids accumulated on streets has been measured in a number of studies. Despite differences between sites, it has been observed that the main fraction of the solids ranges between 20 μm and 1 mm. Some examples are given in Figure 2.1.4.

The approximate size for the median diameter d_{50} is 300–400 μm in Figure 2.1.4. This is also confirmed by Butler and Clark (1995) who suggest $d_{50} = 400$ μm. It is important to appreciate that sampling of these solids is not straightforward. Most studies utilise vacuum techniques, whereas others attempt to wash off the solids by water flushing. Neither approach will sample all of the solids.

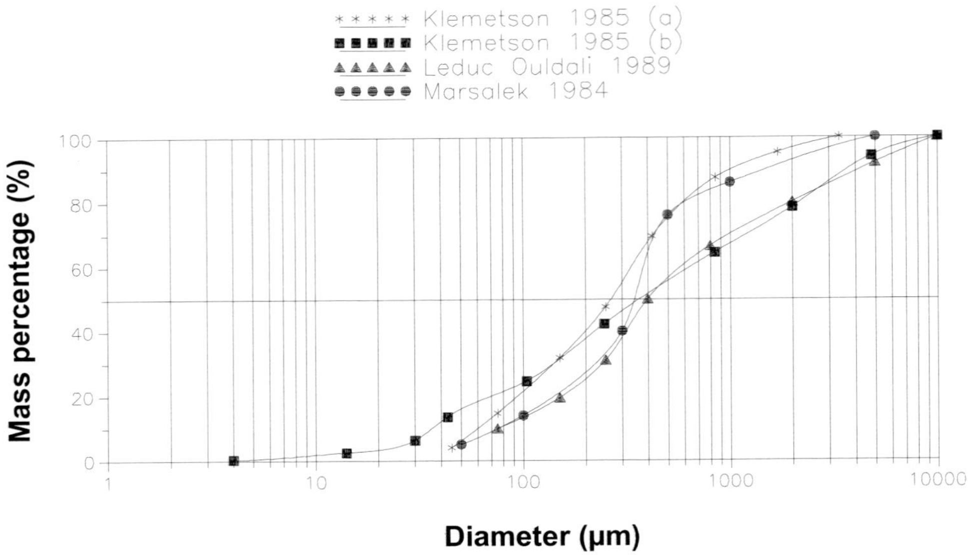

Figure 2.1.4. Examples of size distributions of solids accumulated on streets (Bertrand-Krajewski, 1991).

Not all of the particles are easily washed off during storm events (Pitt, 1987). Artières (1987) distinguished three particle classes (Table 2.1.7).

Table 2.1.7. Propensity of street solids to be transported during storm events (after Artières, 1987)

d < 200 µm	Particles easily transported during storm events by suspension or saltation
200 < d < 2 mm	Particles transported mainly as bed load, only for high intensity rainfall
d > 2 mm	Particles very rarely transported with runoff

Runoff entering the sewer system therefore contains a higher fraction of fine particles than in the accumulated deposits. Usually, the d_{50} at the street inlet is only about 30–40 µm (see also Table 2.2.3 in Section 2.2).

2.1.3.2.6 Organic fraction

Street particles are mainly mineral, with an organic fraction that usually ranges from 10 to 20 %, with relative densities varying between 2.10 and 2.51 (Butler *et al.*, 1992). The density is higher for the fine particles, as shown in Table 2.1.8. Many authors (Chebbo, 1992; Michelbach and Wöhrle, 1992; Stahre and Urbonas, 1990) observed also that the coarse particles are more organic.

Table 2.1.8. Organic fraction and densities of street solids (Chebbo, 1992)

Particle	d < 100 µm	d > 100 µm	all particles	organic fraction all particles
Relative density	2.5	1.83	2.33	10–36 %

2.1.3.2.7 Concentrations and loads

The suspended solids (SS) concentrations and loads in runoff vary widely, depending on site and rainfall characteristics. Some orders of magnitude are presented in Table 2.1.9. The wide ranges in measurements are due to a number of factors, and it has been observed that for the same site, the mean concentration may vary by an order of magnitude between events. This is because of the variability in rainfall characteristics, the antecedent dry weather period, the state of the drainage system, the nature of the street surface, the street cleaning regime, the season, etc. Specific measurements for each catchment during different representative storm events are therefore essential in order to produce a reliable assessment of local SS concentrations and loads.

Table 2.1.9. Ranges of SS concentrations and loads in urban runoff (after Bertrand-Krajewski, 1993)

Suspended solids concentration	10–1200 mg/l
Suspended solids load	350–2300 kg/ha/year

2.1.3.2.8 Potential pollutants associated with the solids

It is well established that a diverse range of potential pollutants may be attached to the solids, and that the particulate fraction includes the largest proportion of potential pollution load (Chebbo, 1992; Pitt *et al.,* 1995) (Table 2.1.10).

The COD/BOD_5 ratio, which is about 2–3 in raw domestic sewage, is usually higher than 4–5 in street runoff, indicating that the organic material is less readily biodegradable (Section 2.2). It should also be emphasised that the finer the particles, the higher the associated proportionate pollutant load. This is observed for many pollutants, and especially for heavy metals as shown in Tables 2.1.11 and 2.1.12.

Table 2.1.10. Percentage of the particulate fraction for different pollutants (after Bertrand-Krajewski, 1993)

Pollutant	Particulate fraction
COD	83–90 %
BOD_5 (biological oxygen demand in 5 days)	77–95 %
TKN (total Kjeldahl nitrogen)	57–82 %
Cd	> 95 %
Pb	68–96 %
Zn	> 95 %
Hydrocarbons	80–90 %
PAHs	79–97 %
PCBs	90–93 %

Table 2.1.11. Particulate fraction of metals depending on the size of street solids (Sartor *et al.*, 1974)

Parameters	< 43 μm		43–246 μm	> 246 μm
Street solids	5.9		37.5	56.5
All metals		51.2		48.7

Table 2.1.12. Metals concentrations associated with highway solids (Ellis and Revitt, 1982)

Site	Size category (μm)	Pb (μg/g)	Cd (μg/g)	Mn (μg/g)	Zn (μg/g)	Cu (μg/g)	Fe (μg/g)
Motorway gutter	> 250	1690	3.8	866	1542	161	-
	< 250	2100	6.0	844	1876	410	-
Motorway dust sample	> 250	895	3.8	1240	1760	212	-
	< 250	1604	6.8	844	2020	640	-
Motorway mid-gutter	> 250	1104	4.2	718	1010	185	22850
	< 250	2296	6.3	766	1212	386	22700
Watford Way (A41)	> 250	723	1.7	694	245	110	14400
	< 250	1826	5.3	792	695	280	15130
Sunnyside Gardens	> 250	261	3.6	792	1715	35.1	8780
	< 250	978	6.0	816	2133	90.7	15280
Country lane	> 250	40	0.72	728	91.6	27.2	4195
	< 250	111	1.3	855	119	42.6	6780

Particles finer than 250 µm are the most contaminated. A more detailed analysis shows that about 70 % of the metallic pollution is attached to particles finer than 100 µm, which represent less than 15 % of the mass of solids accumulated on streets (Ellis and Revitt, 1982, see Table 2.1.12). Similar orders of magnitude are given by other authors (Kern *et al.*, 1992, Kari and Herrmann, 1989; Ball *et al.*, 1996) who conclude that the main fraction of the metals is attached to particles finer than 60 µm.

Hydrocarbons are also mainly found in the particulate phase, especially associated with particles finer than 100 µm (Hoffman *et al.*, 1984; Kari and Herrmann, 1989). As combustion residues are a major source of PAHs, their concentration on street particles is closely related to the traffic intensity as shown by Figure 2.1.5 for benzo(a)pyrene. The profiles are similar for other PAHs (Xanthopoulos, 1992).

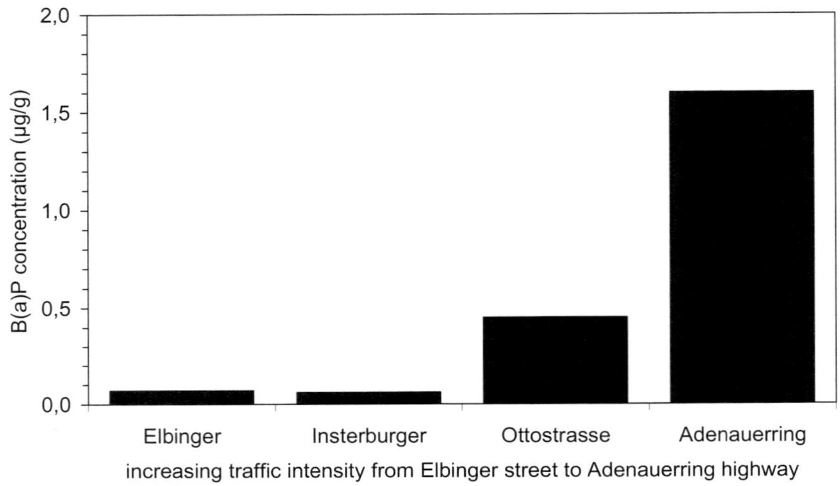

Figure 2.1.5. Example of PAHs concentration: benzo(a)pyrene B(a)P concentration on street solids according to automobile traffic (data from Xanthopoulos, 1992).

The particulate associated fraction of PCBs is typically higher than 90 % (Marsalek, 1986), and 65 % of the mass is attached to particles finer than 20 µm (Kari and Herrmann, 1989) as shown in Figure 2.1.6. Table 2.1.13 gives some complementary information on pollutants associated with surface solids depending on the degree of urbanisation.

Table 2.1.13. Orders of magnitude of pollutant concentrations associated with solids accumulated on surfaces for different degrees of urbanisation (after Bertrand-Krajewski, 1993)

Pollutant	Residential area	City centre	Commercial area	Industrial area	Parking lot	Street	Highway
Cd (µg/g)	0.04–10.7	2.6–7.0	0.02–1.06	0.70–3.40	1.0–14.6	0.22–3.90	0.60–4.30
Cu (µg/g)	14.9–221.0	143–390	10.40	228	206	22–200	90–281
Pb (µg/g)	120–1000	1880–2550	160–220	488–1410	2000–15000	-	130–4800
Zn (µg/g)	47–1170	470–534	53–1065	655–1445	1600	44–480	250–336
HC* (mg/g)	15.7–59.8	8.75–51.80	16.40–34.00	61.90–507.0	-	-	-
PAHs (µg/g)		-	-	-	-	0.2–20	-

*HC: hydrocarbons

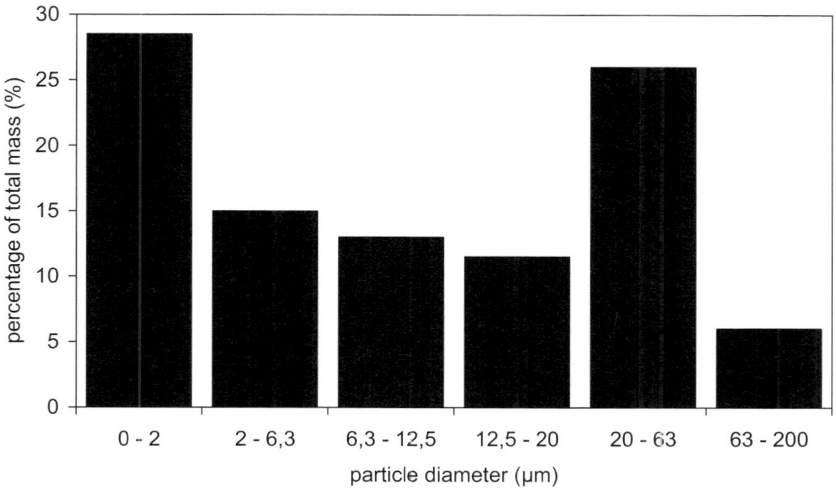

Figure 2.1.6. Percentage of the total load of Clophen A 60 (typical PCB compound) according to the size of street particles (data from Kari and Herrmann, 1989).

2.1.3.3 Solids from gullies

Stormwater sewerage systems have street inlets to allow the ingress of rainfall-runoff, principally from roads and other paved areas. In many countries, these consist of gratings and gully pots (Ristenpart *et al.*, 1995) as illustrated in Figure 2.1.7 (and Figures 4.1.4a–d). Without gully pots (or catchbasins as they are known in the USA), the quality of the stormwater entering the sewer network approximates that of the direct washoff. The inclusion of selective pots has a significant effect on the quality of the washoff into the sewerage system, particularly with respect to solids particle size distribution. It is these altered characteristics which are discussed in the next section.

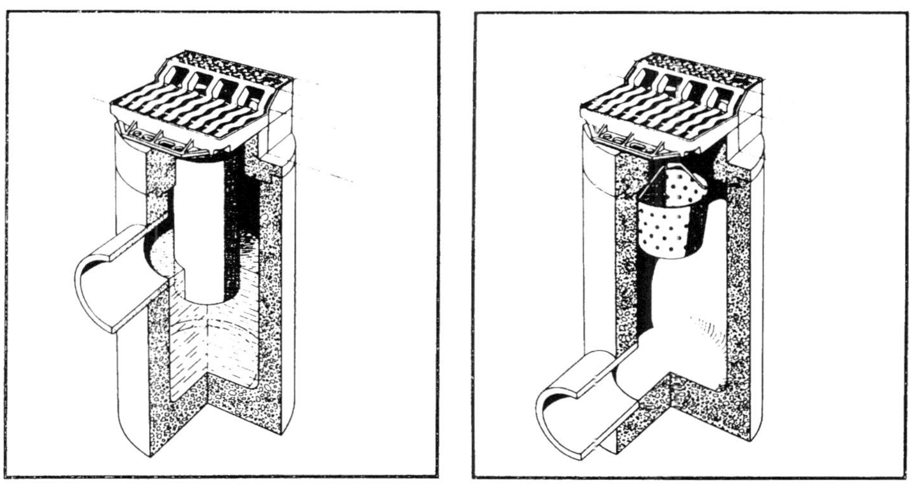

(a) (trapped) with sump (b) without sump and with basket

Figure 2.1.7. Example of gullies (catchbasins) from Pont-à-Mousson, France (source: commercial documentation from Saint-Gobain, France).

The configuration of the inlet also affects the nature of what enters the system. Figure 2.1.8 shows a typical UK gully inlet. This has not been properly cleaned for some time and the gully on the left is blocked with solids. In contrast, Figure 2.1.9 shows an inlet in a busy area of Athens, Greece. There is no trap and any type of solids can easily find its way into the sewer beneath.

Figure 2.1.8. Traditional grated gully inlet and consequences of not cleaning (photograph Richard Ashley).

Figure 2.1.9. Street gully inlet in Athens (photograph Richard Ashley).

2.1.3.3.1 Deposits

The deposits found to some extent in all gullies incorporating sumps or pots, consist of particulate, inorganic sediment with substantial quantities of organic material and larger, non-particulate matter (NPM), together with interstitial liquor. Pratt and Adams (1981, 1984) estimated the bulk density of the bed, in residential area pots in the UK, from samples taken from gully emptying equipment, as 1490 kg/m^3, with a mean particle size of 1.75 mm. Pratt *et al.* (1987) collected sediment from the base of several gully pots for analysis and the particle size distribution of these samples is shown in Figure 2.1.10. Grottker (1990) collected the contents of basal sediment in eight gully pots in Hannover, Germany, and divided them into samples for analysis. The particle size distribution was assessed by wet sieving (see Figure 2.1.10).

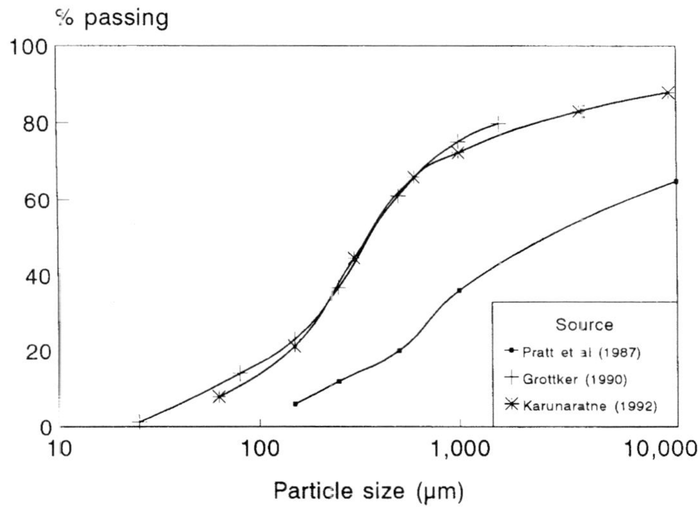

Figure 2.1.10. Particle size distribution of gully basal deposits (Butler and Clark, 1995). Reproduced with permission of CIRIA.

A series of gully bed grab samples were taken by Karunaratne (1992) from locations in inner London, UK, and from emptying gully cleaning tankers. A range of saturated bulk densities was measured (1100–1700 kg/m^3) with an average of 1350 kg/m^3. The proportion classified as NPM varied from 1.1 to 26.6 % (by mass) and the proportion of organic material ranged from 2.9 to 21.6 %. NPM was found to be directly proportional to organic content and inversely proportional to bulk density. Particle size distribution results are similar to Grottker's data and Figure 2.1.10 shows that about 20 % (by mass) of the basal sediments are particles greater than 1.0 mm and 20 % or less are smaller than 0.1 mm with a mean particle size of 350 μm. The material has distinct similarities to class A sewer sediment (Crabtree, 1989) (Section 2.2).

It has been postulated that the basal deposits undergo a process of organic build-up or 'ageing'. Sonification tests have shown that organics are released from gully sediments in two stages: a large release at low energies followed by a more gradual release as the energy input is increased. Clegg *et al.* (1992) have interpreted this to mean that there are two types of solids associated with the sediment: a loosely associated material, probably interstitial solids, and a biofilm-type material that is more tightly bound. In field tests, COD values were generally found to increase with time in undisturbed beds, but not in pre-prepared clean sand beds. Inspection of the cleaned bed after the test period showed little evidence of bio-coverage, with clean sand clearly visible amongst the accumulated black deposits (Clegg *et al.*, 1993).

2.1.3.3.2 Outflow

During rainfall, stormwater is directed into the pot via a grating or inlet and typically falls vertically to impact the surface of the pot reservoir below. This results in considerable turbulence, particularly at the water surface, hence a proportion of the incoming solids will be swept through into the sewer, whilst some of the solids will settle (despite the turbulence). The existing pot liquor will be displaced by the incoming runoff and a fraction of the previously deposited solids may be re-entrained.

Full-scale laboratory experiments by Fletcher and Pratt (1981) and Karunaratne (1992) have established that, in salt removal tests in a gully pot, effluent concentrations arising from liquor discharges follow an exponential decay curve characteristic of the outflow from a completely mixed reactor. It was estimated that such liquors are responsible on average for 11 % suspended solids, 21 % dissolved solids, 14 % BOD_5, 21 % COD, 32 % NH_4-N and 16 % NO_3-N.

Fletcher *et al.* (1978) studied solids erosion in a laboratory based pot with a 50 mm bed of field sediment in the pot invert. They reported that re-entrainment of basal deposits depended on the water flow rate, accumulated sediment bed depth and the mass of sediment 'available for release'. For flow rates of up to 1 l/s, only a maximum of 0.2 % of the basal sediment was washed out.

Results from a simulation model indicated that for the majority of storms monitored, the suspended solids contribution was derived from the material already in suspension rather than by re-suspension of bottom sediments (Fletcher and Pratt, 1981). In Butler and Karunaratne's (1995) laboratory study, significant bed erosion was found to occur on a previously disturbed bed of quartz sand only for a short period at the beginning of inflow. For a flow of 1.5 l/s, a particle specific gravity of 2.58, mean diameter of 68 μm and bed depth of 400 mm, continuous monitoring identified a relatively high (150 mg/l) transient peak during the first 30 s of inflow followed by a continuous erosion output of only 8 mg/l, similarly to the results from modelling foul flushes in laboratory pipe studies.

In the field, Sartor and Boyd (1972) found that for US catchbasins, previously retained solids (both in the liquor and on the bed) were being liberated, even after an hour of flushing with clean water. However, it was noted that only about 1 % of the initial sediment load was removed. Pratt and Adams (1984) noted that in 44 % of the catchment events monitored, the effluent suspended solids concentration was higher than the influent value. The quality was always worsened at high flows in excess of 10 l/s (Pratt *et al.*, 1987). Morrison *et al.* (1988) demonstrated that solids are removed from the deposited bed only by large storms having both high volumes and intensities. For example, they observed that an inflow of 4.5 l/s into a 40 l pot caused an outflow suspended solids concentration of 2145 mg/l, suggesting significant erosion of the bed and mixing in the pot reservoir. Sediment mobilisation was not anticipated for low flow storm events which were characterised by effluent solids concentrations of less than 50 mg/l. In a long-term study of gully sediment bed build-up, Butler *et al.* (1995) never observed wholesale removal of the bed and reductions in bed depth (other than through cleaning) were generally rare and small (≤ 20 mm per month).

Details on particle characteristics *leaving* gully pots are sparse. Most data have been obtained from in-sewer sampling and care is required in their interpretation. A study in the USA by Randall *et al.* (1982) of stormwater suspended solids, found that 89 % of particles by mass passing through catchbasins were smaller than 35 μm in size.

Field studies in Dundee (Ashley and Crabtree, 1992) illustrate how pollutants are washed out from the gully (Figure 2.1.11). As might be expected, the effluent volatile solids fraction was found to be highest at the start of the storm (80 %) declining to 12 %. COD levels were independent of antecedent dry period, being influenced by rainfall intensity and contributing catchment land use. BOD_5 and NH_4-N levels were low (< 25 mg/l and < 1 mg/l, respectively).

Substantial quantities of heavy metals are also flushed out by the incoming runoff. Metals in solution show significant early contributions, coming from the pot liquor and sediment interstitial liquor. Although there appears to be no net rise in soluble metals through the pot, the outflow metals are in a more bio-available form (Morrison *et al.*, 1988). Particulate associated metals can also be significantly enhanced with passage through the pot (Figure 2.1.12).

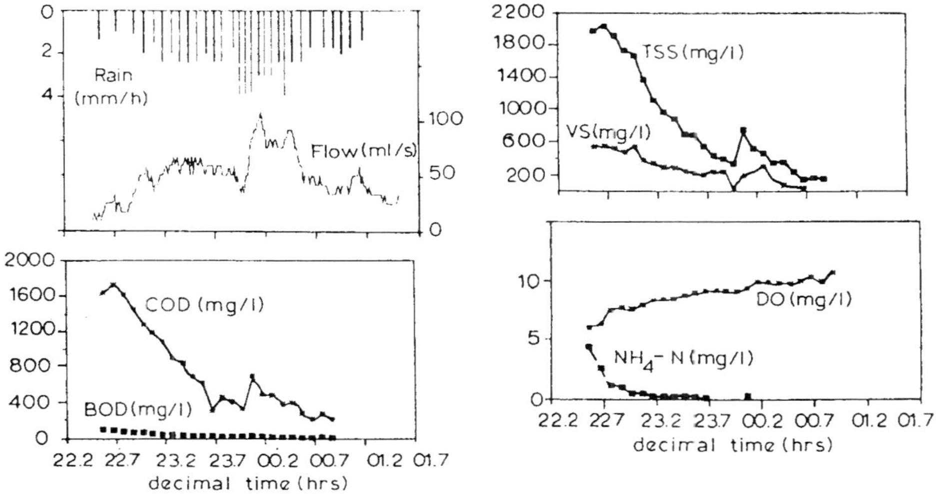

Figure 2.1.11. Quality of gully outflow on a heavily trafficked road (Ashley and Crabtree, 1992).

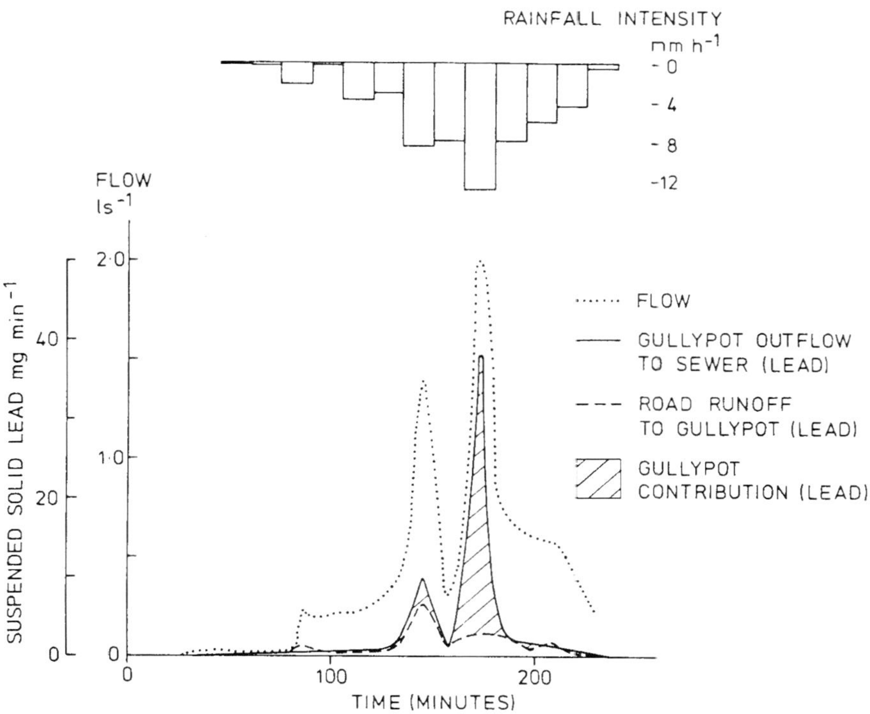

Figure 2.1.12. Contributions of road runoff and gully pot contents to suspended solids Pb loadings in gully pot outflow (Morrison *et al.*, 1990).

2.1.3.3.3 Modelling of solids transfer through gully pots

The storage and erosion phenomena in gully pots depend on runoff, particle characteristics, and antecedent dry weather period. Some modelling approaches are presented below. After experimental studies, Grottker proposed a model for both dry gullies (i.e. without any retention volume and with direct connection to sewer pipes) and wet gullies (i.e. with a volume of trapped water), which predicts the 'passed load' (the load which cannot be retained by the gully pot) as a function of the two main parameters identified: the pollutant load and the flow rate through street inlets (Grottker, 1990; Grottker and Hurlebusch, 1987). This model is written:

$$Mp = Me \cdot a \cdot Q^b \tag{2.1.15}$$

where Mp is mass of particles passing through the gully pot (kg)
Me is mass of particles washed off by rainfall (kg)
Q is flow rate through the gully pot (l/s)
a, b are numerical coefficients depending on the particle diameter ($0.046 < a < 0.791$ and $0.121 < b < 0.737$).

Grottker showed also that this relation was time independent. Equation 2.1.15 gives results with errors in a range of 10 to 30 %. However, the model must be calibrated for use at other sites.

Fletcher and Pratt (1981) proposed a model to reproduce the flushing mechanism of accumulated sediments in gully pots by wet weather. This flushing out is due to the stirring action of the inflow water. Their model was tested and validated using Japanese data by Wada *et al.* (Wada and Miura, 1988; Wada *et al.*, 1987).

Two phenomena can be distinguished:

(1) For sediments already in suspension in the gully water, the concentration is given by the following relation:

$$C = Co \cdot \exp\left(\frac{-t.Q.P}{100.V}\right) \tag{2.1.16}$$

where C is suspended sediment concentration in the outflow (mg/l)
Co is initial sediment concentration in the gully pot (mg/l)
t is time (s)
Q is inflow (l/s)
V is gully pot volume (l)
P is percentage of gully pot fluid mixed (%).

if $Q < 0.12$ l/s, then $P = 664.Q + 19.7$
if $Q \geq 0.12$ l/s, then $P = 100$.

(2) For sediments resuspended from basal deposited sediments, with M total mass of released sediment (kg) and Kr rate of release (mg/s), the relations are:

if $0 < t < M/Kr$ (re-suspension continues):

$$C = \frac{Kr}{Q}\left(1 - \exp\left(\frac{-t.Q}{V}\right)\right) \tag{2.1.17}$$

if $M/Kr < t < \infty$ (re-suspension does not occur):

$$C = \frac{Kr}{Q}\left(1 - \exp\left(\frac{-M.Q}{Kr.V}\right)\right) \cdot \exp\left(\frac{-t.Q}{V} + \frac{M.Q}{Kr.V}\right) \tag{2.1.18}$$

Wada and Miura determined Kr and M with the following empirical equations:

$$Kr = (1.78 \cdot Q + 0.22)\, Mt \tag{2.1.19}$$

$$M = (57.1 \cdot Q + 0.83)\, Mt \tag{2.1.20}$$

where Mt is the total mass of sediment in the bottom of the gully pot.

These numerical coefficients were experimentally determined and the results presented by the authors showed a good agreement with field data. More recently, Butler and Karunaratne (1995) undertook a laboratory study to establish the solids retention performance of a 450 mm diameter, 80 litres volume gully pot. They defined trap efficiency $E=1-Mp/Me$ (as previously defined) and found:

$$E = \frac{1}{1 + \dfrac{72Q\nu}{\alpha\pi g d^2 D^2 (s-1)}} \tag{2.1.21}$$

where Q is flowrate (m^3/s)
ν is kinematic viscosity (m^2/s)
α is turbulence correction factor ($\alpha \approx 0.6$)
g is acceleration due to gravity (m/s^2)
d is particle diameter (m)
D is pot diameter (m)
s is particle specific gravity.

The results are shown in Figure 2.1.13. The trapping performance was found to be dependent (inversely) on the flow rate through the pot and the particle size. The depth of sediment in the pot had a less significant effect. Studies of sediment wash-in to drainage systems and in-pot processes by Memon (2000) and reported by Butler and Memon (1999), demonstrated the application of the concept of the pot as a single completely-mixed reactor. Good model representation of the washout of solids to the downstream drainage system was found by assuming two classes of solids in the pot, one slowly settleable, with a range of 80 ± 160 µm, and all solids above this range assumed to be readily settleable. Only the slowly settleable fraction was assumed to be re-erodible at the sediment liquor interface.

The model of sediment wash-off from paved surfaces developed by Deletic *et al.* (1997) has been extended to predict fine granular sediment inputs into drainage systems via roadside gully pots (Deletic *et al.*, 2000). The model comprises three separate components:

- a model of sediment accumulation on paved surfaces;
- a model of sediment wash-off from these surfaces; and
- a model of sediment deposition and erosion in gully-pots.

The first two model components had been previously successfully tested in Belgrade and Lund (equations 2.1.5 to 2.1.9), whereas the third component of the model has been developed and applied to two small catchments in Dundee, Scotland using continuous simulation periods of up to 8 months.

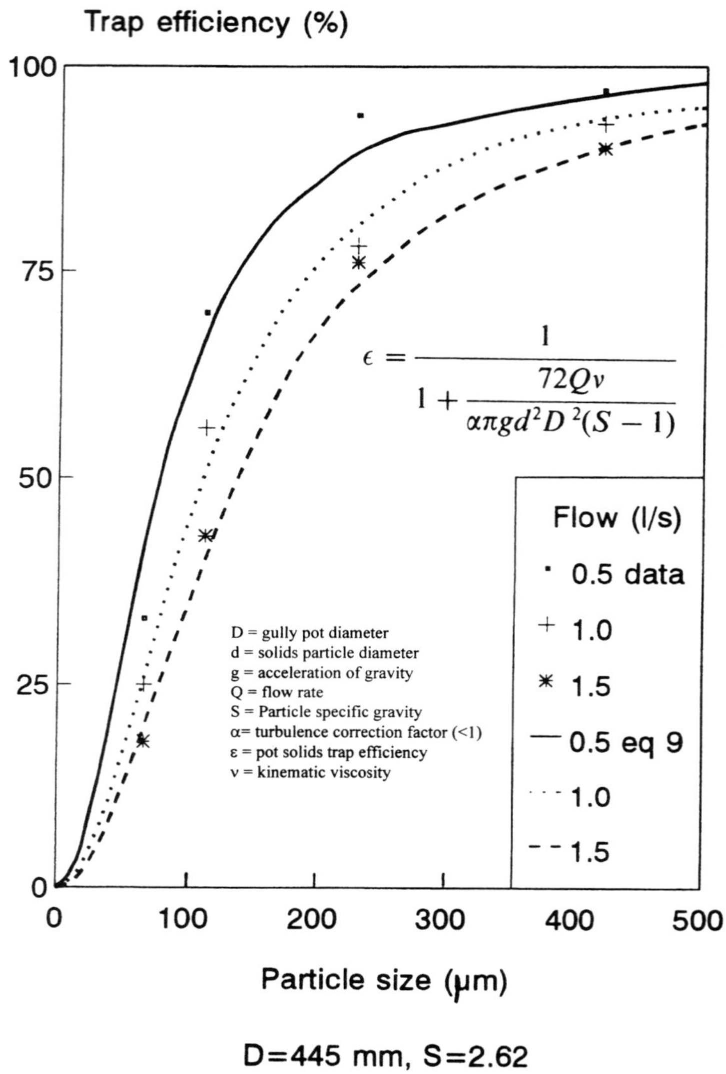

Figure 2.1.13. Efficiency of a gully pot at trapping solids (from Butler and Karunaratne, 1995).

Once a particle conveyed into a gully pot becomes a part of the gully bed it could become eroded during high flow rates (re-suspended into the gully liquor), and eventually carried out of the pot. The concentration of particles in the effluent (and in the gully liquor), C (mg/l), can therefore be modelled as in Equation 2.1.22.

$$\frac{dC(t)}{dt}=\begin{cases}\frac{Q}{V}\left(C_{in}-C\right)\frac{Q}{Q_{mix}} & \text{for } Q<Q_{mix} \quad \text{deposition}\\ \frac{Q}{V}\left(C_{in}-C\right)+\frac{K_g}{V} & \text{for } Q\geq Q_{mix} \text{ and } M_{gb}>0 \quad \text{erosion}\\ \frac{Q}{V}\left(C_{in}-C\right) & \text{for } Q\geq Q_{mix} \text{ and } M_{gb}=0 \quad \text{no bed sediment}\end{cases} \tag{2.1.22}$$

where C_{in} is concentration of particles in the inflow (mg/l)
Q is flow rate (l/s)
V is gully-pot volume (l)
Q_{mix} is flow rate at which all gully liquid is fully mixed and erosion begins (l/s)
K_g is erosion rate (mg/s)
M_{gb} is mass of fine sediment at the gully base (mg).

Thence, a mass balance approach leads to the following expression for M_{gb}

$$\frac{dM_{gb}(t)}{dt}=\begin{cases}\left(C_{in}-C\right)\left(1-\frac{Q}{Q_{mix}}\right) & \text{for } Q<Q_{mix} \quad \text{deposition}\\ K_g & \text{for } Q\geq Q_{mix} \text{ and } M_{gb}>0 \quad \text{erosion}\\ 0 & \text{for } Q\geq Q_{mix} \text{ and } M_{gb}=0 \quad \text{no bed sediment}\end{cases} \tag{2.1.23}$$

The erosion rate, K_g, is calculated according to the linear regression curve found in the laboratory study by Fletcher and Pratt (1981) as described above, with the flow rate for the triggering of erosion adopted from the same study as Q_{mix}= 0.12 l/s.

When runoff ceases, sediment suspended in the gully liquor will deposit. The model predicts the concentration of sediment particles in the gully during dry weather as:

$$C(t_d)=\begin{cases}C_p\left(1-\frac{V_s t_d}{H_g}\right) & t_d<\frac{H_g}{V_s}\\ 0 & t_d\geq\frac{H_g}{V_s}\end{cases} \tag{2.1.24}$$

where C_p concentration of sediments at the end of the previous event (mg/l)
V_s is Stokes' particle settling velocity (m/days)
H_g is gully height (m).

It was concluded that only two calibration parameters are required to model the input of fine granular sediment into storm drainage systems via gully pots, with calibration required only for the build-up component. A typical result is shown in Figure 2.1.14.

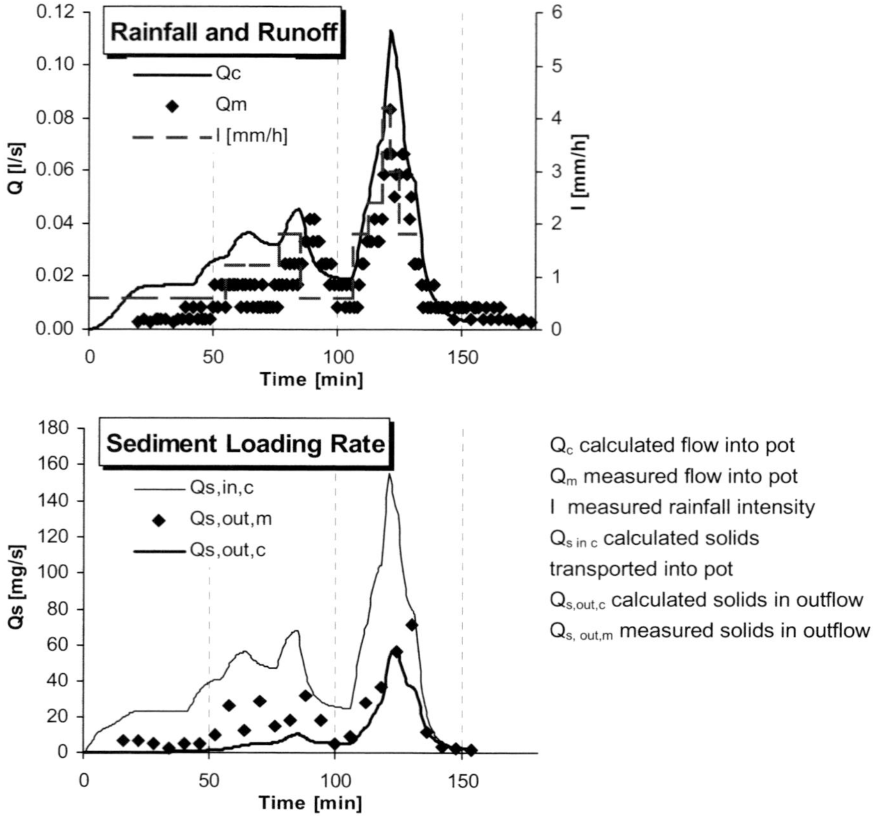

Figure 2.1.14. Modelled compared with measured flow and solids concentrations into a sewer from a gully pot in Dundee (Deletic *et al.* 2000).

2.1.3.4 Solids resulting from de-icing

The declining rate of build-up in solids accumulating on road surfaces, with eventual equilibrium being reached due to rainfall and other factors, does not occur when there are heavy snowfalls or ice formation. Snow and ice continue to accumulate solids and pollutants from all sources whilst freezing conditions prevail. Once thawing begins, however, the release of the accumulated pollutants can result in the highest observed flush loadings of solids and pollutants into the receiving drainage system.

The spreading of de-icing agents to assist with the snowmelt process, or to alleviate icing, is believed to be the single largest contributor to the sediments and pollutant loads entering sewer and drainage systems in temperate regions. As yet, however, there have been no quantified studies to justify this hypothesis. The problems arising from the application of salt to urban surfaces have been known about for some time (e.g. Field, 1974). Another potential problem is the storage of the agents, with careless stockpiling leading to localised solids inputs to drainage systems and high chloride concentrations in receiving waters. De-icing is important for roads, sidewalks and airfields, where both the runways and the aircraft are de-iced, with pronounced seasonal effects. Not all highways are treated: in the UK for example, it is only major highway routes, with some 10 % of minor roads, which are regularly treated with de-icing agents. The solid materials used for de-icing vary, and may include rock salt, or closely graded grit (1–3 mm). In many Scandinavian countries, application of de-icing agents can occur for prolonged periods, e.g. for more than 150 days a year in Denmark.

Butler and Clark (1995) assume that for salt application at the recommended UK precautionary rate of 10–15 g/m^2, with the maximum allowable non-solubility of 7.5 %, the solids residual is about 1 g/m^2. When this is compared with the typical sediment supply rate from other sources of 2 g/m^2, it indicates that even on this very simple balance, the insoluble solids represent some 33 % of the total available for wash-off into drainage systems.

Studies reported in the UK (Colwill *et al.*, 1985; Hedley and Lockley, 1975; Bellinger, 1982) considered motorway runoff quality. Outfall pipe data revealed, for a one year study, that winter salting was the major factor influencing the concentration and total mass of suspended and dissolved pollutants discharged. The monitored flow passed through regularly cleaned gullies and sediment traps, which were only partially successful in abating the level of pollutants. The rock salt used in each case contained between 5 and 10 % by weight of insoluble solids, which then appeared in storm runoff some time after the washoff by rainfall of the soluble chlorions. The results indicated that, on average, some 25–30 % of annual solids loads could be attributed to the application of rock salt.

The rate of salt application was considered excessive and significantly beyond the 14-56 g/m^2 then recommended by the UK Department of Environment. Salt application was considered to be responsible for increasing vehicular and highway surface corrosion, and also for encouraging the release of volatile organic oils and greases. The other studies showed insolubility as within the same range, and with some 80 % of the insoluble solids being grit, the overall effect is to provide some 25 % of the total winter period suspended solids in runoff. These solids and associated pollutants provided some 90 % of the total soluble pollutants during a winter season. It is possible that the salt and other substances from the 'rock' salt encourage toxic metals such as mercury, and lead to be released from sediment adsorption and become more mobile. No attempts have been made to review the relative merits of street sweeping (see Section 4.1) to remove residual salt once weather conditions have improved; this may be a means of helping to minimise the amount of material that eventually finds its way into sewer systems.

Attempts to quantify the relative amounts and effects of de-icing operations on the sediments found in the sewers in Dundee (Scotland) have been made by monitoring the wash-in via road gullies, the quality of the sewage flows, sediment build-up and nature in the sewers. There are a number of effects:

- in-sewer deposits which (by mineralogical examination) can be confirmed as coming from road salting (physical effects);
- changes in the quality of sewage flows in dry as well as wet weather when compared with summer or non-gritted winter conditions (physical and chemical effects) – due to vehicular ‘spraying’;
- changes to the gully sediments and liquor (physical and chemical effects);
- increases in surface deposits on roads and changes in chemical regimes (physical and chemical effects).

Flows and sediments were monitored in the main sewer and in a trunk contributing sewer in the Dundee studies. Significant summer to winter differences were observed. The suspended solids peak levels increased from 195 mg/l during ‘unsalted’ winter dry weather to 395 mg/l after salting. Average solids levels also rise from 80 mg/l to 195 mg/l. Suspended solids concentrations monitored during winter storms show a considerable increase; in one case with a peak concentration of nearly 2000 mg/l. This supports the earlier findings of Hedley and Lockley (1975) that the soluble solids, as measured in the increased DWF (dry weather flow) ammonia levels, are rapidly removed, even by melt water washoff, and that the heavier particulates in the rock salt remain on the road surface until the next rainfall. Examples of the data from Dundee are given in Table 2.1.14 and Table 2.1.15 for the averaged DWF daily variations in sewage characteristics for summer and winter seasons, in the latter case, for conditions when salt has been applied to surfaces and when it has not. On average, salt is applied for some 64 winter days (35 % of the time) in the City of Dundee at a rate of 10-20 g/m^2. The salt has an insolubility of 0.5–5 %, with some 85 % of the residue not being volatile, and a low ammonia concentration (0.000434 mg/g). The characteristics of the salt applied are given in Table 2.1.16.

Table 2.1.14. Average DWF sewage flow quality – Dundee at two sites

	Main interceptor sewer		Contributing trunk sewer	
Season	Summer	Winter (de-icing)	Summer	Winter (de-icing)
SS (mg/l)	173	195	33	116
N-NH_3 (mg/l)	21	91	3.0	11.2
Flow (l/cap-day)	315	258	220	620*

*probable infiltration

Table 2.1.15. Sewer sediment quality averages – main Dundee interceptor sewer

Parameter	Summer	Winter	Gully pots (winter)
VS (%)	5.5	2.9	95.7
COD (mg/kg)*	8534	12093	3793
N-NH_3 (mg/kg)*	88	132	6.9
BOD (mg/kg)*	5821	-	375

* of dry solids

Table 2.1.16. Characteristics of the de-icing salt used in Dundee (UK)

Constituent		Concentration
NaCl		97.4 %
CaSO4		2.00 %
Anti-caking agent		
Blue prussiate of potash (potassium carbonate)		100 ppm
Particulate size	> 10 mm	nil
	< 6.3 mm	84 %
	< 2.36 mm	52 %
	< 300 μm	9 %

2.1.4 Solids from domestic wastewater

Domestic wastewater sources can be categorised in a number of ways, depending of the objectives of their study. In this report they are categorised as:

- fine faecal and other organic particles (sanitary solids);
- large faecal and other organic matter (gross solids);
- paper, rags and miscellaneous sewage litter (sanitary refuse, also generally included as gross solids);
- kitchen sink organics (fine and large).

Few reliable published data exist and the contributions to total sewer solids loads are very different according to water supply, cultural habits, level of development and way of life. However, some data are summarised in Table 2.1.17.

Table 2.1.17. Some per capita contributions of sanitary solids

Nature of solids	Contribution rate	Origin of data
Total suspended solids	60–80 g/capita/day	Europe
Faeces	100–130 g/capita/day	Europe and North America
Toilet paper	6.4 g/capita/day	UK
Gross solids (screeened at WWTPs)	1 g/capita/day	UK

2.1.4.1 Sanitary solids

Sanitary solids are defined as those fine particles derived from domestic sources which are normally assumed to be transported mostly in suspension (Section 2.3) in foul and combined sewers (Ashley and Crabtree, 1992, Verbanck *et al.*, 1994). They are derived from degraded faecal stools, vegetable matter, etc.

Information related to sewage particle characteristics is limited and should be treated with caution for the following reasons:

- most data are collected from wastewater in sewers and not directly from the source of the sewage (e.g. domestic appliances);
- most data are from combined sewers under dry-weather flow conditions rather than from separate foul (sanitary) sewers;
- sewage particles are non-conservative in nature and change temporally and spatially;
- the test procedures used may modify the characteristics being measured.

Detailed studies in France (Chebbo *et al.*, 1990) have shown that fine particles ($30 < d_{50} < 38$ µm) predominate (25 % by mass) in suspended solids during dry weather flow. Work in Brussels by Verbanck *et al.* (1990) tends to support this finding and also shows that approximately one quarter is mineral in content (Figure 2.1.15). It was also noted that 25 % was above 125 µm (90 % of which was volatile) consisting of light organic material. Total loss on ignition ranges from 70–85 % (Verbanck *et al.*, 1994).

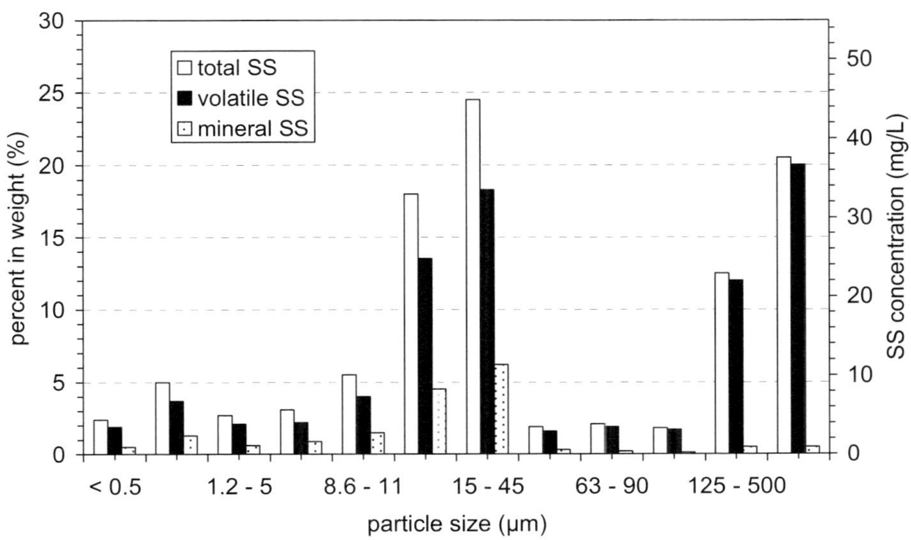

Figure 2.1.15. Particle size distribution for sanitary solids (adapted from Verbanck *et al.*, 1990).

Fair *et al.* (1966) state that specific gravities (SG) of sanitary solids range from less than 1.0 to 1.2 as dry mass, but may be as low as 1.001 when wet. Chebbo *et al.* (1990) found the SG to be 1.6 for particles less than 100 μm and 1.4 above that size. Typical settling velocities are reported in the range 1.0 to 2.5 m/h (Peavy *et al.*, 1986). Chebbo *et al.* (1990) found the settling velocities of particles smaller than 50 μm to be 0.34 m/h and 30 m/h for larger particles, and Michelbach and Wöhrle (1992) found the median settling velocity to be approximately 15 m/h for sanitary sewage. The median velocity reported in a USEPA study (Dalrymple *et al.*, 1975) for sanitary sewage was 1.9 m/h, and results were shown to be time dependent. Verbanck *et al.* (1994) suggest settling velocities lie in the range 1.4–1.8 m/h, but due to the variety of apparatus and procedures used in measurement, care should be taken when comparing the various reported ranges (see Appendix).

The concentration of sanitary solids in sewage is widely reported in the standard texts, although the location or representativeness of the sample is not normally specified (i.e. source, in-sewer or at the sewage treatment works). Similarly, these are assumed to be mean values, representative of the whole flow. Table 2.1.18 summarises various 'typical' international values.

Table 2.1.18. Average reported pollutant concentrations in domestic sewage

Location		SS (mg/l)	BOD_5 (mg/l)	COD (mg/l)	NH_3-N (mg/l)
Abu Dhabi [a]		198	228	600	35
Brussels (Belgium) [d]		290	325	670	35
Brazil (NE) [a]		392	240	570	38
Denmark [e]		120–450	150–350	210–740	12–50
France [f]		150–500	100–400	300–1000	20–80
Germany[g]		325	300–500	600	40–100 (total N)
Jordan [a]		900	770	1830	100
Kenya [a]		520	520	1120	33
USA [b]	weak	100	110	250	12
	medium	220	220	500	25
	strong	350	400	1000	50
UK[c]		80–195	143	40–517	20–90

[a] Horan (1990); [b] Metcalf and Eddy (1991); [c] Crabtree *et al.* (1991); [d] Verbanck (1989); [e] Henze *et al.* (1995); [f] Bertrand-Krajewski (1993); [g] averaged data from a range of sources.

Sewage flow and composition vary on a variety of timescales, including hourly, daily and seasonally, as well as being heavily influenced by socio-economic factors and the extent of industrial input. Routinely, the biggest variation occurs during daily flow variations, with suspended solids concentrations (for example) falling to almost zero during the early hours of the morning (Figure 2.1.16). More detailed studies in the UK (Butler *et al.*, 1995; Friedler and Butler, 1995) derive at-source daily pollutographs based on components of individual domestic appliances.

2.1.4.2 Gross solids and sanitary refuse

Only limited knowledge exists concerning the nature, quantity or variability of gross solids and sanitary refuse. Indeed, these items are not clearly and universally defined. Jefferies and Ashley (1994) follow the convention that gross solids consist of sewage derived material greater than 6 mm in any two dimensions with a specific gravity close to unity. Friedler *et al.* (1996) define 'WC derived solids' to include faecal stools, toilet paper and sanitary refuse. Sanitary refuse is an encompassing term to describe items such as female sanitary products, condoms and other bathroom litter. The majority of items flushed are in this latter category. This material can be a significant problem and is responsible for so-called 'aesthetic' pollution when it escapes from combined sewer overflows into watercourses. Figures 2.1.17 and 2.1.18 illustrate the material in sewers in France and Belgium.

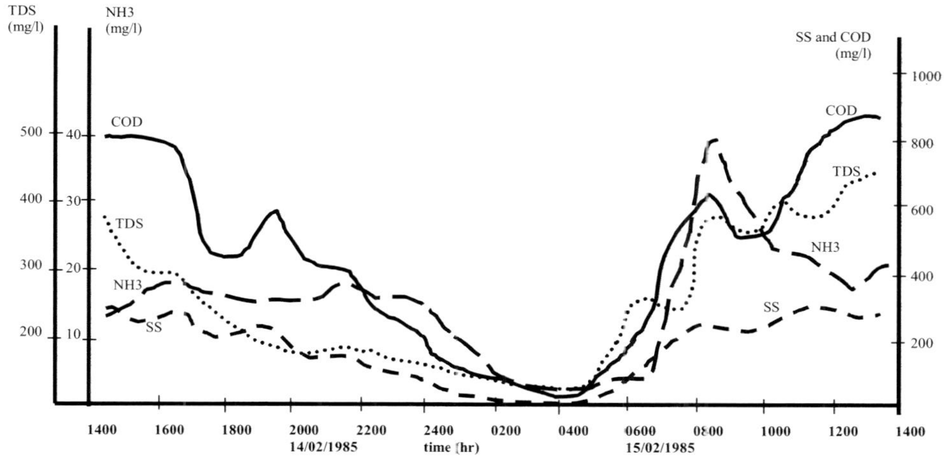

Figure 2.1.16. Diurnal pattern of typical dry weather flow composition (adapted from Thornton and Saul, 1986).

At the start of the 1990s, O'Sullivan (1990) found 'a dearth of information relating to the quantities of gross debris discharged' and a 'lack of information ... for overflow structures in terms of retention of gross solids'.

The use of the WC in the developed world is virtually ubiquitous. However, not all countries with WCs use them in the same way. A questionnaire survey undertaken of 44 countries (Ashley and Souter, 1999) in which the WC is used produced the results shown in Table 2.1.19 for the commonest items. Whilst 13 % of respondents claimed that sanitary items (female hygiene products) were regularly flushed, of those binning, some 62 % of the 26 % in Table 2.1.19 claim that binning is their only disposal route. Whilst a number of countries admitted flushing sanitary items, the practice appears to be most prevalent in the UK, certain UK commonwealth countries and the USA. Little evidence exists for 'disposable' nappies being flushed, although nappy wipes are very evident. In 26 % of the countries responding to the international survey, toilet paper was frequently reportedly placed directly into the solid waste disposal route. Segregation of the respondents into developed, developing and transitional countries revealed that the developed and transitional countries have similar attitudes to WC flushing of items, whereas in the developing world, where flush toilets are less prevalent, disposal via the waterborne route is much less frequent (typically 15 % less than for developed countries for those items flushed). However, it is probable that the respondents contacted in the developing world were those with a lifestyle more akin to the developed countries, and thus were not necessarily representative of the majority of the population.

Table 2.1.19. Reported disposal habits for commonest domestic sanitary waste items for countries with WCs

Numbers disposing via	Sanitary items	Condoms	Nappies	Toilet paper	Cotton buds	Disposable razors
Flushing	13	13	2	25	9	1
Binning	26	22	28	9	26	28

Figure 2.1.17. Floating solids in 'Vieille du Temple Sewer' (rectangular section (60 cm × 60 cm), invert slope 0.06%) in 'Le Marais' catchment (Paris) as a result of no inlet control at street level (photograph Mathieu Ahyerre).

Figure 2.1.18. Gross solids left on a banquette in Brussels following a storm (photograph Michel Verbanck).

In a toilet usage survey within households, asking respondents to record their habits, (Friedler *et al.*, 1996), 29 items were identified, of which tampons were found to be the most commonly flushed, at 23 % of all items (being flushed by 11 % of the households surveyed). In the survey, no respondents admitted to flushing sanitary towels. The survey also indicated that females flushed 1.6 times more 'sanitary refuse' than males. Only once was the flushing of condoms admitted. A number of studies have been carried out at outlet points in larger sewers, where CSOs may be located, or where treatment plants receive all of the flow. Screenings collected at treatment plants, are reportedly some 0.01 m^3 to 0.03 m^3 per cubic metre of sewage in a day (Sidwick, 1984). Some 90 % of these were rags and paper and only some 5 % plastic. During storms these volumes rose by up to seven times. The extreme variability of both amounts and constituents has been reported by a number of studies.

Europeans and North Americans produce an average of 100–130 g/capita/day of faeces, with an average frequency of one defecation a day (Laak, 1974; Feachem *et al.*, 1993). In a detailed survey in the UK (Friedler *et al.*, 1996), weekday defecation was determined at a daily rate of 0.87, but with a discernible diurnal pattern being established (see Figure 2.1.19). The most significant feature of this pattern is the lack of the evening peak which is characteristic of WC usage. In the same study the daily amount of toilet paper used was also established as 11.7 sheets/capita amounting to 6.4 g/capita/day. Outline calculations based on manufacturers' data in the UK result in a figure of 19.4 g/capita/day (National Bag It and Bin It Campaign). A diurnal pattern was present (Figure 2.1.20).

Such data are very dependent upon cultural and institutional practice, and representative only of people in countries with a particular WC usage paradigm. In many parts of Greece for example, any attempt to dispose of large sanitary solids via the WC would result in blockage of the drainage system, and even toilet paper disposal via this route is discouraged. In South Asia, the use of sand for washing dishes results in the input of this into drainage systems together with more 'gross' organic (vegetable) waste than occurs in occidental countries, via kitchen disposal units.

In view of the need to transport the large solids put into WCs, countries like the UK and USA also utilise large volumes of WC flush water (typically 9 litres). In the UK, the water used in this way comprises typically some 35 % of all domestic water supplied. In the USA, the figure is a lower percentage at about 30 % (Field *et al.*, 2000); however, overall water consumption is much higher, and thus WC flushing represents a similar volume to the UK. Given that most WC flush water is currently supplied as fully potable, it has undergone expensive and unnecessary treatment, simply to be used as a transport medium for many large solids which could be disposed of directly into the solid waste route. For example, Swaffield and Wakelin (1995) describe studies which demonstrate that with minor re-design of WCs (low flush toilets), flushes could be reduced to 4 litres, with no risk of downstream blockage of drains. Where standard WC outlets (100 mm diameter gravity) are not feasible, *sanipump* macerators may be utilised. These pump macerated solids via smaller pipes (typically 32 mm), and can collect wastewater from a range of domestic fittings (sinks, baths etc). Such units are not recommended for flushing 'hard' objects containing cotton wool fabrics, e.g. condoms, tampons and wet wipes. No consistent studies have been reported for wastewater systems utilising macerators.

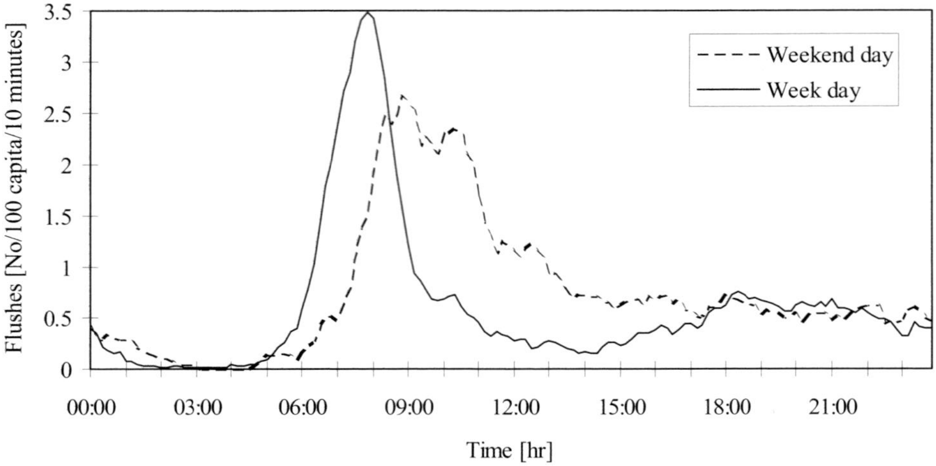

Figure 2.1.19. Diurnal pattern of WC faecal related flushes (Friedler *et al.*, 1996).

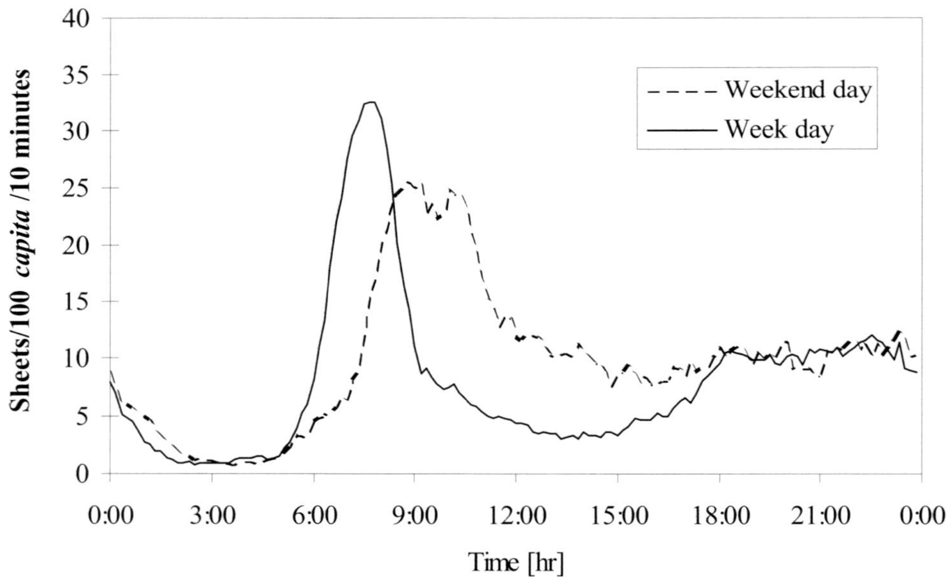

Figure 2.1.20. Diurnal pattern of toilet paper disposal via the WC (Friedler *et al.*, 1996).

Approximately 1 g/capita/day of solid waste material (other than faeces) is screened at UK wastewater treatment plants (Haddon, 1995). Sidwick (1984) reports a screening production rate of 0.01–0.03 m^3/1000 capita/day which can be interpreted as 0.05–0.15 sanitary items/capita/day. Data from Jefferies and Ashley's (1994) study of gross solids discharge in combined sewers can be interpreted to give a rate of 0.05 visible items /capita/day. The average disposal rate reported by Friedler *et al.* (1996) was 0.15 refuse items/capita/day, 72 % of which was due to female WC usage. The most common item of refuse (23 % of those reported) was the tampon. Milne *et al.* (1996) report that the volatile fraction of gross solids retrieved from combined sewers varies from 24 to 81 %. It is significant that adding these to sewer systems reduces their volatility from a level approaching 100 %, and also increases their positive settling velocity. Using laboratory studies, Saul *et al.* (1993) determined the settling velocity of clean sanitary refuse items, and found that of the major items, condoms, sanitary towels and cotton buds (plastic) had a mean rise velocity of up to 19 cm/s. It was concluded from this that CSO structures require screens to prevent such solids being discharged with spill flows. However, field studies show that the behaviour of these solids may be quite different due to absorption of the finer particles in suspension, with a preponderance of gross solids being observed in the lower part of the flow. A study was carried out in Montreal, Canada on the vertical velocity of floatable solids in CSOs (mainly plastics, cigarette butts, sanitary applicators, condoms, etc.) (Gagné *et al.*, 1998). The vertical upward velocities for 51 experimental classes of floatables ranged from 0.01 to 0.51 m/s, with a median value of 0.08 m/s. In Montreal, 80 % of the sampled CSO floatables had upward velocities above 0.07 m/s. In the UK, Saul *et al.* (1998), report on studies carried out categorising gross solids from the outlet of a large sewer entering Wigan sewage works, for which the retention time is several hours. Of the solids collected, some 50 % was found to be 'mush', being of small size and generally unidentifiable, 18 % of the solids were identifiable particles less than 6 mm in size, and the remaining solids (32 %), gross sanitary items and vegetable matter etc. The most prevalent was vegetable matter, followed by stringy material, tissue paper, tampons and toilet paper. Fractionation of the latter into 41 different categories, allowed particle size and settling velocity classes to be determined. Settling velocities ranged from -0.33 m/s to 0.66 m/s.

Gross solids and sanitary refuse are one of the main sources of aesthetic pollution of watercourses due to combined sewer overflows during storm events, rather than a real contribution to chemical, biological and ecological pollution (Gujer and Krejci, 1987). However, as the public is more and more sensitive to the aspect and to the amenity of watercourses, a number of studies have been undertaken in order to improve knowledge about solids control for aesthetic reasons, including their production, characteristics and transport along sewer systems (e.g. Palfrey and Green, 1994). Currently there is a lot of work directed toward the efficient retention of these solids in sewers, at points of discharge, such as CSOs (Section 3.5.1.3).

As well as being environmentally unsightly, gross solids are also responsible for sewer blockages and obstructions. For example, the 56,000 tons of plastic and other sanitary items flushed in the UK every year are held to be responsible for 75 % of all sewer blockages (Section 3.2.1).

For use in countries with limited water supplies and developing countries and areas where main sewers are impractical, a variety of alternative sewerage systems have been developed which utilise solids removal systems prior to the sewer (Section 4.1). For example, in the USA, Small Diameter Gravity Sewers (SGDS) of only 50 mm are used (Oris, 1986). In Africa, settled sewerage systems are based on a similar idea using septic tanks to remove the solids. In such countries, sewers may in fact be open channels, and also therefore susceptible to the ingress of many more large gross and other solids (Section 3).

2.1.4.3 Kitchen waste inputs

In the USA and certain other countries, kitchen sink grinders (macerators) allow the introduction of more organic solids into the sewer system than is usual in countries like Denmark or France, where these are not used. An example of the effects of garbage grinders is given in Table 2.1.20, as summarised from USEPA (1992).

Table 2.1.20 illustrates the importance of kitchen grinders, as these generate typically the highest solids loads per capita and the greatest concentration of solids in the flow due to the relatively low water volume discharged concurrently (8 l/cap/day in USA). Garbage grinders are now becoming popular in other developed countries, and supposedly will take a range of wastes, including plastics. A review of garbage disposers and the consequent particles introduced into the sewer, indicates that the garbage contributes some 21–49 % of suspended solids and 22–77 % of BOD loads (Galil and Shpiner, 2001). Particle sizes larger than 1 mm and with relative densities greater than 2, would be expected to deposit in conventional designed 'self-cleansing' sewers.

Table 2.1.20. Pollutant loads from residential sources discharging to sewer (adapted from USEPA, 1992)

Parameter	Garbage grinders		Toilets		Basins, sinks, appliances	
	(g/cap/day)	(mg/l)	(g/cap/day)	(mg/l)	(g/cap/day)	(mg/l)
BOD_5	11–31	2380	7–24	260	25–39	260
Suspended solids	16–44	3500	13–37	450	11–23	160
Nitrogen	0.2–0.9	79	4.1–16.8	140	1.1–2.0	17
Phosphorus	0.1	13	0.6–1.6	20	2.2–3 4	26

In Europe, one discouragement to putting kitchen organic waste, and certain other sanitary solids, into the solid waste route, the new Directive on Landfill, sets targets limiting the amount of 'organic' waste which can be disposed to landfill sites. It is possible that in order to reach the specified targets, waste operators may encourage the public to dispose of more of such waste into the sewerage system. Danish building standards specify that preliminary treatment, typically using kitchen outlet 'fat traps', can be required for household discharges where they exceed: 300 mg/l SS; 50 mg/l fats; 30 mg/l oils. However, these standards are typically not applied now that all areas have advanced wastewater treatment plants.

2.1.5 Solids from infiltration and sewer decay or degradation and wrong connections

Infiltration along the length of sewers can introduce solids from the surrounding soil. The solids characteristics depend on the nature of the ground. Usually, this source of solids is not constant with time, as blockage of the ingress can occur periodically, or because of the variation in groundwater levels. For example, separate sanitary sewers serve a large portion of the sewered population in the United States. These sewers are of smaller diameter than combined or storm sewers, and serve residential, commercial and industrial areas. Whilst sanitary systems are not specifically designed to carry stormwater per se, stormwater and groundwater do enter these systems and is a common and complicated problem for sewer owners. So common in fact, that the design of sanitary sewers in the US must include capacity for Inflow and Infiltration (I/I), which may actually exceed pure sanitary flowrates. 'Inflow' is defined as wrongly connected inputs, whilst infiltration comes from groundwater sources. Without I/I, peak wastewater flows in sanitary sewers have a low peaking factor since this use is relatively constant. It is estimated that I/I accounts for almost one half of the average flows to WTPs in the United States (Pitt *et al.*, 2000). Excessive I/I can be traced to poor construction techniques and materials and/or poor enforcement of policies regarding illicit connections. Methods for determining I/I are given in WEF/ASCE (1994).

In a recent survey in the UK, White *et al.* (1997) found that of some 1646 catchments surveyed, 28 % reportedly had more than 25 % of the flow coming to treatment as arising from infiltration sources, whereas, 9 % of the catchments had some 50 % of the flow deriving from infiltration. When looking at 38 catchments in detail, it was found that 26 of these had medium levels of infiltration and 12 had 'high' levels of infiltration.

The infiltration was identified as responsible for overloading the WTP at times for thirty-two of these catchments. The commonest sources of infiltration (in order) were:

- normal and high groundwater levels
- tidal waters
- watercourses
- leaking water mains.

Unfortunately, little quantitative information exists regarding the solids associated with infiltration.

Both types of wrong connections (storm flows connected to sanitary sewers, sanitary inputs connected to storm sewers) can have significant consequences: pollution of receiving waters, flooding, overloading of treatment plants, unplanned in-sewer sediment problems, increased operation and maintenance costs. Common non-stormwater entries include: sanitary wastewater; automobile maintenance and operation waste products; laundry wastewater; household toxic substances and pollutants; accident and spill waste streams; runoff from excessive irrigation; and industrial cooling water, rinse water, and other process wastewater. These connections are so far largely uninvestigated. In a recent case, a 'sustainable' system had been built as part of a housing area comprising some 200 houses in the city of Deventer in the eastern part of the Netherlands. It comprised the reuse of storm water for toilet flushing, garden watering and as a main source of water for washing. Since occupation, at least one case of cross-contamination was found due to wrong connections. The real wastewater (to be more precise, the toilet flushes) had been discharging directly into the collection/reuse cistern. As a result, at least one household washed clothing using faecal-contaminated water for longer than 6 months. This was not noticed due to the use of automatic washing machines and the quality of modern washing powders (Schaart, 2003).

Another source of solids is the decay of the sewer itself, when the fabric is progressively eroded or chemically attacked. The decay of the sewer fabric is generally of the fine mineral leading to small particles. Progressive decay will result in larger solids being eroded, causing deposits and blockages.

The following factors are important:

- *chemical attack*
  H_2S generation due to anaerobic conditions (see Section 2.2);
  high temperatures and/or acid/base trade effluents;
- *structural problems*
  very high flow velocities, collapse, breaking or cracking of some parts of the sewer system due to soil imposed loads;
  movements, traffic, accidents, poor material quality or construction.

2.1.6 Sediment ingress from natural water bodies

Where sewer discharges are hydraulically 'free', the natural water body receiving the discharge cannot contribute to any in-sewer sediment. However, where outlet drowning occurs, hydraulic effects may change upstream conditions (backwater influence, etc., see Section 2.3.2). If the sewer has an open outlet, then it may be possible for reverse flows to occur, in which case solids can be introduced from the 'natural' water body into the sewer. This is particularly apparent where there are marine outfalls, for which solids, marine debris and animals can be found in the lower parts of the outfall, either due to settlement of solids from the upstream flow, or brought in by intrusion of sea water (Wilkinson, 1984, 1992 ; Burrows *et al.*, 1996). The widespread use of appropriately designed devices (such as valves preventing reverse flow and outfall diffusers) which disconnect the sewer network from receiving waters when the water stage in the latter is high is a good means of reducing this problem (Brombach 1990; see also Section 4.3). Where the sewer terminates in a designed outfall, with diffusers, salt water intrusion is virtually inevitable and marine debris deposition is a problem. In one study, by the UK Water Research Centre (Ricketts and McNamara, 1987), particles encountered were 0.20–3.35 mm in size, with settling velocities of 2.4-36.2 cm/s, where velocities in the outfall pipe ranged from 0.6 to 5.0 m/s. The problem is usually addressed using periodic flushing cycles (Ali *et al.*, 1995). For these outfalls, the practice of flooding prior to commissioning also leaves a residue of unflushed sediments, despite purging attempts (Wood *et al.*, 1993). In Europe the need to extend the many existing short sea outfalls to comply with Directives will make these problems more apparent, as so far there are no means of avoidance other than flushing.

When sediment ingress from natural water bodies does occur, it may be significant, with large, effectively unlimited amounts of sediments being introduced into the sewer network. The generally fine-grained, cohesive nature of the materials typically available in natural water bodies, makes transport into the sewer system very easy. With a water level imbalance between receiving water and sewer, deposition of these fine riverine/coastal suspensions can occur within the drainage system as a result of the very low flow velocities (actually dropping to zero, and reversing in some instances). For combined sewers, this problem is exacerbated by the encouragement for deposition of fats and greases at the same time (Section 3).

Fine sediments found in sewers as a result of such river inputs have notably been experienced for years in medium-sized cities along the river Scheldt in northern Belgium. Rivers and estuaries such as the Scheldt which experience a strong tidal influence appear to be responsible for the most severe documented cases.

A large part of the drainage system of Buenos Aires, Argentina, has been reported to suffer from severe blockage due to introduction of solids from the nearby mouth of the La Plata river, and the recently constructed long sea outfalls in Hong Kong, even with duckbill diffusers, have suffered from severe in-sewer sedimentation in the diffuser pipe sections. Analysis of the sediment deposits in sewers in the ancient part of Hamburg city, which suffer from tidal ingress, showed that accretion of the material was enhanced by the successive intermixed layers of sewer grit and fine, riverine cohesive material. For both Hamburg (Blumberg and Bauer, 1984) and Buenos Aires (Botzung, 1996) new dredging proprietary devices had to be designed in order to cope with these large amounts of intermixed, concrete-like sediments (see Figure 4.2.23).

2.1.7 Commercial, industrial and construction activities

Depending on the type of activity or industry, the solids from these sources can be very different and should be measured for each specific site. The differences concern all parameters: size, density, organic fraction, associated pollutants, concentrations, etc. However, Table 2.1.21 provides illustrative values of suspended solids concentrations and/or loads for various activities. Some values for sizes and attached pollutants concentrations have also been given in the above paragraphs.

Construction activities can significantly influence the solids loads introduced into sewer systems. For example, Broeker (1984) indicated that surface washoff loadings increased up to 300 % because of construction sites. In the USA, erosion losses from uncontrolled construction sites have been measured to be in the range of 30 to more than 100 kg/ha/yr.

Table 2.1.21. Values of suspended solids concentrations and loads for various industrial activities (from Blanic, 1973 and Ramade, 1998)

Industry			TSS in effluents	Water consumption
Agro-alimentary industries	Slaughter house	Bovines and large carcases	9–15 g/kg of carcase	6–9 l/kg of carcase
		Porcines carcases	6–13 g/kg of carcase	5–11 l/kg of carcase
		Poultry carcases	3.5–5.5 g/kg of carcase	7–9 l/kg of carcase
		General mean concentrations	500–1500 mg/l	
	Tanneries and taweries		75–150 kg/ton of skin	2—120 m^3/ton of skin
		Milk and yoghurts	50 mg/l	1–25 l/l of milk
		Cheese	100–500 mg/l	100 l/l of milk
		Milk powder and butter	30 mg/l	1–3 l/l of milk
	Breweries		500–1000 mg/l	200–700 l/hl of beer
	Potatoes industry		20–200 kg/ton of potatoes	2–6 m^3/ton of potatoes
	Vegetable canning industry		200–1000 mg/l	
	Distillery		1500–5000 mg/l	
Textile industries	Wood pretreatment industry		20–30 kg/ton of raw wood	3–7 m^3/ton of raw wood
	Acrylic		30–400 mg/l	35 m^3/ton
	Wood		30–400 mg/l	70 m^3/ton
	Cotton		30–400 mg/l	100 m^3/ton
	Sponge		30–400 mg/l	200 m^3/ton
Paper industry			10–40 kg/ton of pulp	10–100 m^3/ ton of pulp
			2–25 kg/ton of paper	3–30 m^3/ton of paper
	General mean concentrations		200–3000 mg/l	
Siderurgy	Cold laminating		0.2 g/l of effluents	1 l/ton of coke
	General mean concentrations		1000–3000 mg/l	
Petrol refinery	General mean concentrations		50–100 mg/l	

The nature of these solids depends both upon the materials used for the construction and on the nature of the ground (silt, clay, gravel, etc.). The solids are usually predominantly mineral, with low or reduced associated pollutant loadings. However, they can significantly contribute locally to the solids load entering into sewer systems and to the sewer sediment build-up.

The relative importance of construction activities as a source of solids also depends on construction conditions and habits:

- protection of the construction site (in many countries, especially in densely urbanised areas, geotextile 'skirts' are used to control washoff of solids from sites, Section 4.1);
- storage arrangements for construction and excavated materials;
- protection arrangements if any, for nearby gullies;
- pumped groundwater directly discharged into the sewer system or stored and settled before discharge, etc.

In addition, elevated sediment discharges occur for several years after development has been completed as the landscaping matures.

2.1.8 Developing countries

Little consistent data have been published about sewer solids in developing countries. However, some data exist in unpublished literature, especially in consulting reports. Nevertheless, many difficulties remain:

- data are difficult to obtain.
- measurement procedures and analytical standards are not always known or reliable.
- variability from one study to another is very great; this reflects very different local urban conditions, including the density of the sewer system, the structural state of roads and sewer systems, the percentage of asphalted or impervious road, the highly variable population density, the proportion of inhabitants connected to the sewer system, the percentage of pervious and/or rural surfaces connected to the sewer system.
- notionally separate systems (typically open storm drains) are invariably rife with wrong connections and are used for refuse disposal (Kolsky, 1998).

As an example, Table 2.1.22 gives sewage quality data for different experimental catchments in the area around São Paulo, Brasil. Dry weather values appear very low in comparison with data from Table 2.1.18, whilst storm loads and or concentrations are similar or significantly higher in the case of the Vila Calu catchment.

Table 2.1.22. Example of mean solids concentrations and loads measured at the outlet of different catchments around São Paulo, Brasil (from PRIME, 1998)

Catchment	Urbanisation	Dry weather	Storm flow	
		Concentration (mg/l)	Concentration (mg/l)	Load (kg/ha/year)
Sem Nome	urban, totally urbanised,	10	186	1732
Rio das Pedras	urban totally urbanised	80–200	863	8035
Vila Calu	accelatated urbanisation, few pervious roads and sewer systems	100–200	1771	12719
Ribeirao Itaim	semi urban	15	501	203
Rio Embu Mirim	semi urban	12	20	105
Golfe Club	rural	10	124	495
Ribeirao Itaquaxiara	rural	12	74	294

Figure 2.1.21 illustrates how the Indian data, from Kolsky (1998), reveal that the solids encountered in countries like India are considerably coarser than those found both in Europe and Malaysia (Ab Ghani *et al.*, 2000). This is also illustrated in Figure 2.1.22. Clearly, open drains (which this shows for India) allow easier ingress of the larger solids, but other factors include: the poor garbage handling facilities; higher rates of road surface erosion and more construction debris in India. In Malaysia, which is more developed than India, the data averaged for five cities shows a greater similarity to European solids, despite being collected on open (storm) drains with many wrong connections. In Malaysia, dumping of refuse in storm drains is virtually non-existent.

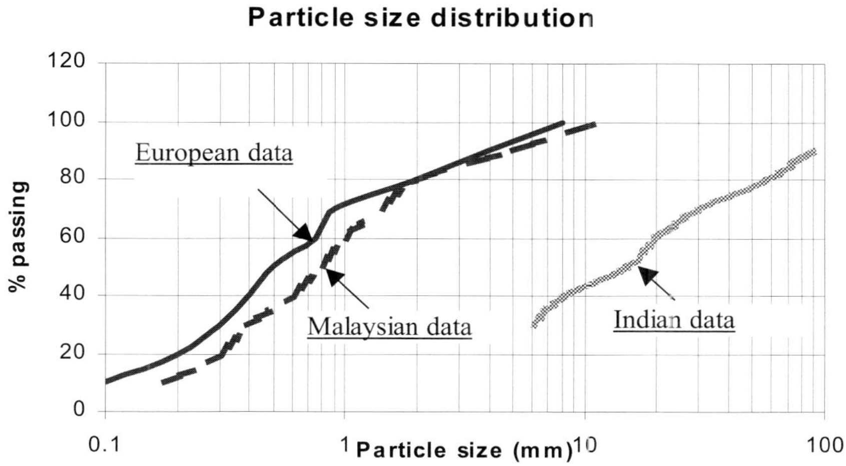

Figure 2.1.21. Drain solids size distribution comparisons between Indore (India) (Kolsky, 1998), Malaysia (Ab Ghani *et al.*, 2000) and Europe.

These results illustrate the necessity for specific measurement campaigns, since the wide variability in solids in drainage systems in the developing world mean that mean values cannot be applied.

Extreme events can also cause the greatest sewer solids impacts. Whilst it is likely that events such as flooding (from whatever cause) also bring solids into sewers and other drainage systems, these are specific to the site and the catastrophic event. Figure 2.1.23 illustrates the result of one such event and of not cleansing in Algeria, where the City of Bab El Oued was severely damaged by upland flood waters. The sewer is being cleaned by workers from Marseille.

Figure 2.1.22. Open storm drains are sewers in Mumbai, India, hence litter is the most predominant solids (photograph Richard Ashley).

Figure 2.1.23. Solids in a sewer in Bab el Oued (Alger, Algeria) following the catastrophic floods of 10 November 2001 (photograph Dominique Laplace).

2.1.9 References

Aalderink, R.H., Van Duin, E.H.S., Peels, C.E. and Scholten, M.J.M. (1990) Some characteristics of run-off quality from a separated sewer system in Lelystad, The Netherlands. *Proceedings of the 5th International Conference on Urban Storm Drainage*, Osaka, Japan, July 1990, 427–432.

Ab Ghani, A., Zakaria, N.A., Kassim, M. and Nasir, B A. (2000) Sediment characteristics of urban drains in Malaysian Cities. *Urban Water*, 2(4), 335–341.

Ali, K.H.M., Burrows, R. and Wose, A.E. (1995) Sedimentation in long sea outfalls. *Wat. Sci. Tech.*, 32(2), 123–132.

Alley, W.M. (1981) Estimation of impervious area washoff parameters. *Water Resources Research*, 17(4), 1161–1166.

Alley, W.M. and Smith, P.E. (1981) Estimation of accumulation parameters for urban runoff quality modelling. *Water Resources Research*, 17(6), 1657–1664.

APWA (1969) *Water Pollution Aspects of Urban Runoff.* American Public Works Association, Water Pollution Control Research Series WP-20-15, Federal Water Pollution Control Administration, January 1969.

Armitage, N. and Roseboom, A. (2000) The removal of urban litter from stormwater conduits and streams: Paper 1 – The quantities involved and catchment litter management options. *Water SA*, 26(2), 181–187.

Artières, O. (1987) *Les dépôts en réseau d'assainissement: origine, caractéristiques pollution, transport.* Thèse de doctorat, Université Louis Pasteur, Strasbourg, 214 pp.

Ashley, R.M. and Crabtree, R.W. (1992) Sediment origins, deposition and build-up in combined sewer systems. *Wat. Sci. Tech.*, 25(8), 1–12.

Ashley, R.M. and Souter, N. (1999) Urban Drainage – The Human Dimension. Keynote paper to 8th International Conference on Urban Drainage, Sydney, Australia.

Ball, J., Jenks, R. and Aubourg, D. (1996) Dry weather build-up of constituents on road surfaces. *Proceedings of 7th International Conference on Urban Storm Drainage*, Hannover, Germany, 9-13 September 1996, 2, 785–790.

Bellinger, E.G. (1982) The character and disposal of motorway runoff water. *Water Pollution Control*, 8(3), 372–389.

Bertrand-Krajewski, J.-L. (1991) Modélisation des débits et du transport solide en réseau d'assainissement – Etude bibliographique. Rapport ENITRTS/Lyonnaise des Eaux-Dumez, Strasbourg, France, April 1991, 207 pp.

Bertrand-Krajewski, J.-L. (1993) Pollution des rejets urbains par temps de pluie. Synthèse générale. Rapport *CIRSEE Lyonnaise des Eaux*, Le Pecq, November 1993, 137 pp.

Bertrand-Krajewski, J.-L., Briat, P.P. and Scrivener, O. (1993) Sewer sediment production and transport modelling: a literature review. *Journal of Hydraulic Research,* 31(4), 435–460.

Blanic, R. (1973) *Eaux urbaines et eaux industrielles – Traitement et épuration.* Supplément à la revue Industrie Minérale, March 1973, 352 pp.

Blumberg, D. and Bauer, W. (1984) Beseitigung von Ablagerungen in Abwasserkanälen grosser Durchmesser. *Korrespondenz Abwasser*, 12/84, 1063–1072

Botzung, P. (1996) Scour power. *Water Quality International*, Sept.–Oct., 10–11.

Briat, P. (1995) Développement du modèle HYDROPOL. *Rapport Interne Lyonnaise des Eaux*, Bordeaux.

Broeker, H.W. (1984) Impact of deposition on sewer operation. *Proceedings of 3rd International Conference on Urban Storm Drainage*, Göteborg, Sweden, 3, 879–887.

Brombach, H.H. (1990) Equipment and instrumentation for CSO control. US Engg. Foundation Conf. Proc. Urban Stormwater quality enhancement: source control, retrofitting and combined sewer technology. Davos, Switzerland, 459–478.

Bujon, G. (1988) *Prévision des débits et des flux polluants transités par les réseaux d'égouts en temps de pluie.* Le modèle FLUPOL. La Houille Blanche, 1, 11–23.

Burrows, R., Ali, K.H.M., Davies, P.A. and Wose, A.E. (1996) Purging studies on a model sea outfall diffuser. *Proc. UK Inst. Civ. Engrs, Water Maritime and Energy*, 118.

Butler, D. and Clark, R.B. (1995) *Sediment management in urban drainage catchments.* CIRIA Report no. 134, CIRIA, London.

Butler, D. and Karunaratne, S.H.P. (1995) The suspended solids trap efficiency of the roadside gully pot. *Water Research*, 29(2), 719–729.

Butler, D. and Memon, F. (1999) Dynamic modelling of roadside gully pots during wet weather. *Water Research*, 33(15), 3364–3372.

Butler, D., Thedchanamoorthy, S. and Payne, J.A. (1992) Aspects of surface sediment characteristics on an urban catchment in London. *Wat. Sci. Tech.*, 25(8), 13–19.

Butler, D., Xiao, Y., Karunaratne, S.H.P. and Thedchanamoorthy, S. (1995) The gully pot as a physical, chemical and biological reactor. *Wat. Sci. Tech.*, 31(7), 13–24.

Chamberlain, A.C. (1986) Deposition of gases and particles on vegetation and soils. *Air pollutants and their effects on terrestrial ecosystems*, edited by Legge, A.H. and Krupa, S.V., J. Wiley, New York, 189–210.

Chebbo, G. (1992) *Solides des rejets pluviaux urbains: caractérisation et traitabilité.* Thèse de doctorat, ENPC, Paris, March 1992, 450 pp.

Chebbo, G., Musquère, P., Milisic, V. and Bachoc, A. (1990) Characterization of solids transferred into sewer trunks during wet weather. *Wat. Sci. Tech.*, 22(10/11), 231–238.

Clegg, S., Forster, C.F. and Crabtree, R.W. (1992) Sewer sediment characteristics and their variation with time. *Environmental Technology*, 13, 561–569.

Clegg, S., Forster, C.F. and Crabtree, R.W. (1993) An examination of the ageing of gully pot sediments. *Environmental Technology*, 14, 453–461.

Collwill, D.M., Peters, C.J. and Perry, R. (1985) *Motorway runoff: the effect of drainage systems on water quality.* TRRL Research Report no. 37, UK.

Crabtree, R.W. (1989) Sediment in sewers. *Journal of the Institution of Water and Environmental Management*, 3(6), 569–578.

Crabtree, R.W., Ashley, R.M. and Saul, A.J. (1991) *Review of Research into Sediments in Sewers and Ancillary Structures*. FWR Report no. FR205, Foundation for Water Research, UK.

Dalrymple, R.J., Hodd, S.L. and Morin, D.C. (1975) *Physical and Settling Characteristics of Particulates in Storm and Sanitary Wastewaters.* US EPA Report no. 670/2-75-011, Cincinnatti, Ohio, USA, April 1975, 41 pp.

Deletic, A. (2001) Modelling of water and sediment transport over grassed areas. *Journal of Hydrology*, 248, 168–182

Deletic, A., Ashley, R.M. and Rest, D. (2000) Modelling input of fine granular sediment into storm drainage systems. *Water Research*, 34(15), 3836–3844.

Deletic, A., Maksimovic, C. and Ivetic, M. (1997) Modelling of storm washoff of suspended solids from impervious surfaces. *Journal of Hydraulic Research*, 35(1), 99–118.

Deutsch, J.-C. and Grange, D. (1986) Qualité des eaux de pluie: comparaison des flux polluants avant et après ruissellement. *Actes des 19èmes Journées de l'Hydraulique, rapport II-2, Société Hydrotechnique de France*, Paris, 5 pp.

Ellis, J.B. (1986) *Pollutional Aspects of Urban Runoff. Urban Runoff Pollution*, NATO ASI Series vol. G10, edited by H.C. Torno, J. Marsalek and M. Desbordes, Springer-Verlag, Berlin, 1–38.

Ellis, J.B. and Revitt, D.M. (1982) Incidence of heavy metals in street surface sediments: solubility and grain size studies. *Water, Air, and Soil Pollution*, 17, 87–100.

EPA (1977) *Control of Reentrained Dust from Paved Streets*. US EPA Report no 905/9-77-007, Kansas City, Missouri, USA.

Fair, G.M., Geyer, J.C. and Okun, D.A. (1966) *Water and Wastewater Engineering. Volume 2: Water purification and wastewater treatment and disposal.* Wiley, USA.

Feachem, R.G., Bradley, D.J., Garelick, H. and Mara, D.D. (1993) Elements and health risks of excreta and wastewater. *Sanitation and Disease: Health aspects of excreta and wastewater management*, edited by R.G. Feachem, D.J. Bradley, H. Garelick and D.D. Mara, World Bank Studies in Water Supply Sanitation 3, John Wiley, New York, USA, 3–22.

Field, R. (1974) Design of a combined sewer overflow regulator/concentrator. *Journ. WPCF*,.46(7), 1722–1741.

Field, R., Heaney, J P. and Pitt, R. (2000) *Innovative Urban Wet-Weather Flow Management Systems.* Technomic Pub. Lancaster Pa. ISBN 1-56676-914-0.

Fletcher, I.J. and Pratt, C.J. (1981) Mathematical simulation of pollutant contributions to urban runoff from roadside gully pots. *Proceedings of 2nd International Conference on Urban Storm Drainage*, Urbana, Illinois, USA, 116–124.

Fletcher, I.J., Pratt, C.J. and Elliot, G.P. (1978) An assessment of the importance of roadside gully pots in determining the quality of storm runoff. *Urban Storm Drainage*, edited by Helliwell, Pentech Press, 586–602.

Förster, J. (1990) Roof runoff: a source of pollutants in urban storm drainage systems. *Proceedings of 5th International Conference on Urban Storm Drainage*, Osaka, Japan, July 1990, 469–474.

Förster, J. (1996) Heavy metal and ion pollution patterns in roof runoff. *Proceedings of 7th International Conference on Urban Storm Drainage*, Hannover, Germany, 9–13 September 1996, 1, 241–246.

Friedler, E., Brown, D.M. and Butler, D. (1996) A study of WC derived solids. *Wat. Sci. Tech.*, 33(9), 17–24.

Friedler, E. and Butler, D. (1995) Quantifying the inherent uncertainty in the quantity and quality of domestic wastewater. *Proceedings of the International Symposium on Uncertainty, Risk and Transient Pollution Control. Acute risks to the Environment*, Exeter, UK, July 1995.

Gagné B., Cigana, J., Couture, M., Meunier, C. and Comeau, Y. (1998) Determination of the vertical velocity distribution of floatables in CSOs. *Proceedings of Novatech 98*, Lyon, France, 1, 355-360.

Galil, N. and Shpiner, R. (2001) Additional pollutants and deposition potential from garbage disposers. *J. CIWEM*, 15, March.

Göttle, A. (1978a) Atmospheric contaminants, fallout and their effects on storm water quality. *Progress in Water Technology*, 10(5/6), 455–467.

Göttle, A. (1978b) *Ursachen und Mechanismen der Regenwasserverschmutzung. Ein Beitrag zur Modellierung der Abflussbeschaffenheit in städtischen Gebieten.* Berichte der Technischen Universität München, Heft nr 23, Munich, Germany, 313 pp.

Grottker, M. (1990) Pollutant removal by gully pots in different catchment areas. *The Science of the Total Environment*, 93, 515–522.

Grottker, M. and Hurlebusch, R. (1987) Mitigation of stormwater pollution by gully pots. *Proceedings of 4th International Conference on Urban Storm Drainage*, Lausanne, Switzerland, 66–67.

Gujer, W. and Krejci, V. (1987) Urban storm drainage and receiving waters ecology. *Proceedings of 4th International Conference on Urban Storm Drainage*, Lausanne, Switzerland.

Haddon, M. (1995) Out of sight? Out of your mind. *Water Bulletin*, 642, 8–10.

Heaney, J.P. and Huber, W.C. (1973) *Stormwater Management Model. Refinements, testing and decision-making.* Department of Environmental Engineering and Sciences, University of Florida, Gainsville, Florida, USA.

Hedley, G. and Lockley, J.C. (1975) Quality of water discharged from an urban motorway. *Journal of Water Pollution Control Federation*, 74(6).

Hémain, J.-C. (1985) La pollution du ruissellement pluvial urbain. Campagne de mesure expérimentale française 1980–1982. *Rapport de synthèse. STU (Service Technique de l'Urbanisme)*, Paris, August 1985, 80 pp.

Henderson, R.J. and Moys, G.D. (1987) Development of a sewer quality model for United Kingdom. *Proceedings of 4th International Conference on Urban Storm Drainage*, Lausanne, Switzerland, 201–207.

Henze, M., Harremoës, P., Jansen, J.L.C. and Arvin, E. (1995) *Wastewater Treatment: Biological and chemical processes*. Springer-Verlag, Berlin, Germany (ISBN 3-540-58816-7).

Hoffman, E.J., Mills, G.L., Latimer, J.S. and Quinn, J.G. (1984) Urban runoff as a source of polycyclic aromatic hydrocabons to coastal waters. *Environmental Science and Technology*, 18(8), 580–587.

Horan, N.J. (1990) *Biological Wastewater Treatment Systems. Theory and operation.* John Wiley, New York, USA.

Hough, A.M. (1988) Atmospheric chemistry at elevated sites. A discussion of the processes involved and the limit of the current understanding. *Acid Deposition at High Elevation Sites*, edited by M.H. Unsworth and D. Fowler, NATO ASI Series, Kluwer, Dordrecht, 1–45.

Huber, W.C (1986) *Deterministic Modeling of Urban Runoff Quality. Urban runoff pollution*, NATO ASI Series vol. G10, edited by H.C. Torno, J. Marsalek and M. Desbordes, Springer-Verlag, Berlin, 167–242.

Huber, W.C. and Dickinson, R.E. (1992) *Storm Water Management Model, version 4: user's manual.* US EPA Report no EPA-600/3-88/001a (NTIS PB88-236641/AS), Athens, Georgia, USA, October 1992, 569 pp.

Huber, W.C., Heaney, J.P., Aggidis, D.A., Dickinson, R.E. and Wallace, R.W. (1981) *Urban Rainfall-Runoff-quality Database.* US EPA Report no EPA-600/2-81-238 (NTIS PB82-221094), Cincinnatti, Ohio, USA.

Jefferies, C.J. and Ashley, R.M. (1994) The behaviour of gross solids in sewer systems. *Journal of the European Water Pollution Control Association,* 4(5), 11–17.

Jewell, T.K. and Adrian, D.D. (1978) SWMM stormwater pollutant washoff functions. *Journal of the Environmental Engineering Division*, 104(5), 1036–1040.

Kari, F.G. and Herrmann, R. (1989) Abspülung von organischen Spurenschadstoffen und Schwer-metallen aus einem städtischen Einzugsgebiet: Ganglinienanalyse, Korngrössenzuordnung und Metallspeziesauftrennung. *Deutsche Gewässerkundliche Mitteilungen*, 33(5/6), 172–183.

Karunaratne, S.H.P. (1992) *The Influence of Gully Pot Performance on the Entry of Sediment into Sewers.* Unpublished PhD thesis, South Bank University, London.

Kern, U., Wüst, W., Daub, J., Striebel, T. and Herrmann, R. (1992) Abspülverhalten von Schwermetallen und organischen Mikroschadstoffen im Strassenabfluss. *GWF Wasser Abwasser*, 133(11), 567–574.

Klemetson, S.L. (1985) Factors affecting stream transport of combined sewer overflow sediments. *JWPCF*, 57(5), 390–397.

Kolsky, P. (1998) *Storm Drainage: An engineering guide to the low-cost evaluation of system performance*, IT Publisher, London.

Laak, R. (1974) Relative pollution strength of undiluted waste materials discharged in households and the dilution waters used for it. *Manual of grey water treatment practice*, edited by J.H.T. Winneberger, Ann Arbor Science, Michigan, USA, 68–78.

Leduc, R. and Ouldali S. (1989) Apports en sédiments et envasement des lacs de rétention des eaux de drainage urbain. *Sciences et Techniques de l'Eau*, 22(4), 309–314.

Marsalek, J. (1984) Caractérisation du ruissellement de surface issu d'une zone urbaine commerciale. *Sciences et Techniques de l'Eau*, 17(2), 163–167.

Marsalek, J. (1986) Toxic contaminants in urban runoff: a case study. *Urban Runoff Pollution,* NATO ASI Series vol. G10, edited by H.C. Torno, J. Marsalek and M. Desbordes, Springer-Verlag, Berlin, 39–57.

McElroy, A.D., Chiu, S.Y., Nebgen, J.W., Aleti, A. and Bennett, F.W. (1976) *Loading Functions for Assessment of Water Pollution from Nonpoint Sources.* US EPA Report no EPA-600/2-76-151, Washington D.C., USA.

Memon, FA. (2000) Simulating the influence of roadside gully pots on runoff quality. PhD thesis, Imperial College, London.

Metcalf and Eddy Inc (1991) *Wastewater Engineering: Treatment, Disposal and Reuse.* 3rd edition, McGraw-Hill, New York, USA.

Michelbach, S. and Wöhrle, C. (1992) Settleable solids in a combined sewer system: measurement, quantity, characteristics. *Wat. Sci. Tech.,* 25(8), 181–188.

Milne, D.A., Jefferies, C. and Ashley, R.M. (1996) Pollutional aspects of gross solids and their interaction with sewer sediments. *Wat. Sci. Tech.*, 33(9), 31–37.

Morrison, G.M., Revitt, D.M. and Ellis, J.B. (1990) Metal speciation in separate stormwater systems. *Wat. Sci. Tech*., 22(10/11), 53–60.

Morrison, G.M.P., Revitt, D.M., Ellis, J.B., Svensson, G. and Balmer, P. (1988) Transport mechanisms and processes for metal species in a gullypot system. *Water Research,* 22(11), 1417–1427.

Novotny, V. and Olem, H. (1994) *Water Quality: Preventions, identification and management of diffuse pollution*. Van Nostrand-Reinhold Publishing, New York.

Novotny, V., Sung, H.M., Bannerman, R. and Baum, K. (1985) Estimating nonpoint pollution from small urban watersheds. *Journal of Water Pollution Control Federation*, 57(4), 339–348.

Oris, R.J. (1986) *Small Diameter Gravity Sewers: An alternative wastewater collection method for unsewered communities*. US EPA Report no EPA/600/S2/86/022, EPA, Cincinnatti, Ohio, USA.

O'Sullivan, P. (1990) *A Review of the Performance of Storm Sewage Overflow Structures with Respect to Aesthetic Criteria*. WRc report UM1037, Swindon, UK.

Palfrey, E.A. and Green, M. (1994) *Final Summary Report on Aesthetic Pollution.* Report no FR0440, Foundation for Water Research, UK.

Payne, J. and Osborne, M. (1996) Control of pollution from highway drainage discharges. *Proceedings of 7th International Conference on Urban Storm Drainage*, Hannover, Germany, 9–13 September 1996, 2, 791–796.

Peavy, H.S., Rowe, D.R. and Tchobanoglous, G. (1986) *Environmental Engineering.* McGraw-Hill, New York, USA.

Philippe, J.P. and Ranchet, J. (1987) *Pollution des eaux de ruissellement pluvial en zone urbaine. Synthèse des mesures sur dix bassins versants en région parisienne*. Rapport de recherche LCPC n° 142, Laboratoire Central des Ponts et Chaussées, Paris, February 1987, 76 pp.

Pitt, R. (1979) *Demonstration of Nonpoint Pollution Abatement through Improved Street Cleaning Practices.* US EPA Report no EPA-600/2-79-161, Cincinnatti, Ohio, USA, August 1979.

Pitt, R. (1983) *Urban Bacteria Sources and Control by Street Cleaning in the Lower Rideau River Watershed.* Rideau river stormwater management study technical report. Ontario Ministry of the Environment and Environment Canada, Ottawa, August 1983, 165 pp.

Pitt, R. (1984) *Characterization, Sources and Control of Urban Runoff by Street and Sewerage Cleaning.* Contract no R-80597012, US EPA Report, Cincinnatti, Ohio, USA.

Pitt, R. (1987) *Small Storm Urban Flow and Particulate Washoff Contributions to Outfall Discharges.* PhD thesis, Civil and Environmental Engineering Department, University of Wisconsin, Madison, Wisconsin, USA, November 1987.

Pitt, R., Field, R., Lalor, M. and Brown, M. (1995) Urban stormwater toxic pollutants: assessment, sources and treatability. *Water Environment Research,* 67(3), 260–275.

Pitt, R. and McLean, J. (1986) *Toronto Area Watershed Management Strategy Study. Humber River Pilot Watershed Project.* Ontario Ministry of Environment, Toronto, Ontario, June 1986.

Pitt, R. and Shawley, G. (1982) *A Demonstration of Non Point Source Pollution Management on Castro Valley Creek.* Alameda County Flood Control and Water Conservation District (Hayward, CA) for the Nationwide Urban Runoff Program, US EPA, Washington, D.C., USA, June 1982.

Pitt, R. and Sutherland, R. (1982) *Washoe County Urban Stormwater Management Program. Volume 2: Street particulate data collection and analysis.* Washoe Council of Governments, Reno, Nevada, USA, August 1982.

Pratt, C.J. and Adams, J.W.R. (1981) Sediment washoff into roadside gullies. *Proceedings of 2nd International Conference on Urban Storm Drainage*, Urbana, Illinois, USA, 174–183.

Pratt, C.J. and Adams, J.W.R. (1984) Sediment supply and transmission via roadside gully pots. *The Science of the Total Environment,* 33, 213–234.

Pratt, C.J., Elliott, G.E.P. and Fulcher, G.A. (1987) Suspended solids discharge from highway gully pots in a residential catchment. *The Science of the Total Environment*, 59, 355–364.

PRIME (1998) Relatório síntese: avaliação da poluição por fontes difusas afluente ao reservatório Guarapiranga. *PRIME Engenharia relatorio para SMA/CETESB*, São Paulo, Brasil, May 1998, 46 pp.

Qing, X. (1984) A mathematical model for studying pollutant flows in urban runoff. *Proceedings of 3rd International Conference on Urban Storm Drainage*, Göteborg, Sweden, 3, 957–967.

Ramade, F. (1998) *Dictionnaire encyclopédique des sciences de l'eau.* Editions Ediscience International, Paris, France, 786 pp., ISBN 2-84074-152-0.

Randall C.W., Ellis K., Grizzard, T.J. and Knocke, W.R. (1982) Urban runoff pollutant removal by sedimentation. *Proceedings of the ASCE Engineering Foundation Conference*, ASCE, USA.

Randall C.W., Helsel D.R., Grizzard T.J. and Hoehn R.C. (1978) The impact of atmospheric contaminants on storm water quality in an urban area. *Progress in Water Technology*, 10(5/6), 417–431.

Raudkivi, A.J. (1990) *Loose boundary hydraulics*. Pergamon Press, Oxford, UK.

Ricketts, D.J. and McNamara, M. (1987) *Field Study to Evaluate the Hydraulic Performance of Sea Outfalls. No. 2: Langley Point outfall.* WRc external report No. ER275E.

Ristenpart, E., Ashley, R.M. and Uhl, M. (1995) Organic near-bed fluid and particulate transport in combined sewers. *Wat. Sci. Tech.*, 31(7), 61–68.

SAFEGE (1996) *Etude des flux de pollution par temps de pluie sur un bassin versant test du département des Hauts de Seine*. Rapport d'avancement n° 3, validation de MOUSETRAP, Safege, Nanterre, France, 44 pp.

Sakakibara, T. (1996) Roof runoff storm water quality. *Proceedings of 7th International Conference on Urban Storm Drainage*, Hannover, Germany, 9–13 September 1996, 1, 157–162.

Sartor, J.D. and Boyd, G.B. (1972) *Water pollution aspects of street surface contamination.* US EPA Report no R2-72-081, Washington D.C., USA, November 1972.

Sartor, J.D., Boyd, G.B. and Agardy, F.J. (1974) Water pollution aspects of street surface contaminants. *Journal of Water Pollution Control Federation*, 46(3), 458–467.

Saul, A.J., Gupta, K. and Stovin, V.R. (1998) Gross solids characteristics and an event-based CSO efficiency prediction methodology. *Proceedings of International Conference on Developments in Urban Drainage Modelling*, London, UK, 21–24 September 1998, 1, 373–380.

Saul, A.J., Ruff, S.J., Walsh, A.M. and Green, M.J. (1993) *Laboratory studies of CSO performances.* WRc report No. UM 1412, Swindon, UK.

Schaart, N. (2003) *Sustainable Urban Drainage and Public Hygiene: A risk analysis.* (In Dutch. Orig. title: Duurzame waterketen en volksgezondheid; een risicoanalyse) MSc thesis, Delft University of Technology, section sanitary engineering, Delft, January 2003.

Schnitzlein, M., Blumberg, D. and Bauer, W. (1984) Voruntersuchungen zur Entschlammung des innerstädtischen Mischsielnetzes Hamburgs. *Korrespondenz Abwasser*. 12/84, 1058–1062.

Servat, E. (1984) *Contribution à l'étude des matières en suspension du ruissellement pluvial à l'échelle d'un petit bassin versant urbain.* Thèse de doctorat de l'Université des Sciences et Techniques du Languedoc, Montpellier, 182 pp.

Shaheen, D.G. (1975) *Contributions for Urban Roadway Usage to Water Pollution.* US EPA Report no 600/2-75-004, US EPA, Washington D.C., USA, April 1975.

Shinoda, T. (1990) Comparative study on surface runoff by stormwater infiltration facilities. *Proceedings of 5th International Conference on Urban Storm Drainage*, Osaka, Japan, July 1990. 783–788.

Shivalingaiah, B. and James, W. (1984) Algorithms for build-up, washoff and routing pollutants in urban runoff. *Proceedings of 3rd International Conference on Urban Storm Drainage*, Göteborg, Sweden, 4, 1445–1455.

Sidwick, J.M. (1984) *Screenings and Grit in Sewage: Removal, treatment and disposal.* Preliminary report. CIRIA/WRc Technical Note no 119, CIRIA, London, UK.

Stahre, P. and Urbonas, B. (1990) *Stormwater detention for drainage, water quality and CSO management.* Prentice-Hall, Englewood Cliffs, New Jersey, USA, 338 pp.

Stockhausen, E.M. and Souter, N. (1996) *International Survey of Sanitary Waste Disposal Practices.* Wastewater Technology Centre, University of Abertay Dundee, Scotland.

Swaffield, J.A. and Wakelin, R.H.M. (1995) Water conservation: the impact of design, development and site appraisal of a low volume flush wc. *Proceedings of International Conference on Low Cost Sewerage*, July 1995, Leeds, UK.

Thornton, R.C. and Saul, A.J. (1986) Some quality characteristics of combined sewer overflows. *The Public Health Engineer*, 14(3), 35–38.

Uchimura, K., Nakamura, E. and Fujita, S. (1996) Characteristics of stormwater runoff and its control in Japan. *Proceedings of 7th International Conference on Urban Storm Drainage*, Hannover, Germany, 9–13 September 1996, 1, 55–60.

USEPA (1992) *Manual for Wastewater Treatment/Disposal for Small Communities.* EPA/625/R-92/005.

Verbanck, M. (1989) L'épuration des eaux à Bruxelles: caractérisation des charges polluantes. *Actes du Forum 'Quelles solutions pour les eaux usées à Bruxelles?', Forum A.Ir.Br.-V.Ir.Br.*, Bruxelles, Belgique, 31 May 1989, 15–26.

Verbanck, M. (1995) *Transferts de la charge particulaire dans l'égout principal de la ville de Bruxelles.* PhD thesis, Université Libre de Bruxelles, May 1995, 193pp.

Verbanck, M., Ashley, R.M. and Bachoc, A. (1994) International workshop on origin, occurence and behaviour of sediments in sewer systems: summary of conclusions. *Water Research*, 28(1), 187-194.

Verbanck, M., Vanderborgh,t J.P. and Wollast, P. (1990) Size distributions of suspended particles in combined sewers during dry and wet weather. *Proceedings of 5th International Conference on Urban Storm Drainage*, Osaka, Japan, 891–896.

Wada, Y. and Miura, H. (1988) *Characteristics of accumulated loads in the street inlets in urban area and its flushing*. Technology Reports of Kensai University, 30, 153–166.

Wada, Y., Miura, H. and Hasegawa, K. (1987) Model building and analysis of runoff water quality of flush from the street gully pots. *Proceedings of 4th International Conference on Urban Storm Drainage*, Lausanne, Switzerland, 60–65.

Waller, D. and Hart, W.C. (1986) Solids, nutrients and chlorides in urban runoff. *Urban Runoff Pollution*, NATO ASI Series vol. G10, edited by H.C. Torno, J. Marsalek and M. Desbordes, Springer-Verlag, Berlin, 59–85.

WEF/ASCE (1994) Existing sewer evalution and rehabilitation. WEF Manual of Practice FD-6. ASCE Manuals and Reports on Engineering Practice No. 62. ISBN 1-881369-92-7.

White, M., Johnson, H., Anderson, G., Misstear, B. (1997) *Control of Infiltration to Sewers*. CIRIA report 175.

Wilkinson, D. L. (1984) Purging of saline wedges from ocean ourtfalls. *J. Hydr. Engg.*, 110 (12).

Wilkinson, D. L. (1992) Model studies of outfall riser hydraulics. *J. Hydr. Res.* 30(5).

Wood, I.R., Bell, R.G. and Wilkinson, D.L. (1993*) Ocean Disposal of Wastewater. Advanced series on ocean engineering*, 8. World Science Publishers, Singapore.

Xanthopoulos, C. (1992) Beschaffenheit und Behandlung des Regenabflusses von Strassen und Autobahnen. *ATV Seminar für die Abwasserpraxis – Aktuelles zur Regenwasserbehandlung*, Bremen, Deutschland, April 1992, 16 pp.

Yaziz, M.I., Gunting, H., Sapari, N. and Ghazali, A.W. (1989) Variations in rainwater quality from roof catchments. *Water Research*, 23(6), 761–765.

Young, R.A. and Wiersma, J.L. (1973) The role of rainfall impact in soil detachment and transport. *Water Resources Research*, 9(6), 1629–1636.

Zug, M., Phan, L., Bellefleur, D. and Scrivener, O. (1998) Pollution washoff modelling in sewer networks: calibration, validation, transposition. *Proceedings of Novatech 98,* Lyon, France, 1, 131–138.

2.2 CHARACTERISTICS AND TRANSFORMATION PROCESSES

2.2.1 Processes in sewers – a general overview

A sewer system is an integral part of the urban wastewater system and it interacts with wastewater treatment plants as well as the local receiving waters: wastewater is transported through the sewers to treatment, and overflow, during rain events, of untreated wastewater into receiving waters takes place. During transportation of wastewater in the sewers both soluble and particulate substances undergo transformation processes including degradation. These solids accumulate, which affects design as well as operation of the sewer and hence is of importance for interactions within the entire urban wastewater system. Physical, chemical and biological processes take place in the water phase, in the biofilms and in the sewer sediments. The exchange of soluble as well as solid matter between these phases is therefore also important, as shown conceptually in Figure 2.2.1.

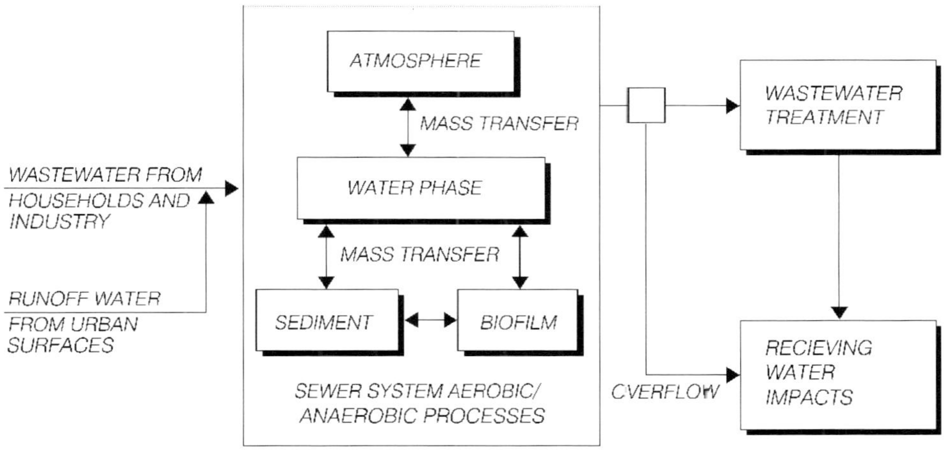

Figure 2.2.1. Transport and transformation of wastewater in a sewer system.

Within this context, three types of sewer solids may be defined:

- suspended solids and colloids
- sewer sediments, i.e. deposits and near-bed material
- biofilm.

The characteristics and the physical, chemical and biological processes associated with each of these types of solids are different. This is of fundamental importance for the definition of the characteristics and nature of sewer solids. Figure 2.2.2 shows in outline the processes and occurrence of solids in sewers.

In comparison with current knowledge about receiving waters, and what is known about the transformation processes for pollutants in wastewater treatment processes, relatively little is known about organic matter and other pollutant transformation and accumulation processes in sewer systems. Quantification of the processes of transport and transformation of wastewater within the holistic urban wastewater system requires that processes in sewers are properly understood. This is important when dealing with the function of the sewer itself and when integrating and establishing boundary conditions to wastewater treatment plants and receiving waters.

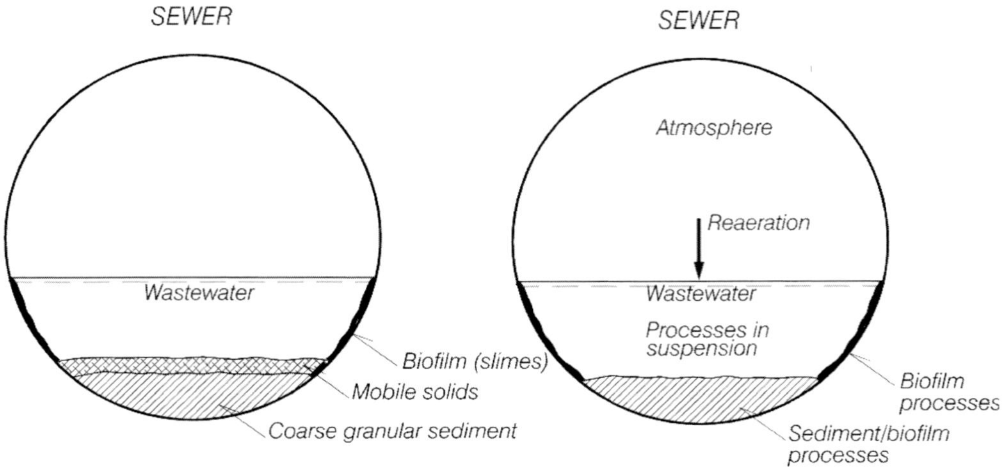

Figure 2.2.2. Diagrammatic illustration of the processes and occurrence of solids in sewers.

A process concept for a sewer which includes the physical, chemical and biological processes is outlined in Figure 2.2.3. The following interpretation of the process concept emphasises the microbial transformation of wastewater organic matter and related transport phenomena. Despite the overall objective of this section, existing knowledge is limited, hence it is not possible to be entirely comprehensive when considering the characteristics and processes to be included in this review. As an example, heterotrophic biomass is considered homogeneous despite the knowledge that biomass consists of both cell biomass and extracellular polymers, each having different properties (Jahn and Nielsen, 1997).

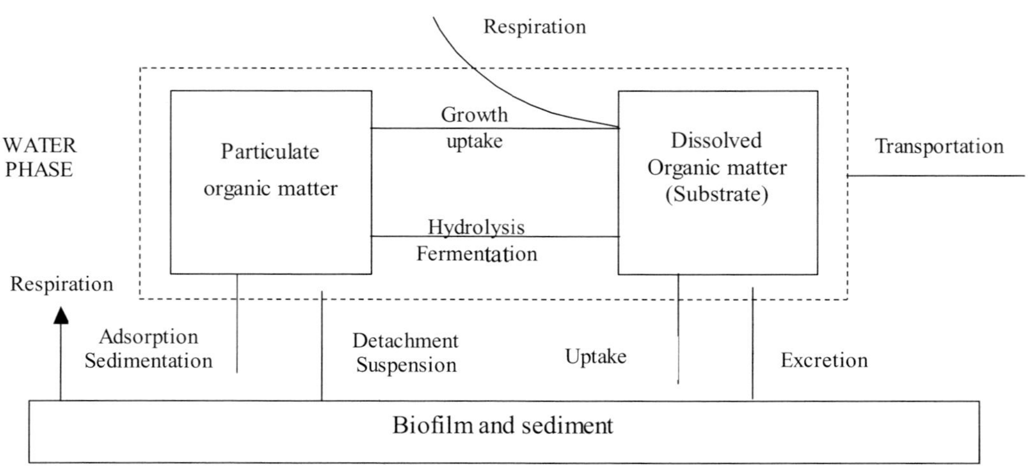

Figure 2.2.3. Major components and processes that influence the concentration of dissolved and suspended organic matter under transportation in sewage (after Raunkjær *et al.*, 1995).

Organic matter in solid as well as soluble forms plays an essential role in sewer processes, for example, in relation to the availability of dissolved oxygen determining not just whether aerobic or anaerobic conditions in the sewer prevail but also the major characteristics of the three types of sewer solids. Organic matter includes living biomass and is needed as substrate for the microbial processes in the sewer. The organic matter exists in dissolved forms as well as in particulate fractions subdivided into biomass and particles of different size and biodegradability. The transformations between dissolved and particulate fractions are determined by a number of processes, e.g. hydrolysis, fermentation, decay, and substrate uptake and growth. Furthermore, dissolved organic matter uptake to the biofilm and into the sediments causes growth followed by detachment of biofilm into the bulk water. Microbial processes taking place in the biofilm may influence the quality of the bulk water. Other processes which influence the particulate and dissolved fractions are adsorption, sedimentation and resuspension. Sedimentation takes place during dry weather periods and resuspension occurs mainly during rain or other high flow velocity periods, as illustrated in Figure 2.2.3.

2.2.1.1 Characteristics of sewer solids in suspension

The characteristics of the particles in suspension are defined by their physical, chemical and biological properties.

2.2.1.1.1 Characterisation of suspended solids

Suspended solids (SS or total suspended solids, TSS) concentrations in wastewater can be determined as described in APHA *et al.* (1995), taking due account of the specific problems arising from analyses of wastewater samples; for example, the 'exclusion of large floating particles, or submerged agglomerates' for filtration, but subsequently accounting for these larger solids in the total mass. Thus the constituents associated with the suspended solids phase must be determined based on analysis of both the filtered and pre-segregated wastewater sample. An alternative or complementary method for separation of suspended solids is centrifugation.

2.2.1.1.2 Organic fractions in wastewater suspended solids

Characterisation of wastewater organic matter in terms of particle size distribution has been provided by Levine *et al.* (1985), Ødegaard (1987), Logan and Jiang (1990) and Levine *et al.* (1991), as shown in Figure 2.2.4. Several early studies related particle size fraction to measurement of specific organic groups in wastewater (Heukelekian and Balmat, 1959; Painter and Viney, 1959; Hunter and Heukelekian, 1965; Rickert and Hunter, 1971). These early methods for characterisation of specific organic fractions especially concerning carbohydrate, protein, lipid and humic substances have recently been supplemented with new analytical methods and techniques according to Raunkjær (1993), Raunkjær *et al.* (1994) and Frølund *et al.* (1996). Measurement of the specific fractions of carbohydrate, protein, lipid and volatile fatty acids (VFAs) is useful because the measurement of the biodegradability of these fractions is a way of studying changes in wastewater organic matter composition during transport in a sewer. Some of these fractions may undergo considerable transformations in terms of COD or BOD value, (Raunkjær *et al.*, 1994; Hvitved-Jacobsen *et al.*, 1995a).

The best method of measurement for carbohydrate in sewage is the anthrone method originally introduced by Dreywood (1946). For lipids, an infrared method modified from Standard Methods (1995) may be utilised; both methods are described by Raunkjær *et al.* (1994). Protein is measured according to Lowry *et al.* (1951) with modifications concerning measurement of humic substances as well (Frølund *et al.*, 1996). VFA is measured using ion chromatography as described by Hvitved-Jacobsen *et al.* (1995a).

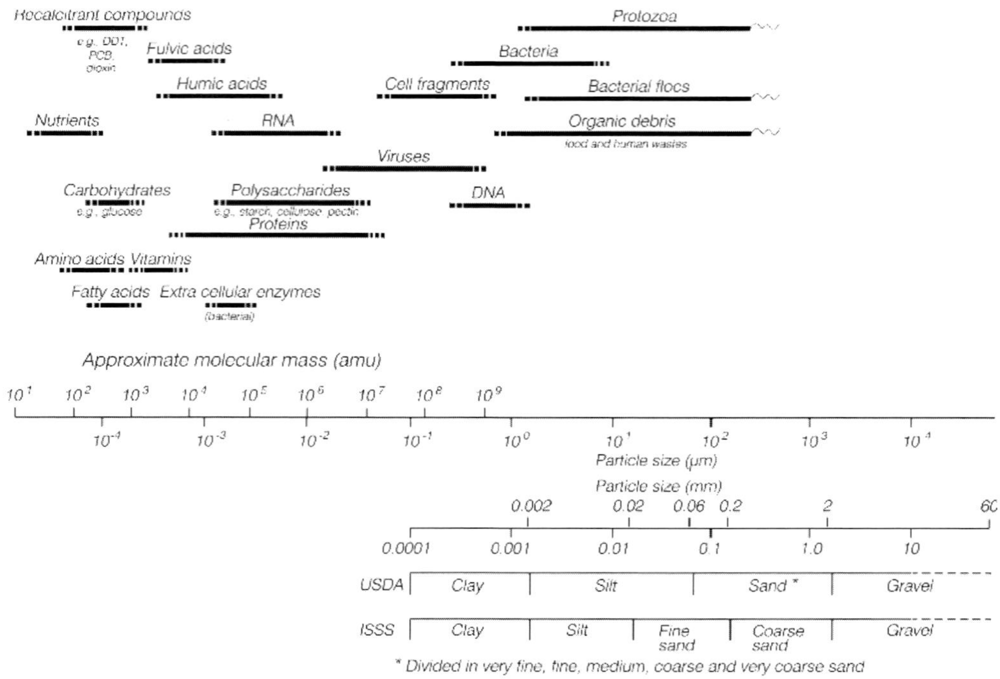

Figure 2.2.4. Particle size of typical organic constituents in settled municipal sewage (adapted from Levine *et al.*, 1985). Information on size characteristics of inorganic material originates from two of the most commonly used standards: US Department of Agriculture (USDA) and the International Soil Science Society (ISSS)

A widely used method for characterisation of wastewater organic matter is the COD fractionation related to biodegradability as used in the IAWQ model Nos 1 and 2 for activated sludge (Henze *et al.*, 1987, 1995). Based on this concept, the three main organic fractions of wastewater suspended solids are:

X_S slowly biodegradable hydrolysible substrate (g/m^3)
X_I inert, non-biodegradable organics (g/m^3)
X_{Bw} heterotrophic biomass (g/m^3).

These COD fractions can be determined using a combination of respiration measurements and model calibration techniques (Henze *et al.*, 1995). The suspended fractions typically account for about 2/3 of the total COD in raw domestic wastewater and even more during a rain event caused by erosion of sediment and biofilm. Sollfrank and Gujer (1991) divided the particulate slowly biodegradable fraction into rapidly hydrolysible and slowly hydrolysible. This type of COD fractionation is also useful for sewer solids, (Bjerre *et al.*, 1995; Hvitved-Jacobsen *et al.*, 1998).

Characteristic values for the COD components in wastewater are given in Table 2.2.1. (Other illustrative characteristics of dry weather sewage are given in Tables 2.1.14 and 2.1.18).

Table 2.2.1. COD-components and corresponding dissolved oxygen for wastewater at the head of a sewer

Component (includes soluble (S) and particulate (X) fractions)		Typical value wastewater	Diluted***	Unit
X_{Bw}	heterotrophic active biomass in water phase	20–100	40	$gCOD/m^3$
X_{Bf}	heterotrophic active biomass in biofilm	-	-	$gCOD/m^2$
X_{S1}	hydrolysible substrate, fast biodegradable	50–100	40	$gCOD/m^3$
X_{S2}	hydrolysible substrate, slowly biodegradable*	300–450	210	$gCOD/m^3$
S_S	readily biodegradable substrate	0-50	30**	$gCOD/m^3$
S_F	fermentable substrate	variable	20	$gCOD/m^3$
S_A	fermentation products (i.e. VFAs)	variable	10	$gCOD/m^3$
S_O	dissolved oxygen	0–4	2	gO_2/m^3
COD	total COD	≈600	320	$gCOD/m^3$

* includes very slowly biodegradable and inert organic matter
** $S_S = S_F + S_A$
*** diluted with e.g. groundwater or infiltration

The fractions of total protein, carbohydrate, lipid and VFA of the total COD at the inlet to four wastewater treatment plants were measured 4 times during a day and night. The measured figures were found to remain constant with average values of 28 %, 18 %, 31 % and 6 % respectively (Raunkjær *et al.*, 1994). These fractions together accounted for 83 % of the total inlet COD. The suspended fractions as a percentage of the total concentration for COD, protein, carbohydrate and lipid averaged 53 %, 56%, 76 % and 100 % respectively (lipids are defined as suspended). The protein fraction may, however, include humic substances due to interference of the Lowry method as described by Frølund *et al.* (1996).

2.2.1.2 Characteristics of suspended solids in urban wet weather discharges

2.2.1.2.1 Introduction

During dry weather conditions, suspended solids in wastewater may settle and accumulate as sewer deposits and biofilms may form on sewer walls as a result of microbial processes. During a successive period of increased flow, e.g. due to urban wet weather runoff into combined sewers, erosion of both sewer deposits and biofilm may occur. In addition to the suspended solids which are introduced from surface runoff, a relative increase in suspended solids caused by erosion of bed deposits and biofilms is expected when compared with the dry flow solids concentrations. Corresponding quality changes in terms of the physical, chemical and biological characteristics of the suspended solids is also expected.

The local catchment and sewer conditions and the runoff characteristics vary considerably in time and space. It is therefore not surprising that there is significant variability in the suspended solids characteristics during runoff events. Even at a specific site, the standard deviation of the characteristics of the suspended solids fraction is often of the same order of magnitude as the mean value. This must be taken into account when evaluating specific sewer solids characteristics and consequent impacts caused by wet weather discharges.

2.2.1.2.2 Pollutants attached to suspended solids in urban wet weather discharges

Table 2.2.2 shows that a significant pollutant load is associated with the suspended solids fraction during wet weather in developed urban areas (Section 2.1.3). This is even more pronounced for the micropollutants attached to the solids, for example, 93 % of PCBs

(Marsalek, 1985); 90 % PAHs, 97 % of benzopyrene and 85 % of fluoranthene (Hermann and Kari, 1990).

Table 2.2.2. Pollutants associated with combined sewage suspended solids during wet weather for particles larger than 0.45 μm (Chebbo, 1992)

Sites in France	Parameters (% of the total pollutant concentration)				
	COD	BOD_5	TKN	Total hydrocarbon	Pb
Béquigneaux (storm sewer, Bordeaux)	83–90	77–95	67–82	86–87	95
Trunk no. 13 (combined sewer, Marseilles)	83–92	91	70–80	82–99	99.5
La Molette (CSO, Seine-Saint-Denis)	88	83	48		99

The percentages indicated in Table 2.2.2 would be lower for large combined sewers which receive relatively low fractions of dry weather flow mixed with storm runoff. It is probable that at the start of an event the sewer flow is composed principally of dry weather flow which is already in the network when the runoff begins. This is observed for example, in a number of large sewers in Paris, France and at the downstream ends of trunk sewers in Clichy and Asnières in the Paris suburbs (see also Section 3.4).

2.2.1.2.3 Grain size, density and settling characteristics for suspended solids in urban wet weather discharges

Based largely on extensive French results, as indicated in Table 2.2.3 and Figure 2.2.5, the grain size characteristics for suspended solids in wet weather discharges can be generalised as follows (Chebbo, 1992; Chebbo and Bachoc, 1992; Chebbo *et al.*, 1995):

- Fine particles (< 100 μm) dominate in the suspended phase in the downstream sections of a sewer network. Particles smaller than 100 μm represent between 66 and 85 % of the total mass with a median diameter (D_{50}) ranging from 25 to 44 μm. In Tables 2.2.3 and 2.2.4, D_{10} and D_{90} correspond to meshes through which 10 % and 90 %, respectively, of the mass of solids would pass.
- Fine particles (< 100 μm) have a tendency to agglomerate. Table 2.2.4 shows that after ultrasonic dispersion, the average diameter of the particles is considerably reduced. The sizes return to almost the original value of D_{50} after 5 minutes without mixing, indicating that the ultra-sound waves do not permanently separate the particles.
- Results of grain size measurements are similar for a range of sewer systems and also for combined or separate systems.
- Solids deposited in man-entry size sewers (< 1 m) are generally coarser than the solids transported in suspension in storm or dry weather flows.

Table 2.2.3. Values of the grain size indicators D_{10}, D_{50}, D_{90} and % < 100 mm for suspended solids in urban wet weather discharges

Type of network	D_{10} (μm)	D_{50} (μm)	D_{90} (μm)	% < 100 μm
Storm sewer networks	6–9	29–38	265–1375	76–85
Combined networks	4–13	25–44	183–497	66–82

Table 2.2.4. Influence of test procedure on particle size characteristics (particle size fraction < 100 mm)

Site	Test procedure*	Grain size characteristics		
		D_{10} (μm)	D_{50} (μm)	D_{90} (μm)
Béquigneaux	1	3.9	22	53
(storm sewer, Bordeaux)	2	1.8	13	36
	3	2.5	18	37
La Molette	1	3.2	20	62
(CSO, Saint-Seine-Denis)	2	1.6	12	42
	3	2.2	23	52

*Mode of procedure:
1 with mixer and no ultrasonic dispersion
2 with mixer and increasing duration of ultrasonic dispersion until stabilisation of the grain size curve
3 as 2 but taking the measurement 5 minutes after exposing the sample to ultrasonic dispersion.

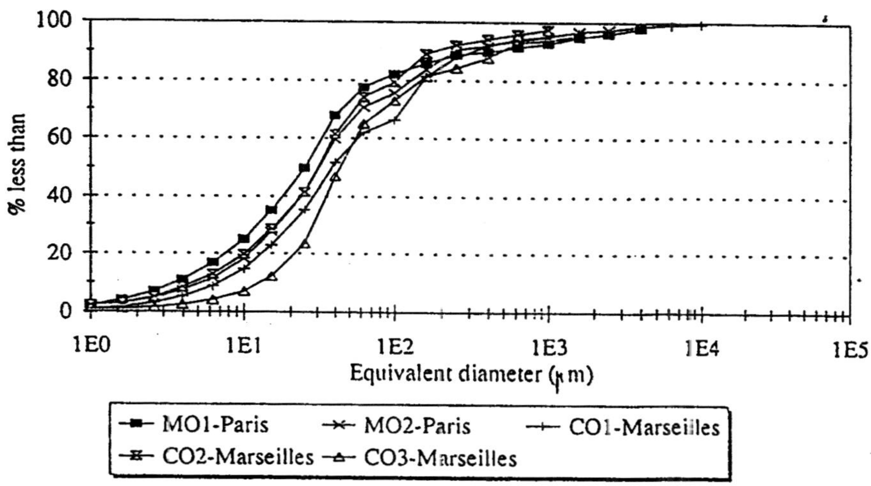

Figure 2.2.5. Cumulative grain size distribution for suspended solids in a range of catchments with combined sewerage systems (Chebbo, 1992).

Typically, the largest particles have the lowest specific gravity in wastewater systems. During wet weather, and in combined networks, solids are less dense (1700 to 2200 kg/m^3) than the solids found in separate storm sewers (> 2200 kg/m^3). This demonstrates that there is a major contribution from the deposits already in the sewers (and the wastewater already within the sewer when the runoff starts) to the conveyed pollutant load in combined sewage during wet weather. However, for a given type of network (separate or combined) the density of the particles is found to be reasonably uniform from site to site.

The difficulties of determining representative results for the settlement of sewage particulates are well known (Ashley and Crabtree, 1992; Pisano, 1996; Lucas-Aiguier *et al.*, 1998) and are discussed in the Appendix. For the setting of design standards, settling velocities of 350 samples of sediment, slime and wastewater were measured in a German study (Michelbach and Wöhrle, 1992). These single settling curves of each type of solid, defined as what settles in 2 hours within an Imhoff-cone, were then compared with a characteristic cumulative distribution of settling velocity (Figure 2.2.6).

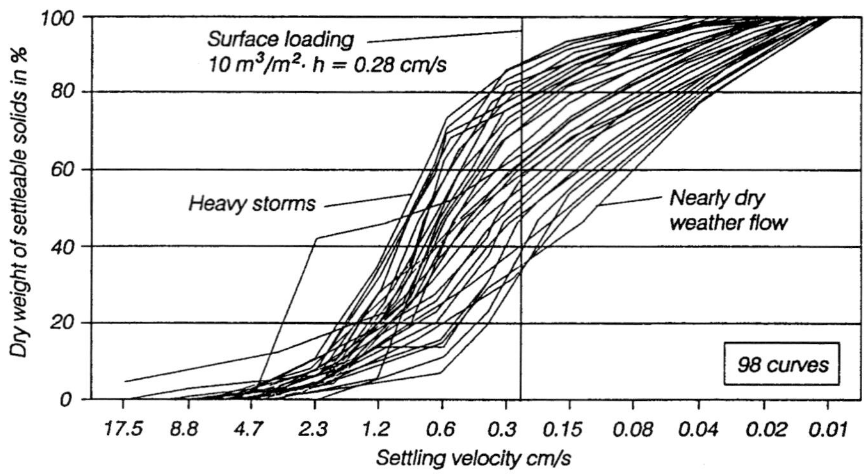

Figure 2.2.6. Cumulative distribution of settling velocity of settleable solids in combined sewage. 'Settleable solids' here refers to what will settle in 2 hours within an Imhoff-cone (Michelbach, 1995).

Most of the curves in Figure 2.2.6 are S-shaped. This is reasonable because there are no solids which settle faster than 175 mm/s and there are only a small percentage of solids which settle slower than 0.1 mm/s. The curves are the result of a natural process and they reach the borders of the range as asymptotes.

A number of researchers have attempted to determine the relationship between the settling velocity and the pollutant concentration of sewer deposits. For example, Michelbach and Wöhrle (1992, 1994) used samples of combined sewage which were fractioned by settling. It was expected that most pollutants would be associated with fine particles of small grain size and low settling velocities. Figure 2.2.7 shows, however, that fractions with a settling velocity of 8 and 4 mm/s had the highest concentration of heavy metals and organic micropollutants. The reason may be that this fraction contained a high amount of eroded wall slime.

Settling velocities for the particles conveyed in suspension in wet weather are quite high with the exception of a small fraction (less than 10%) of fine particles, as shown in Figure 2.2.8a and 2.2.8b. w_{50} (median velocity) is generally between 4 and 11 m/h. w_{90} is smaller in a combined network (46 to 98 m/h) than in a stormwater system (31 to 133 m/h), whilst being of the same order of magnitude. w_{10} and w_{20} are highly variable. This has major implications because the small particles are those which have relatively the highest amount of associated pollutants (Sections 2.1.3 and 2.1.4).

The fastest settling particles are mostly those which are inorganic, whereas the slower settling particles comprise the largest organic fraction. Particles which settle faster than a few cm/s typically comprise in aggregate, less than 5 % of the total organics in suspension. Particles with settling velocities less than a few mm/s typically comprise more than 50 % of the total organic mass in suspension. In terms of biodegradability, the slower settling organic particles are more easily degraded than the organic fraction with the higher settling velocity. Table 2.2.5 (Vollertsen *et al.*, 1998) illustrates that the sewer solids discharged from a combined sewer overflow during rainfall, i.e. those with the lowest settling velocity, are likely to exert the most significant impact in terms of oxygen demand. This contrasts with the presumed effects of discharge of entrained sediment deposits (the particles with the highest settling velocity), which would be less significant.

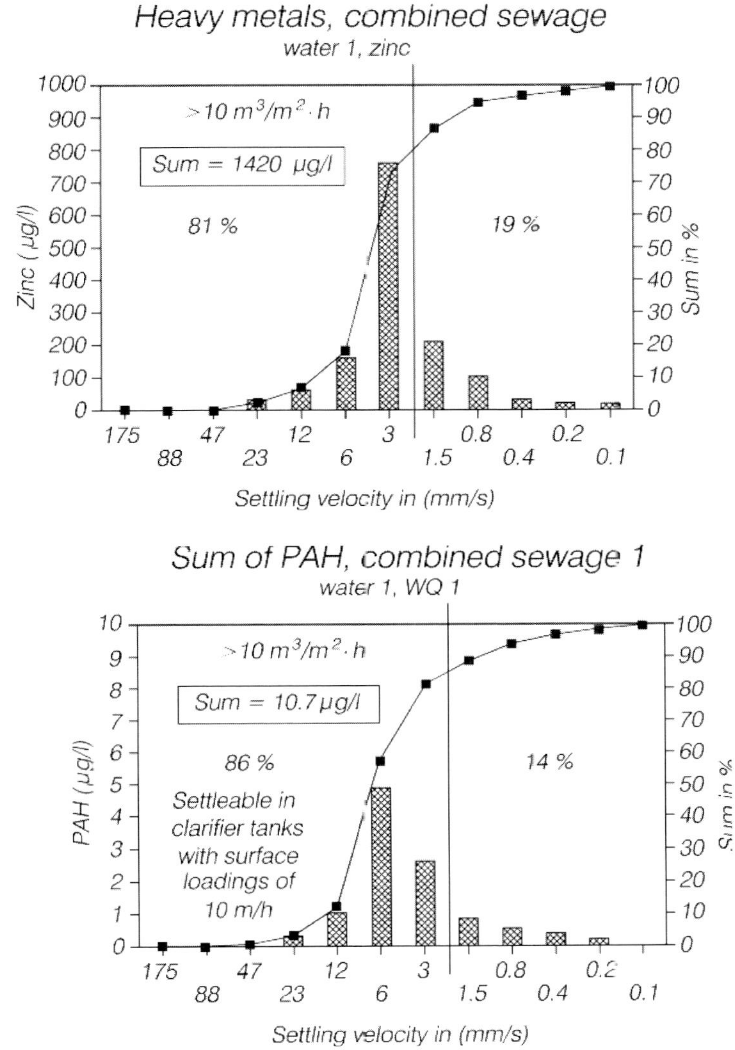

Figure 2.2.7. The relationship between settling velocity and the concentration of zinc and polycyclic aromatic hydrocarbons (PAH) in combined sewage for solids settling in 2 hours within an Imhoff-cone (Michelbach and Wöhrle, 1994).

Table 2.2.5 Typical distribution of organic matter (expressed in terms of COD) in sewer solids in terms of biodegradability (Vollertsen, 1998)

Settling velocity (mm/s)	Xs, fast* (% of COD)	Xs, med* (% of COD)	Xs, slow* (% of COD)	X_{Bw} (% of COD)
>20	4	10	83	3
2–20	6	20	68	6
0.02–2	8	30	56	6

* Rate of hydrolisation of substrate and biomass

** Heterotrophic bacterial biomass.

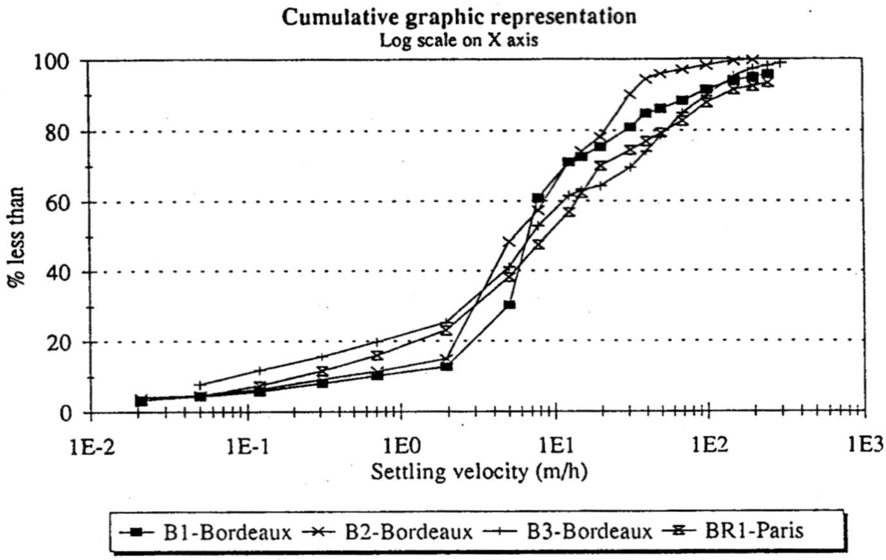

(a) Stormwater.

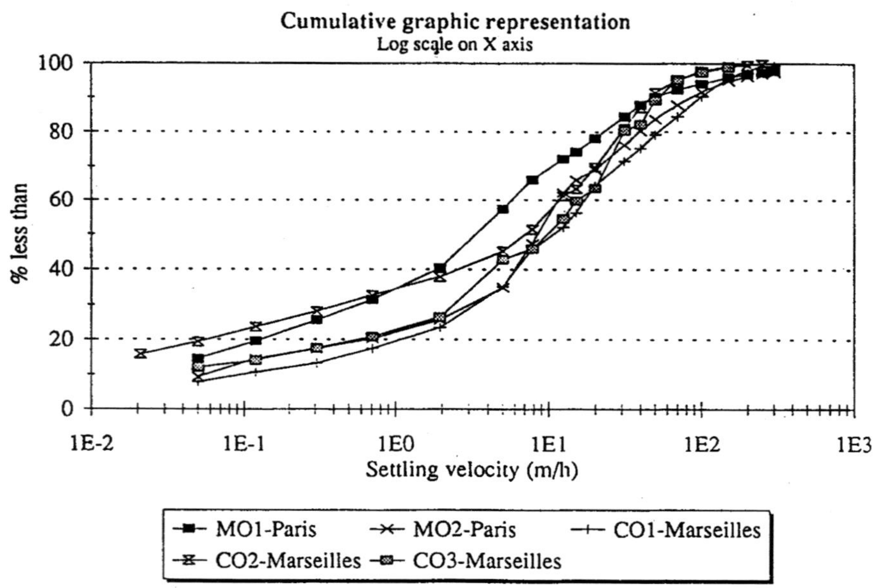

(b) Combined sewage

Figure 2.2.8. Cumulative distributions of settling velocities for suspended solids in stormwater and combined sewer networks (adapted from Chebbo, 1992).

2.2.2 Processes related to suspended solids and colloids

2.2.2.1 Suspended solids and particle definitions – what is a solid?

Traditionally, the definition of suspended solids as used in wastewater transport and urban runoff is defined according to the method used in the analysis of total suspended solids (TSS), e.g. as described in Standard Methods (1995). This definition of suspended solids is usually based on a filtration procedure (Section 2.2.1.1) and is therefore appropriate for the larger particles having a diameter > about 0.5 μm (Figure 2.2.4).

Whilst this definition for determining what are 'solids' is rational as far as physical transport processes in sewers are concerned, when dealing with the chemical and microbial processes for sewer conditions, an extension of the 'suspended solids' definition is required.

Physico-chemical processes, e.g. adsorption, are especially important when dealing with colloidal particles, i.e. particles with diameters from 10^{-3} to 1 μm (Figure 2.2.4). For the microbial processes it is also important as to whether the biodegradable organic matter can be transported without further disintegration through the bacterial cell wall. Particles larger than about 10^{-4} μm cannot be transported through the cell wall and are therefore, from a microbial point of view, to be considered as non-dissolved – and hence by definition, as particles in suspension – which before microbial uptake and utilisation is possible, must be broken into small molecules, i.e. undergo a hydrolysis process initiated by extracellular enzymes.

In order to avoid confusion, these small 'solids in suspension' will be referred to here as colloidal particles and particles which might undergo hydrolysis.

2.2.2.2 Processes in the suspended water phase

Suspended solids and particles, including the associated pollutants, comprise the key components of the processes which are important for sewer design and operation. The resulting properties of the solids are also important for wastewater treatment processes and for impacts on receiving waters. Examples of sewer processes are:

Physical processes, transport

- Re-aeration (air-water oxygen transfer)
- Sedimentation and erosion

Chemical and physico-chemical processes

- Adsorption and desorption, e.g. adsorption of heavy metals and micropollutants. These processes are of specific importance for the amount of pollutants associated with the fine particulate suspended solids fractions (colloidal particles).

Microbial processes

- Especially those processes related to the wastewater carbon transformations in sewer systems. The main processes under aerobic conditions affecting the production and removal of the heterotrophic bacterial biomass (X_{Bw}) and the biodegradable (hydrolysible), organic solids fractions of X_S are (Bjerre *et al.*, 1998a; Hvitved-Jacobsen, *et al.* 1998):

  Growth of heterotrophic bacterial biomass, X_{Bw} (bacterial floc particles) and associated respiration processes.

  Hydrolysis of particulate biodegradable substrate (including different fractions of X_S). This process is initiated by extracellular enzymes.

  Energy requirement for maintenance of the heterotrophic bacterial biomass.

2.2.2.3 Transformation of wastewater organic matter under aerobic conditions

The behaviour of organic matter in sewer systems plays a dominant role as regards the quality of the wastewater. The microbial processes involving organic matter as substrate and heterotrophic biomass are therefore of particular significance. Both dissolved and particulate fractions of wastewater organic matter interact in the production and removal of organic solids in sewers. It is especially important to realise that the growth process includes two basic, consecutive processes:

- a process resulting in removal of dissolved organic substrate and a production of heterotrophic biomass with a yield coefficient Y typically between 0.5 and 0.6 g COD biomass produced per g COD substrate totally consumed;
- an energy producing process using a fraction of (1-Y) of the dissolved organic substrate removed. Under aerobic conditions this energy producing process manifests itself in a corresponding dissolved oxygen (DO) consumption rate – also referred to as respiration rate, oxygen utilisation rate or oxygen uptake rate (OUR).

Under aerobic conditions in a (gravity) sewer, wastewater microbial transformations are primarily caused by growth and maintenance of heterotrophic biomass associated with carbon and oxygen uptake as well as hydrolysis of particulate organic substrate. The driving and governing process for these transformations is the continuous supply of oxygen by reaeration (Jensen and Hvitved-Jacobsen, 1991; Jensen, 1995). These transformations are illustrated in Figure 2.2.9.

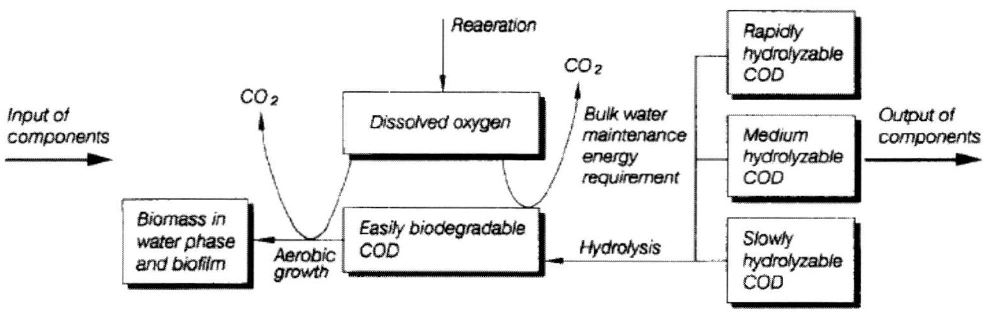

Figure 2.2.9. The simplified aerobic wastewater system under gravity sewer conditions (Hvitved-Jacobsen *et al.*, 1998a)

2.2.2.4 Anaerobic processes

Wastewater transformations under anaerobic conditions in sewers have been investigated by Tanaka and Hvitved-Jacobsen (1998) and have been integrated with the aerobic processes as shown in Figure 2.2.9 and 2.2.10. In contrast to the aerobic heterotrophic transformations, the rate of transformation is relatively low under anaerobic conditions. Volatile fatty acids (VFA's) are produced under anaerobic conditions from fermentable readily biodegradable organic matter (Hvitved-Jacobsen *et al.*, 1995a). It is, therefore, necessary to classify readily biodegradable substrate (S_S) into two fractions: fermentable substrate (S_F) and fermentation products (S_A), i.e. VFAs (Figure 2.2.10). The VFAs may be utilised as substrate for methane producing bacteria.

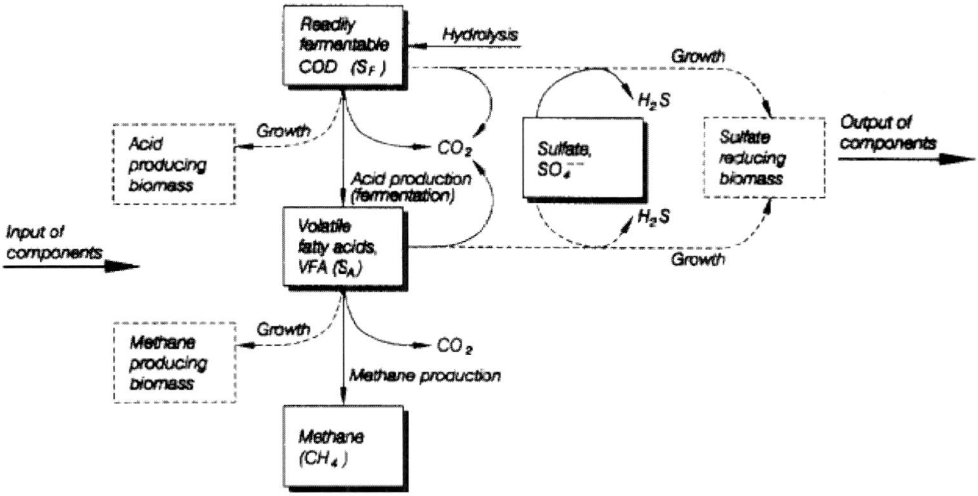

Figure 2.2.10. The simplified anaerobic system for wastewater transformations in sewer integrated with aerobic processes shown in Figure 2.2.9.

The sulphur cycle in terms of the sulphate respiration processes can be integrated with the anaerobic carbon cycle (Figure 2.2.10). Fractionation of readily biodegradable substrate (S_s) into S_F and S_A corresponds with the expectation that it is mostly the S_F which is used by the sulphate reducing biomass in sewer biofilms. By integrating the sulphide formation in a process description, a conceptual approach is obtained instead of the traditional empirical descriptions, (Thistlethwayte, 1972; Boon and Lister, 1975; Pomeroy and Parkhurst, 1977; Hvitved-Jacobsen *et al.*, 1988).

2.2.2.5 Process characteristics

As part of the process of generation of biomass solids, a considerable reduction in dissolved organic substrate may take place during wastewater transportation in a (gravity) sewer under aerobic conditions (Raunkjær *et al.*, 1995; Bjerre *et al.*, 1998b; Hvitved-Jacobsen *et al.*, 1998). Transformation rates (growth of biomass) are generally dependent on the following characteristics (Figure 2.2.9):

- availability of soluble, readily biodegradable organic substrate, i.e. both the nature of the organic matter and its concentration;
- DO concentration, which in addition to the organic substrate, determines the size/location of the aerobic zone for suspended particles and especially within the biofilm;
- heterotrophic biomass characteristics and concentration;
- temperature.

These characteristics may vary considerably in time and space in a sewer system and a corresponding variability in the transformation rates of organic substrate will be seen: e.g. readily biodegradable substrate may be consumed during wastewater transportation and a corresponding increase in the biomass concentration will take place. The primary organic compounds which are removed under aerobic conditions are dissolved substances, e.g. carbohydrate and volatile fatty acids (VFA).

A typical example of the transformation of organic matter, based on non-limiting substrate conditions, i.e. > 3–5 mg COD/l of dissolved, readily biodegradable substrate, and assuming a biomass concentration of 20–30 mg COD/l and a DO concentration of 2–4 mg/l, gives a rate of total transformation of substrate of the order of 15–30 mg COD/(l^3h) at 15 °C. This level of transformation results in a biomass production of 8–18 mg COD/(l^3h) and a corresponding OUR of 6–15 mg O_2/(l^3h). Biofilm processes may increase the total transformation of organics compared with this example; especially in small diameter sewers with a relatively high surface area/volume ratio (Section 2.2.5).

Under substrate limiting conditions the transformation rates will be reduced. If the growth process is controlled by hydrolysis, the total transformation rate of COD in the bulk water phase of a sewer may be reduced from the 15–30 mg COD/(l^3h) above to about 2–10 mg/(l^3h).

Transformations of wastewater organic matter can be monitored based on the variation in the OUR value taking place within a batch experiment (Bjerre *et al.*, 1995) (Figure 2.2.11). The actual OUR value – and in addition to this the DO consumption from the biofilm and sediment microbial processes – will determine if the reaeration rate for the flowing wastewater is sufficient for maintaining aerobic conditions in a gravity sewer. If this is not the case, the water phase in the sewer turns anaerobic. The oxygen balance in a sewer is of major importance for the microbial transformations taking place (Tanaka and Hvitved-Jacobsen, 1998).

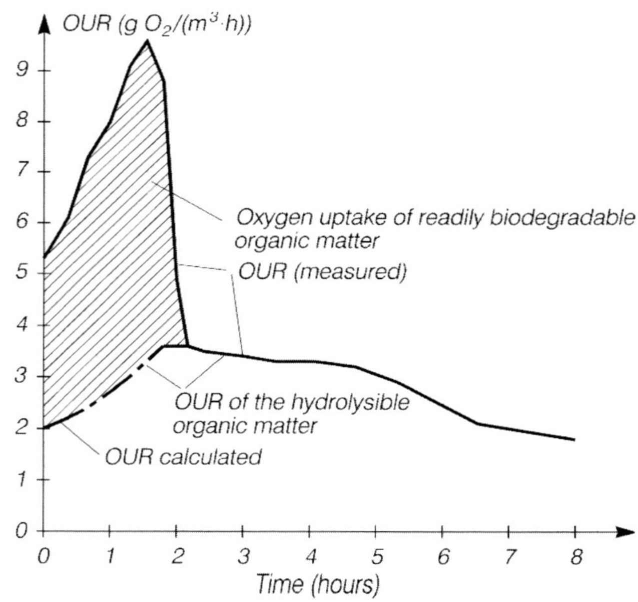

Figure 2.2.11. OUR variation of a wastewater sample in a batch experiment measured at a DO level of 6-8 mg O_2/l.

Unlike flows in gravity sewers under aerobic conditions, there is a low net transformation of organic matter taking place in pressure pipes caused by anaerobic processes, e.g. sulphate reduction and fermentation (Hvitved-Jacobsen *et al.*, 1995a; Tanaka and Hvitved-Jacobsen, 1998). There is a tendency for a net production of VFA and a net removal of dissolved carbohydrate, and overall a net production of 1–2 mg COD/(l^3h) has been observed under anaerobic conditions. No significant net production of biomass occurs under anaerobic conditions. As indicated in Figure 2.2.11, the processes related to the suspended particles in the bulk water phase interfere with the processes in the biofilm and the sediment. This is dealt with in Section 2.2.7.

The increased OUR value in the beginning of the experiment illustrated in Figure 2.2.11 is caused by an increased bacterial biomass concentration produced under non-limiting substrate conditions. When the concentration of the readily biodegradable organic matter becomes limiting for the growth process (in this example after about 2 hours), the OUR value declines to a level determined by hydrolysis.

2.2.2.6 Transformation of organic matter in resuspended sewer sediment

In a similar way to the processes occurring in the water phase, resuspended sewer sediments are transformed by the heterotrophic biomass, and the process results in biomass growth and a demand for energy. The main difference between the water phase and sediment processes is the relatively low biodegradability of the bulk of the organic matter in sewer sediments. Hence, the concept presented in Figure 2.2.9 can also be applied to the aerobic transformation of sewer sediment organic matter with slight modifications in order to focus upon the hydrolysis processes (Vollertsen and Hvitved-Jacobsen, 1998). The transformation processes can be monitored in a similar way to the processes in the water phase, based on the variation in OUR. The time required to monitor OUR of sewer sediment may be large as a consequence of the low biodegradability of the organic matter, as illustrated in Figure 2.2.12.

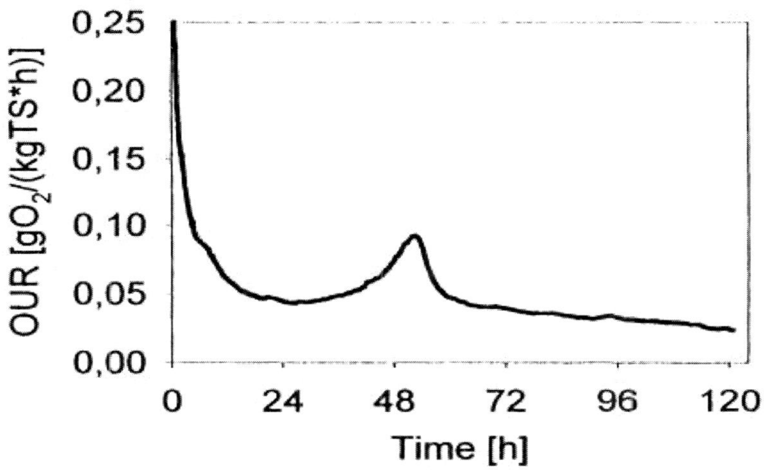

Figure 2.2.12. OUR variation of a sewer sediment sample in resuspension in a batch experiment measured at a DO level of 6-8 mg O_2/l.

2.2.3 Characteristics of bed deposits

2.2.3.1 Introduction

In principle at least, the extent of deposits in a sewer network may be investigated either remotely using closed circuit TV (CCTV) camera inspection, or directly by 'man'-entry.

However, there are a number of reasons why it may be impossible to obtain any useful information by either of the following methods.

- Small sewers are impossible to inspect, particularly if they do contain solid deposits.
- Large sewers may have continuous and deep flows, rendering useless normal equipment, although alternative 'remote sensing' devices which may be floated down the sewer are available for larger sewers (Ashley *et al.*, 1993).
- Surveys will observe only transient conditions – which may vary rapidly in time.
- The common practice of 'surveying' sediments by manhole survey (often from the surface) is misleading due to the important physical discontinuity caused by the manhole.

Bed deposits found in the invert of combined sewer pipes predominantly consist of granular mineral particles. The accumulation of these deposits has been observed in discontinuities of sewers, like for example, depressions and obstacles in pipe inverts, branches and poor pipe joints. The bed-load transport process is primarily responsible for the deposition of sediments (Section 2.3). The filling-up of depressions leads to a smoothing of the grade line and therefore ultimately results in a correction of existing hydraulic irregularities. On the other hand, singularities which are blocking the bed-load transport can cause a severe build-up of sediment bed deposits in the sewer length just upstream of the obstacle.

The most comprehensive and co-ordinated studies of sewer sediments and related phenomena have been those undertaken in Europe, notably in the UK, (Crabtree *et al.*, 1991; Ashley and Crabtree, 1992; Gent *et al.*, 1995), Belgium (Verbanck, 1992; Torfs, 1995), France (Bachoc, 1992; Laplace *et al.*, 1992) and, to a lesser extent, the Netherlands. Early studies carried out in Germany and Scandinavia only reported limited assessments of in-sewer deposit characteristics, usually based on flushing studies of small-sized collector sewers (e.g. Larson *et al.*, 1990). These latter studies dealt primarily with assessments of the extent of deposits within systems and the polluting consequences of storm flow wash-out, rather than the nature of the deposits. Important studies to assess the shear resistance of in-sewer sediments have been reported, however, from Germany, and sediment characteristics and deposition has been studied in detail in Hanover (Ristenpart, 1995a).

Since the last major review of sewer sediment knowledge (Verbanck *et al.*, 1994), considerable advances have been made in the understanding of where, when and how sewer sediments deposit in sewers. Unfortunately the larger organic solids have so far been largely ignored, and whilst their effects at the downstream end of large networks may be insignificant (although even that is unclear), these solids are likely to play a major part in determining the 'sediment' transport and deposition characteristics of a system, particularly for the sewers higher up in the system (Section 2.1.4).

The organic content of the sewer sediment strongly influences the physical as well as the chemical and biological characteristics.

2.2.3.2 Cohesive-like properties

Sewer deposits normally contain a certain amount of organic pollutant substances, which may lead to an organic binding creating a cohesive-like sediment bed. For this reason bed deposits often exhibit high (cohesive) shear resistance and require higher erosional shear stresses than non-cohesive granular material such as sand (Williams *et al.*, 1989). Once the cohesive-like deposits have been entrained and are in movement, they lose these bonding forces. Hence for the deposition and transport processes of solids, the cohesive properties may be ignored (Crabtree, 1989).

With the help of flushing tests, Stotz and Krauth (1986) found that the characteristics of sewer deposits were subject to changes during dry weather periods. The longer the bed deposits remained, the more kinetic energy had to be applied to cause erosion. After a residence time of only 12 hours a greater stability was observed. This was explained by the binding of organic bed particles and a consolidation of the sediment bed by dewatering. According to Crabtree (1989) and Ashley *et al.* (1992a), the organic binding of the single particles is caused by chemical interaction and mainly by biological degradation processes in the sediment bed itself.

The behaviour of cohesive-like sediment beds may be described with the help of rheological parameters. Rheology is defined as the science of the deformation and flow of matter. Parameters such as yield strength are used for the description of the strength of resistance to movement of cohesive materials. Small stresses at the water/solid interface deform the sediments elastically, but when the stress exceeds the yield value, continuous deformation occurs with viscous flowing of the material (Wotherspoon and Ashley, 1992). Applied stress rheometry (Williams *et al.*, 1989) has been used to investigate the elastico-viscous properties of sewer sediments (Section 2.3.4.5).

2.2.3.3 Physical characteristics of sewer sediments

The physical characteristics of sewer deposits can be described in terms of both individual particulates and bulk properties. The hydraulic and structural conditions in the sewer, together with the nature of the inputs, will control the type of material which deposits at a given location.

Crabtree (1989) proposed a working taxonomy based on 4 primary classes A, C, D, E, with a fifth class B, comprising agglutinated or cemented Class A material. The primary characteristics of these classes are given in Table 2.2.6. The large organic fraction found in Classes C and D shows that these sediments differ markedly from traditional 'sediments' found in pipe inverts. The clear distinction implied by these classes is not in fact so apparent in reality, and the taxonomy should be viewed only as a means of rapidly making a pragmatic classification of the solids found in sewer deposits. Sediments are usually mixed between the class types as illustrated in Table 2.2.7, from Ashley *et al.* (1989) and Ashley *et al.* (1992a). There is also some uncertainty as to what constitutes sediments and what constitutes 'bed-load'. The Class C sediment defined above may be a true sediment only in stagnant zones as it is only weakly resistant to shear (see also Section 2.3.4.3). This is illustrated in Figure 2.2.13a.

Table 2.2.6. Sewer sediment taxonomy as proposed for UK usage (Crabtree, 1989)

Sediment type	Description/ where found	Wet density x 10^3 kg/m^3	% by granular particle size (mm) minimum-mean-maximum			Organic content (%)
			< 0.063	0.063–2.0	2.0–50	
A	Coarse, granular bed material – widespread	1.72	1-6-30	3-6-87	3-33-90	7
C	Mobile, fine grained found in slack zones, in isolation and overlying Type A*	1.17	29-45-73	5-55-71	0	50
D	Organic pipe wall slimes	1.21	17-32-52	1-62-83	1-6-20	61
E	Fine grained mineral and organic material found in CSO storage tanks	1.46	1-22-80	1-69-85	4-9-80	22

*Since these results were published, Class C material now also includes larger (organic) particles.

Table 2.2.7. Particle characteristics in Dundee sediments, as perceived with mixed deposits

Particle characteristic		Perceived class			
		1: A	2: A/C	3: C/A	4: C
Vol. solids (%)	Max	3.4	10	17.6	95
	Min	0.2	0.2	0.5	4
D_{10} (mm)	Max	0.4	0.2	0.4	0.2
	Min	0.05	0.01	0.01	< 0.1
D_{50} (mm)	Max	2.1	1.3	0.7	0.9
	Min	0.55	0.2	0.09	< 0.8

Figures 2.2.13a, 2.2.13b, 2.2.13c and 2.2.13d show sediment classes C, organic NBS, A and encrustations respectively. Figure 2.2.21 shows a slime layer at the dry weather flow-air interface.

The sediments found in different parts of a combined sewer network will have different characteristics, as described in Table 2.2.8. In general, the larger particles will be found near the heads of networks, with the finer particles in the downstream parts of the systems (e.g. Bachoc, 1992). The particle sizes for the granular materials found in sewers around Dundee are illustrated in Figure 2.2.14 (Crabtree *et al*., 1991). The Dundee main sewer deposits may be compared with French, German and Belgian results, as illustrated in Figure 2.2.15 (Ashley and Crabtree, 1992).

Figure 2.2.13a. Class C solids overlying bulk (Class A) sediment in Dundee (photograph Richard Ashley)

Figure 2.2.13b. Organic NBS overlying bulk (Class A) sediment in Le Marais catchment (photograph Mathieu Ahyerre).

Figure 2.2.13c. Class A sediment being removed from the main Dundee Interceptor sewer (photograph Neil Watson).

Figure 2.2.13d. Build-up of solids on the walls in a Dutch concrete pipe (these are encrustations and not in the standard classes) (photograph Robin Veldkamp).

Table 2.2.8. Deposition of sediments in combined sewers

Sewer type	Location of deposits	Nature of deposits
Small collector sewers	At discontinuities*, otherwise randomly located in discrete 'lumps'	Large organics, sand and gravel
Trunk sewers	At discontinuities*, otherwise only larger, denser particles deposited	Large granular particles, some intermixed (large) organics
Interceptor sewers	At discontinuities*, otherwise where gradients slack	Fewer large organics than above, plus finer granular particles than above

* Structural or hydraulic discontinuities

The sediment particle size ranges observed for the various French sites (Chebbo *et al.*, 1990; Bachoc, 1992) are summarised in Table 2.2.9, together with sewage particulate data, reported data for Brussels from Verbanck (1990), (1992) and ranges for the UK. It should be noted that the particle size of the smallest fraction (< 63μm) has in each case been determined indirectly by settling velocity determination. In view of the uncertainties surrounding this practice (see Appendix) these results should be treated with caution.

The Belgian data are reportedly for samples with a small (< 5%) volatile fraction, although the proportionate volatile content of each particle size (sieve) class was found to increase with decreasing particle size. The overall frequency distribution of volatile fraction for these samples is shown in Figure 2.2.16.

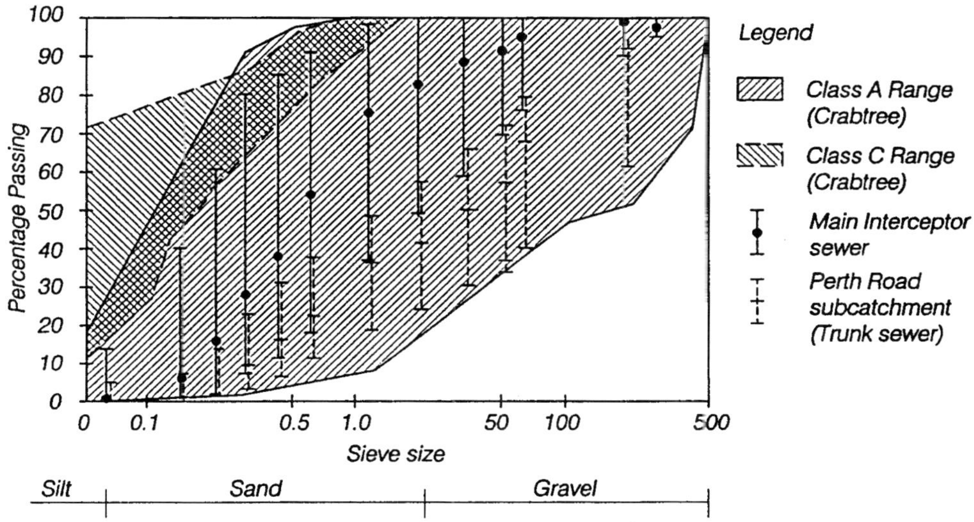

Figure 2.2.14. Sewer sediment particle size distributions – Dundee and UK data (Crabtree *et al.*, 1991).

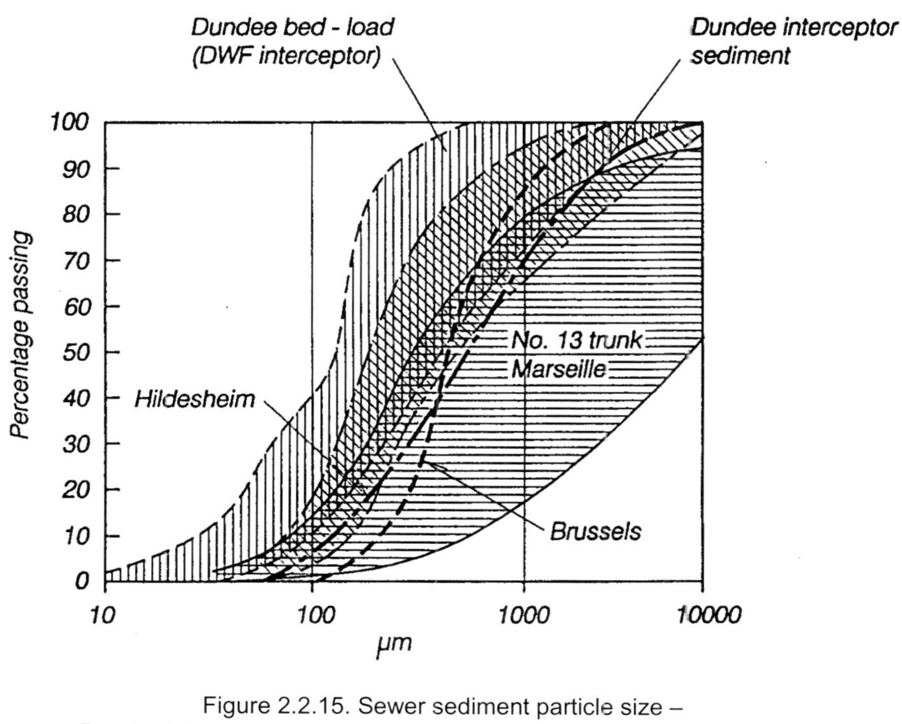

Figure 2.2.15. Sewer sediment particle size – Dundee/Marseille/Hildesheim/Brussels (Ashley and Crabtree, 1992).

Table 2.2.9. Particle size ranges – sewer sediments

Country	Sewer	Particle size range (mm)				
		< 0.1	0.1–0.2	0.2–2	2–20	> 20
France	Man-entry 'upstream' sewers	2	8 – 10	23–60	25–47	7–18
	Trunk sewers	4	5–10	38–55	25–45	8–10
	Interceptor sewer	4.5	<-----35.5	-------->	<-------60	-------->
Belgium	Interceptor sewer	3	17	78	2	NR
	All sewers in Brussels	20	30	<-------43	-------->	9
UK	Interceptor sewer	1 – 20	1–21	32–80	65–20	NR
	Trunk sewer	1 – 8	1	18–41	41–50	NR

NR Not recorded.

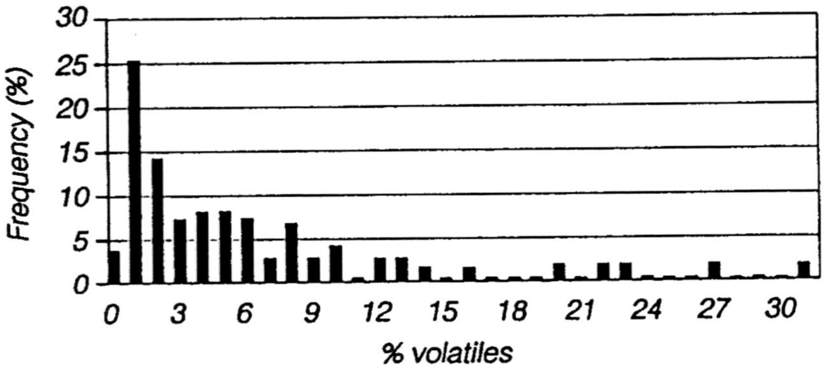

Figure 2.2.16. Sewer sediments in Brussels from a variety of sources – volatile (organic) frequency distribution (adapted from Verbanck, 1992a).

The UK data are for samples with organic contents from about 3 % up to 96 %, depending upon the WRc classification. Generally the Class A sediments are those most granular, and the French and Belgian sediments appear to correspond with this type.

As part of the French programme a continuous study has been carried out of the No. 13 trunk sewer in Marseille which has shown a gradual sediment build-up following cleaning (Laplace *et al.*, 1992). The Marseille data show an 'evolution' of particle sizes temporally and spatially. Initially, the sediments dispersed along the sewer length were of relatively uniform size, but these were observed over the course of a year to become more graded as the bed developed. Near the head of the 2.75 m high sewer, the D_{50} of the bed particles gradually increased from 2 mm up to about 8 mm, with smaller particles (2 mm) predominating near the end of the studied sewer. No significant consolidation of the bed was noted, and throughout the study the bed was found to be composed of relatively large particles, with less than 10 % having a size of 200 μm or less. This observation is consistent with the results for flushing tests carried out in the USA, where the flush wave apparently 'leached' many of the finer particulates from the bed deposits. During storms in Marseille the particles in suspension 30 cm above the bed were sampled and shown consistently to be predominantly (70 %) smaller than 100 μm. These particles had a low D_{50} of 40 μm. Sample preparation using ultrasound dispersion was applied to some of the French samples. Testing following this showed an apparent reduction in particle sizes, presumably because of the breakdown of flocculated particle groups (Table 2.2.4). The Brussels study also indicated that the finer particulates (< 125 μm) were being leached from the sediments by diurnal increases in dry weather flow. The organic content of the sediment throughout the sediment bed, which had a depth of 150 mm, was examined and found on a number of occasions to be highest in the deeper layers. One explanation for this was attributed to 'shielding' by the overlying material preventing washout of the finer (mostly organic) particles.

During sediment studies in Dundee, more than 150 bed samples have been taken from the main Interceptor sewer and contributing trunk and collector sewers. The data are for sieved samples following ashing at 550 °C and represent only the mineral sample fraction. Coulter Counter analyses of the smaller-sized (ashed) fraction were used to further investigate the particulates below 63 μm in size. The D_{50} for the Interceptor sewer sediment samples was found to be in the range 0.2–2.0 mm, whereas the steeper gradient Perth Road trunk sewer samples had a D_{50} of 2–20 mm. The depth of deposits in the Interceptor sewer averaged 100-200 mm and approximately 75 mm in the Perth Road trunk sewer. The trunk sewer deposits had an average volatile fraction of about 3 %, compared with a range up to 20 % in the Interceptor sewer. The average bulk density of the Interceptor deposits was found to be 1580 kg/m^3 compared with 1806 kg/m^3 for the trunk sewer. These figures are comparable with the Brussels one of 1510 kg/m^3.

Results from the Hildesheim sewer in Germany (Ristenpart, 1995a) have shown that the characteristics of sediments vary extensively both spatially and temporally. Of note is the ageing of 'fresh' sediments at a point in the interceptor sewer, illustrated by the increase in bulk density from 1200 kg/m^3 to 1840 kg/m^3 over an (unspecified) period. These observations of coarser, more granular, sediments being found in trunk sewers where gradients are steeper than in interceptors are consistent with the other studies described, and indicative that the finer organic particulates are more likely to deposit and remain in large interceptors with slacker bed gradients rather than local collector or trunk sewers.

The shear strength characteristics of sediment deposits have been examined in limited French studies and in more detail for UK studies in which disturbed samples have been analysed using applied stress rheometrical techniques. These have revealed that deposits can have an apparent yield strength in excess of 2500 N/m^2 (Williams *et al.*, 1989; Wotherspoon and Ashley, 1992).

Figure 2.2.17 illustrates a typical result from creep tests carried out using an applied stress rheometer on a sample of combined sewer sediment. Additional rheological measurements of the yield strength of sediment bed samples from a combined sewer in Hildesheim (Germany) were undertaken by Ristenpart (1995a) to compare the transferability of the UK results. The range of values obtained corresponds to that determined by Wotherspoon (1994), but on average the German sediments showed significantly higher yield strength. The German deposits showed only a weak coherence between shear strength and moisture content, unlike those tested in the UK study.

In addition to the use of these sophisticated specialist rheometric techniques, a range of standard geotechnical tests have been applied to in-situ sediments in Dundee and in France (Laplace *et al.*, 1990). Vane tests using standard soil testing equipment for Dundee, Marseille and Paris sewers have indicated very similar yield strengths for deposits, with up to 2-10 kN/m^2 and 2–20 kN/m^2 being measured respectively in Scotland and France. The development of a cohesive-like structure and resistance to shear stresses in sewer sediments is a complex process involving consolidation, chemical and biological interactions and temporal and spatial changes. Many questions remain regarding the development of a bed structure from depositing suspended or bed-load material, and its transition from a fluid to a weak and gradually strengthening 'solid' bed (Ristenpart, 1995a). Even in the artificial, but controllable, conditions of a laboratory (Nalluri and Alvarez, 1992; Stotz and Krauth, 1986;

Torfs, 1995; Skipworth *et al.*, 1996), the study of changes in cohesive sediment beds in response to imposed flow conditions is difficult. How such studies relate to field conditions is not always clear, but a number of conclusions may be made.

- Sewer sediments can have cohesive like properties, but these may not necessarily be because of cohesive forces in the classical sense, which are caused by electrostatic attraction – in sewers the effect is usually produced by agglutination.
- Sewer sediment structures very quickly recover following deformation – i.e. regain strength rapidly (in hours rather than days).
- Measured bulk yield strength generally exceeds by several orders of magnitude the likely applied shear stresses from sewage flowing over the deposits.
- Yield strengths vary with solid/liquid phase proportions and bulk density – hence any depth layering or consolidation effects are important.

Settlement velocity tests on sewer sediments have been found to be very difficult to carry out due to the sticky nature of the material. However, some success has been achieved using a test developed in France (see Appendix).

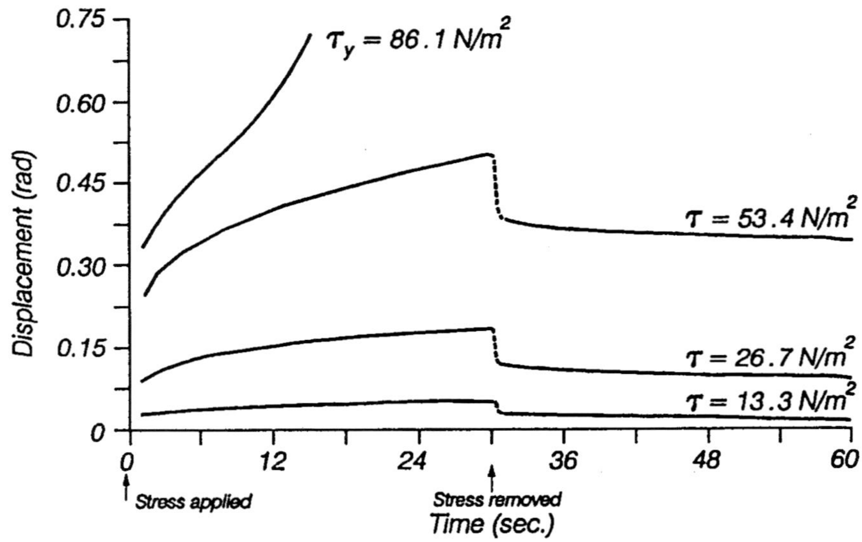

Figure 2.2.17. Creep test result on a sample of sewer sediment from Dundee using an applied stress rheometer (Wotherspoon and Ashley, 1992).

2.2.3.4 Chemical and biochemical properties

Type A sewer sediment material is the most commonly found in combined sewer systems. Numerous studies have been collated to demonstrate the deposit properties given in Table 2.2.10.

Bulk characteristics of sewer sediments are now well established, e.g. Table 2.2.11 (Ashley and Crabtree, 1992) which illustrates results from a wide range of UK sewers. It is clear that these do vary spatially and temporally over short and long time and distance scales, and that there are important national differences.

Table 2.2.10. Chemical and biochemical properties of sewer sediment bed material (type A)

Parameter	Unit	Sampling programme	
		min. – max.	mean
Total solids	(g/kg)	450–650[4]	540[4]
		355–815[3]	691[3]
			734[2]
			796[5]
Volatile solids	(%)	3–30[2]	10[2]
		1.1–18.6[1]	4.4[1]
COD	(g/kg)*	6–80[2]	23[2]
		14.9–69[3]	69[3]
BOD_5	(g/kg)*	1.3–14[2]	4.2[2]
		3.2–89[3]	14.2[3]
BOD (4 hours)	(mg/kg)*	100–700[2]	400[2]
Organic N	(mgN/kg)*	200–1500[2]	800[2]
Ammonia (NH4-N)	(mgN/kg)*	10–300[2]	100[2]

References: 1) Brombach *et al.*, 1992; 2) Crabtree, 1989; 3) Ristenpart, 1995a; 4) Schmitt and Seyfried, 1992; 5) Verbanck, 1990.
* dry weight.

Sewer sediment bed material exhibits wide heterogeneity in terms of all of the characteristics in Table 2.2.10 (Ashley *et al.*, 1990; Ristenpart, 1995a), hence the definition of general properties is not possible. The primary reasons for the non-uniformity and variability of these sediments are:

- different origins of bed deposits;
- varying inlet conditions (gully pots, etc.);
- diverse sewer classes (collectors, trunks, interceptors);
- dynamic transformation processes in the sewer itself;
- the general imprecision of the analytical techniques (problems caused by the nature of the liquid and solid mixture as well as by unusual parameter ranges) (McGregor *et al.*, 1993; Verbanck *et al.*, 1994).

Based on Table 2.2.10, the average total solids content of the bed material is typically between 540 and 724 g/kg. This means that there is a water content in the bed deposits of about 25 % to 50 %. The low organic content (average 10 %) shows that mainly mineral material make up the deposits. The grain size distribution of the bed material (D_{50} from tenths of millimetres up to centimetres) illustrates a larger size range in comparison with the suspended solids and also when compared with bed load (Ristenpart, 1995a). As a consequence, the bed deposits have the highest settling velocities (Ashley and Verbanck, 1996).

Generally, two layers of sediment can be identified. In the upper part of the sediment bed, an increase of total solids content as well as bulk density with depth is typical, whereas in the lower layers, these parameters may decrease downwards or remain constant. The distribution of mineral, inorganic solids content is virtually fixed and equals the total solids content as, typically, more than 90 % of the bed deposits are inorganic.

Table 2.2.11. Combined sewer sediments – chemical characteristics

Measured parameter	Sediment origin / class													
	Dundee interceptor sewer				Dundee Perth Road trunk Sewer		UK Averages							
	A/C		Near bed solids (C)		A/C		A		C		D		E	
	Mean	Range	Mean	Range	Mean	Range	Mean	CV %	Mean	CV %	Mean	CV %	Mean	CV %
Bulk density (kg/m^3)	1559	< 2150	1070	< 1450	1807	474–2111	1720	-	1170	-	1210	-	1460	-
Solids total (%)	56	0.6–82.4	4.7	2.9–50.2	67	48–85	73.4	-	27	-	25.8	-	48	-
Volatile (%)	3	0.2–17.6	76	14.7–97.4	2.9	0.7–10.3	7	77	50	47	61	54	22	77
COD (g/kg)*	16.2	0.6–52.2	1691	277–6353	2.5	< 4	23	63	76	23	193	83	48	70
BOD_5 (g/kg)*	12.4	0.9–43.4	369	73–873	0.7	< 0.6	4.2	91	20	51	103	95	13	61
Amm-N (g/kg)*	0.28	0.007–3.5	7.5	0.1–237	0.03	< 0.065	-	-	-	-	-	-	-	-

* g/kg dry solids

classes:

A - coarse granular material

C - mobile, fine grained, overlying class A

D - organic wall slimes

E - deposits found in tanks

The volatile solids content measured by Ristenpart (1995a) also corresponds with the physical properties, being significantly higher in the near-invert layer of the bed. Loss on ignition in the lower area (8.7–13.3 % on average) was 1.4–2.8 times higher compared with the upper area (3.8–8.2 %). Schmitt and Seyfried (1992) and Verbanck (1990) also found a smaller organic fraction in the upper layers of the bed deposits. The former explained the increase with depth by anaerobic degradation processes near the invert, the latter explained the decrease in the upper area by the wash-out of the finer (organic) particles, due to diurnal shear stress peaks in the sewage flow (elutriation process).

The parameters listed in Table 2.2.11 show that sewer deposits are potentially responsible for high levels of pollution in terms of oxidisable constituents. The relatively low Biochemical Oxygen Demand after a 4 hour period compared with the consumption after 5 days, shows that the oxygen demand exists as slow hydrolysible fractions and not as easily biodegradable organics. The characteristics (especially COD, BOD_5 and NH_4-N) and volatile solids content exhibit very broad variability compared with the physical parameters.

In order to assess the effects which may be caused by erosion of a sediment bed as a potential source of pollutants, it is not the absolute concentrations, but the amount of potential pollutants accumulated per metre pipe length which is relevant. For a typical situation in a pipe (D = 1.5 m and sediment bed depth 0.15 m), the amount of pollutants in the bed deposits as well as in the raw sewage must be taken into account. Crabtree (1989) determined the sediment bed pollution potential as about eight times higher than the sewage flow (BOD_5, based on weight). Similar estimations of Ristenpart (1995a) resulted in even higher relative proportions of potential pollutant deposits in units of COD. For dry weather flow conditions, the pollution potential of the bed material was found to be 29 times higher than for the sewage; for storm flow conditions this was reduced to about 10.

When assessing the pollution potential of the bed deposits, it should be appreciated that the estimates above are based on the potential total pollution which could be released by bed deposits. The actual impact of the pollutant deposit will probably not be so significant, as:

- only a part of the bed deposits will erode during storm events;
- a part of the mobilised particles will settle downstream in the sewer. McGregor *et al.* (1993) determined this in terms of COD to be 15 % of the total amount of pollutants in the sediments (See Section 2.3.4.5.2).

In a study concerning the origin of solids in combined sewage flows, Chebbo and Bachoc (1992) determined mass balances of pollutants for wet weather periods. They concluded that 22 % of the bed deposits contributed to the annual pollutant load for each of the parameters COD and BOD_5.

The aerobic biodegradability of resuspended organic matter from sewer sediments is low compared with suspended wastewater organic matter (Vollertsen and Hvitved-Jacobsen, 1997). Due to the time scale of the processes involved, the degradation of the organic matter taking place in the sewer bulk water phase itself under storm events is small compared with the total amount of degradable organic matter in transport. The time scale of the biodegradation process has been found to be consistent with delayed oxygen consumptions measured after CSO discharges into receiving waters (Hvitved-Jacobsen, 1982).

Recent studies in Denmark have found that about 82 % of the organic matter in sewer sediments is slowly hydrolysible or inert. The biodegradability of the organic matter was not uniformly distributed between a range of sediment fractions with different settling velocities. The fraction of the sediments that settled slower than 17 mm/s contained some 20 % of the organic matter but was found to include only 67 % of the slowly hydrolysible or inert organic matter. Hence it is recommended that the determination of biodegradability is essential when addressing CSO impacts on the DO concentration of receiving waters.

Very few studies have been concerned with heavy metal pollution from bed deposits. Results from two German investigations of bed deposits (Brombach *et al.*, 1992; Ristenpart, 1995a) are shown in Table 2.2.12.

Table 2.2.12. Heavy metals in sewer bed deposits

Material	Sewer bed deposits (study site)		Sludge
	Bad Mergentheim** samples not fractioned (Brombach *et al.*, 1992)	Hildesheim*** fraction D < 0.2 mm (Ristenpart, 1995a)	Thresholds for agricultural use in Germany (1998 data)
Lead (mg/kg)*	14	173	900
Copper (mg/kg)*	26	168	800
Zinc (mg/kg)*	327	14	2500
Nickel (mg/kg)*	14	26	200
Cadmium (mg/kg)*	0.3	1.8	10

Each result is based on 2–3 random samples.
* concentrations based on dry weight
** residential catchment
*** mixed industrial and residential catchment.

Xanthopoulos and Augustin (1992) observed varying distributions of heavy metals on different particle size fractions of bed deposits from combined sewers and storm sewers. Fine particles from storm sewer bed deposits were found to be significantly more polluted than coarser particles. In contrast, combined sewer deposits showed a more uniform distribution of heavy metals. These differences were explained by the organic coating of the latter deposits and the more extensive biological processes taking place in combined sewers.

Brombach *et al.* (1992) found that, due to the small specific surface, the relatively coarse bed deposits were the least polluted material compared with sewage and sewer slime. Chebbo and Bachoc (1992) determined from storm sewage pollutant mass balances that 28 % of the annual load of lead in combined sewage flow originates from the bed deposits.

2.2.4 Processes related to bed deposits

2.2.4.1 Transformation processes

2.2.4.1.1 Introduction

Very few studies have been directly concerned with the transformation processes of sewer deposits, e.g. investigation of sulphate reduction, (Schmitt and Seyfried, 1992; USEPA, 1985). However, in some cases, e.g. concerning microbiological processes, sediments may be compared with biofilms (cf. Section 2.2.5). Even under aerobic conditions in the bulk water phase, only the upper few millimetres of the sediment are aerobic.

The results reported in this section mainly originate from investigations on variations in bed deposit properties. High temporal variations as well as vertical gradients of bed deposit characteristics have been found, particularly for pollutants (COD, BOD and volatile solids content). These variations in time are related to in-sewer processes, transformation processes in the sewer sediment bed, exchange processes between bed deposits and sewage and solid transport processes.

2.2.4.1.2 Temporal fluctuations of deposit properties

Ristenpart (1995a) observed the characteristics of bed deposits in a 1.5 m interceptor sewer at a fixed observation point over a period of 9 months. The physical properties, bulk density and total solids content showed only minor variability over the 9 months (coefficient of variance CV < 10 %). However, the volatile solids content (CV = 28.5 %), COD (CV = 78.1 %) and BOD_5 (CV = 77.1 %) showed much greater variability. The changes were believed to be attributed to in-sewer processes.

A general relationship between the immediate history of deposits before sampling and the pollutant characteristics cannot as yet be established. The hypothesis that the variations correspond to the rainfall-runoff history, or to the period with no disturbance of the sediment bed (period without erosion) was contradicted by Ristenpart (1995a), by continuous measurements of the sediment bed level and a lack of correlation between the erosion 'events' and the bed properties.

The broad ranges (in terms of minima and maxima) of analysis results of deposit sampling programmes published by several authors and summarised in Table 2.2.10 also show wide deviations. These deviations correspond to the range of variations observed by Ristenpart (1995a) and therefore confirm that this variability is widely observed.

The complex fluctuations of BOD_5 and COD are shown in Figure 2.2.18. The maximum BOD_5 (23.0 g O_2/kg) and COD (134.4 g O_2/kg) were measured on 10.05.93 after a very long dry weather period of 3 weeks without any erosion of sediments. On another occasion when the preceding time period was similar, without bed erosion (15.02.94), the pollutant concentration in the deposits was found to be significantly lower. In summary it is apparent that the nature of the bed material characteristics over the prolonged measurement period was very variable for all the parameters measured. No general trends were found from the investigation, and the wide variability of deposit characteristics could not be fully explained.

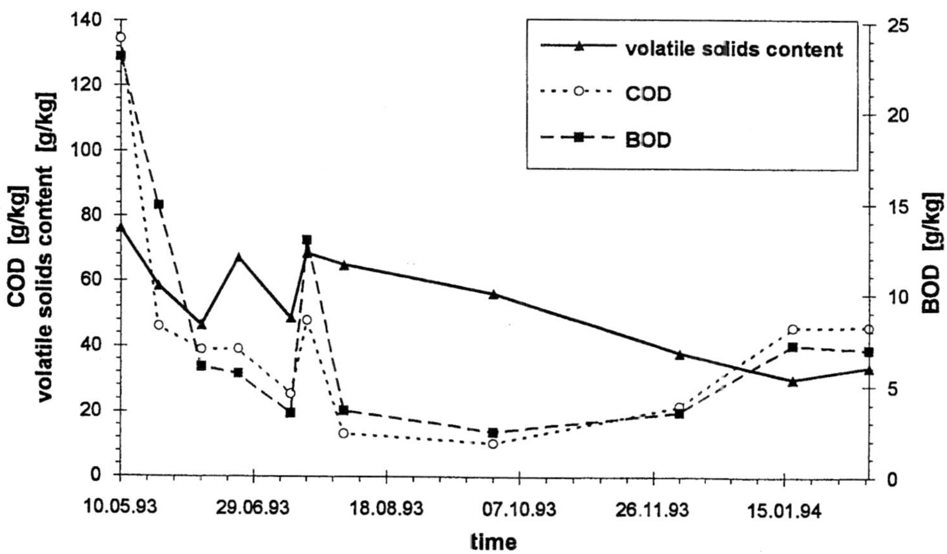

Figure 2.2.18. History of bed deposit characteristics in an interceptor sewer (adapted from Ristenpart, 1995b).

2.2.4.1.3 'Ageing' of deposits

Crabtree (1989) first postulated an expected 'ageing' of deposits which should occur due to interactions between bed deposits and sewage. The characteristics of deposits with different 'ages' have been investigated by Ristenpart (1995b).

Samples were taken at three different points in a large sewer where the residence time of deposits was known. One sample was obtained a few days after cleaning of the whole sewer length and represents the fast reaccumulation of fresh sediments. A second sample was taken from a semi-permanent bed where some erosion and deposition occurred (sediments of middle age). The third sample was extracted from a point where there was an undisturbed sediment bed (old deposits). The results are shown in Table 2.2.13.

Table 2.2.13. Variability of characteristics of bed deposits due to 'ageing' (Ristenpart, 1995b)

Deposits		Residence time		
		Short (newly deposited)	Medium (disturbed from time to time)	Long (consolidated)
Bulk density	(kg/m^3)	1200	1510	1840
Total solids	(g/kg)	355	705	812
Volatile solids	(% dry matter)	27.0	8.8	2.4
pH	(-)	5.68	7.11	7.66
BOD_5	(g/kg wet weight)	31.6	12.5	2.7
COD	(g/kg wet weight)	95.6	55.3	19.0

The 'ageing' of sediments is clearly illustrated in Table 2.2.13, both physical and pollutant characteristics are shown to be altered as a consequence of in-sewer processes. As expected, total solids content and bulk density were found to be low in the newly deposited sediments and high in the old, consolidated deposits. Volatile fractions are directly related to the pollutant concentration and a decrease in organic matter coincides with a reduction in COD and BOD. The lower pH-values for fresh sediments probably indicate anaerobic biodegradation processes, whereas the high pH-value and low organic content for the old deposits may show that most of the biodegradable organic substrate has been consumed. The reduction of pollutant concentration was found to be significant, and the older deposit samples contained only 8.4 % of the BOD_5 and 19.9 % of the COD when compared with the fresh samples. These uniform changes with time corresponded with the results for the semi-permanent sediment bed, for which the characteristics were always found to be between the old and the very fresh deposits.

One of the important conclusions from this study is that fresh deposits are shown to have the highest pollutant potential. Considering that the critical shear stress for erosion is much lower for the fresh deposits, they will be readily entrained, hence the polluting potential of fresh sediments is likely to be much more significant compared with older, consolidated sediments.

Even short time scales can produce significant changes in bed structure. For example, Vollertsen and Hvitved-Jacobsen (1997), used a series of laboratory test flumes with reformed sediment beds, in which methane production was so significant over even just a few days, that the bed volume increased by 20–40 % due to the cavities. This resulted in a corresponding decrease in bed strength. Hence the microbial processes have been shown by this work to be crucial in determining bed strength.

2.2.4.1.4 Vertical variability of bed deposit characteristics

Ristenpart (1995b) investigated the vertical distribution of material characteristics in bed deposits in an interceptor sewer. By fractionated sampling it was possible to investigate vertical profiles of bed deposit properties. An example of the results for a deposit with a depth of 17 cm is given in Figure 2.2.19.

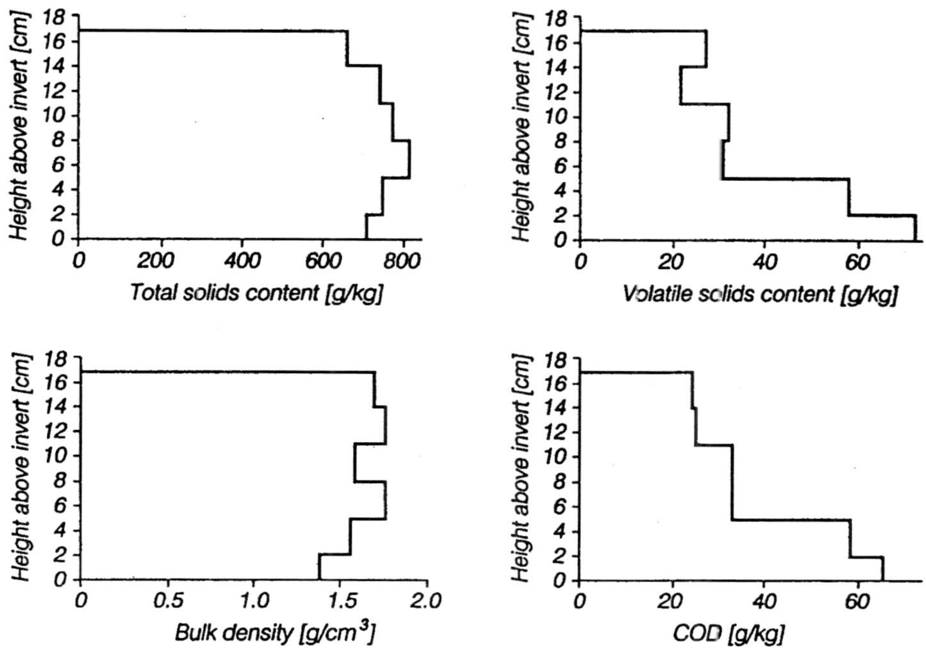

Figure 2.2.19. Vertical distribution of bed deposit properties (adapted from Ristenpart, 1995b).

2.2.4.1.5 Sulphate reduction

Sulphate reduction (or hydrogen sulphide formation) may take place under anaerobic conditions in sewer deposits. Sulphate reduction characteristics for sewer deposits and sewer biofilms are in many respects similar (Section 2.2.5). However, surface production potential for deposits may exceed what is observed for sewer biofilms on walls (Schmitt and Seyfried, 1992). Bjerre *et al.* (1998a) for example, reported values of 60–170 gS/m^2h in biofilms.

Sulphide formation, effects and control are not dealt with in detail in this report. These phenomena are described in a number of publications. However, limited discussion is given in Section 3.3.

2.2.4.1.6 Potential polluting properties

The variation of the potential polluting properties with sediment depth observed by Ristenpart (1995b) may be illustrated for COD. The marked gradient of this parameter is clearly recognisable in Figure 2.2.19. The pollutant concentrations of the lowest layer (58.0-78.7 g O_2/kg wet weight) were found to be 1.4–2.7 times higher than for the upper layer (23.8–52.3 g O_2/kg wet weight). The increase of COD with depth corresponds with a decrease of pH and an increase in volatile solids content. These two factors, together with the black colour of the samples may indicate anaerobic biodegradation processes predominating in the deposits nearest to the sewer invert. In the upper layer the concentrations are lower, because there are exchange processes of organic matter and degradation products with the overlying flowing sewage.

Schmitt and Seyfried (1992) also found an increase of COD as well as organic acids with depth in deposits in combined sewers. The highest concentrations were found in the lowest layer directly above the sewer invert (COD = 6.33 g/l, acetic acid = 1.08 g/l, both analysed from centrifuged liquid samples taken from sediment samples). The high pollutant content near the bottom was explained by biological processes under anaerobic conditions (fermentation) producing energy-rich organic acids, which provide evidence for high loadings of organic substrate. This is confirmed by the distribution of the volatile solids fraction, which was significantly more concentrated in the deeper layers (bottom: 17–27 %, top: 4–7 %). Schmitt and Seyfried (1992) were able to draw parallels from these results between the structure of sewer sediment bed deposits and biofilms.

Investigations at Aalborg University, Denmark, indicate that a considerable part of the organic matter in sewer deposits is biodegradable. A typical oxygen utilisation rate (OUR) for the deposits in suspension was found to vary between 0.1 and 0.2 mg O_2 (g dry deposits per·h), observed to be virtually constant over several days. This accords with the observations of delayed effects of combined sewer overflow discharges on the DO concentration in receiving waters caused by the washout from sewers of bed deposits (Vollertsen and Hvitved-Jacobsen, 1997).

2.2.4.2 Settleable solids, heavy metals and organic micro pollutants

Heavy metals and organic micro pollutants are adsorbed by sewer solids. This is significant for the treatment of combined sewage. By employing a settlement process, not only will the settleable solids be removed, but also many of the associated pollutants. The concentrations of heavy metals and organic micropollutants which are adsorbed on sewer solids correspond with the phases which they are generally associated with (Figure 2.2.20).

As expected, the concentrations of pollutants associated with granular sediments are small. The amount of pollutant associated with the solid particles is most influenced by the surface area available. In Figure 2.2.20, the sediment particles generally have a small specific surface area. Slime 2a taken ‘above the water level’ (a.w.) is a mixture of organic and mineral material which would not be eroded during storms. Slime 2b was taken ‘under the water level’ (u.w.) and was mostly real slime, i.e. microorganisms. This material had the highest concentration of heavy metals and organic micro pollutants. Sample Slime 1 was a mixture of these two phases of slimes.

The combined sewage samples had high concentrations of heavy metals and organic micropollutants. During dry weather flow, pollutants were adsorbed and enriched by the slime organisms. During light rain, some of the pollutants washed from the urban surfaces are adsorbed by the slime layer. During high storm flow, however, slime and associated pollutants will be eroded and high concentrations of pollutants in the combined sewage will be measured. Reiff (1992) reported that slime is easily eroded during storms.

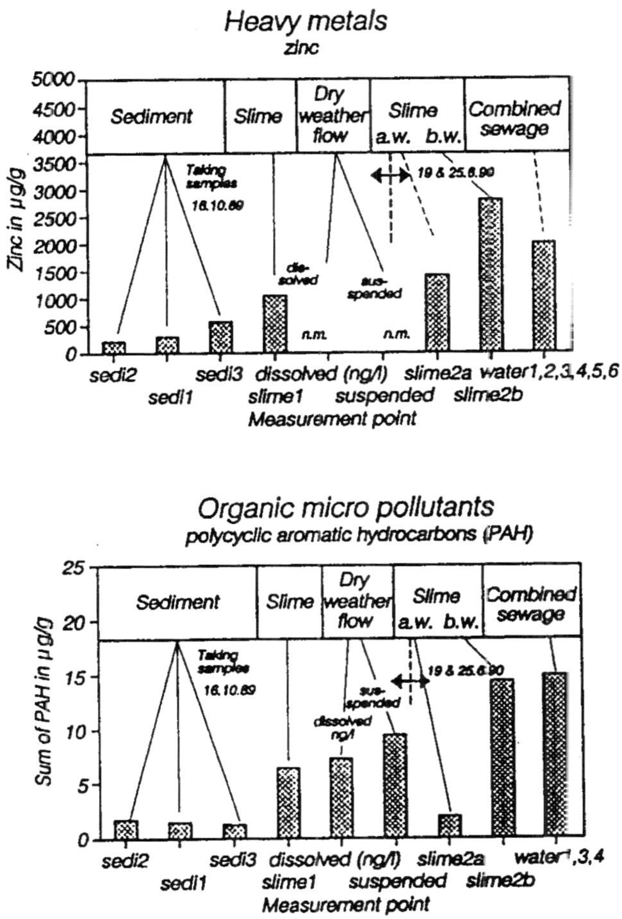

Figure 2.2.20. Pollutant load of different types of sewer solids samples (adapted from Michelbach and Wöhrle 1994).

2.2.5 Biofilms

2.2.5.1 Characteristics

In all types of sewers, biofilms are produced at the surfaces exposed to the water phase, as illustrated in Figure 2.2.21. Biofilms are often referred to as 'slimes' and consist mainly of microorganisms, extracellular polymeric substances (EPS) produced by the microorganisms and adsorbed organic and inorganic compounds from the wastewater. Biofilms in gravity sewers usually have a thickness of a few millimetres, depending mainly on flow conditions in the system. The biofilm thickness in pressure mains is usually lower due to a higher water flow velocity and lower bacterial growth rates during anaerobic conditions. In each case, however, examples may be found of very thick biofilms, where a significant increase in the friction factor has led to a reduced hydraulic capacity of the sewer (Characklis and Marshall, 1990; USEPA, 1985).

Figure 2.2.21. Slime layer on the walls of the Rivoli sewer in 'Le Marais' catchment (Paris) (photograph Mathieu Ahyerre).

The composition of sewer biofilms has not been investigated extensively. Microorganisms are embedded in a matrix of EPS which is presumed to be mostly polysaccharides produced by the bacteria (Characklis and Marshall, 1990). The EPS fraction is the largest organic fraction in biofilms with reported values up to 90 % of the total organic content. Only limited studies of the total composition of biofilms in terms of protein, carbohydrates and humic substances have been undertaken for sewer biofilms (Jahn and Nielsen, 1997). The protein fraction in biofilms taken from various gravity sewers was found to comprise the largest organic 'pool', making up approximately 50 % of all the organics, whilst carbohydrate and humic substances were each responsible for 15–25 % of the organics. No published information exists for biofilms in pressure mains.

2.2.5.2 Processes related to biofilms

Biofilms are responsible for a number of chemical and microbiological processes taking place in a sewer:

- degradation of dissolved and particulate organic compounds;
- production of biomass which detaches into the bulk water;
- consumption of oxygen and nitrate if available and, in some cases, sulphate reduction and sulphide production.

Biofilms are also believed to provide a means of developing bed strength in deposits (Crabtree, 1989), and recent evidence to support this has been produced in Danish studies (Vollertsen and Hvitved-Jacobsen, 1997).

2.2.5.3 Biomass production in biofilms

During the transportation of organic compounds in sewers, a proportion of these are converted to biomass by the various microbial processes. The generated biomass detaches from the biofilm and is transported to the treatment plant or via overflow structures, to receiving waters together with the biomass produced in the bulk water. A significant biomass production can take place (Raunkjær *et al.*, 1995). A simple method to assess the amount of biomass produced is to take an estimate from the consumption of electron acceptors (dissolved oxygen, nitrate and sulphate) in the system. If the actual electron acceptor consumption is known, the biomass production can be calculated using the yield coefficient for the process.

2.2.5.3.1 Oxygen consumption

Oxygen respiration is the main respiration process in aerobic gravity sewers. Heterotrophic growth in sewers usually has a relatively high yield coefficient; 50–60 % of the consumed COD will be used for the production of new biomass. This implies that the biomass production, in terms of COD, is greater than the corresponding oxygen consumption.

The consumption of oxygen is mainly due to microbial oxidation of organic compounds. In addition, there is biological and chemical oxidation of some of the reduced substances diffusing out from deeper parts of the biofilm, e.g. sulphide (Norsker *et al.*, 1995) (Figure 2.2.22). Since sulphate reduction has a low yield coefficient the overall biomass production will be lower than under aerobic conditions. Oxygen utilisation rates (OUR) of sewer biofilms may be very high, depending on the type and amount of organic compounds present, oxygen concentration, flow velocity and biofilm surface structure. Typical values reported from full scale systems are 0.5–1 $gO_2/(m^2h)$, although this may be lower in some cases 0.1–0.3 $gO_2/(m^2h)$ (Bjerre *et al.*, 1998). In biofilm reactors using wastewater comparable to sewer conditions, consumption rates have been reported from 0.5 to 4.0 $O_2/(m^2h)$ at 20 °C. Biofilm kinetics is approximately ½ order for DO lower than 5-8 gO_2/m^3, and for experiments performed without substrate limitation (Bjerre *et al.*, 1998a/b).

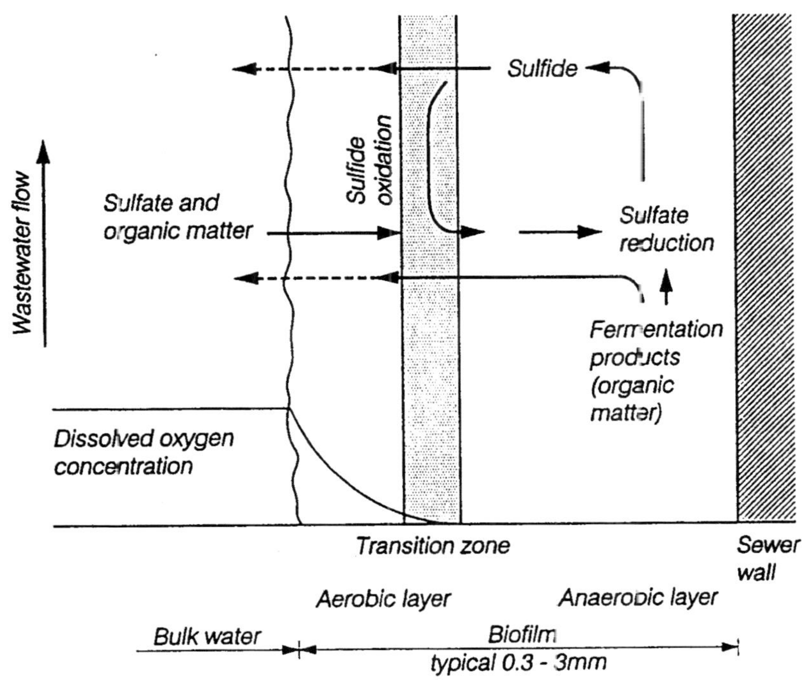

Figure 2.2.22. Aerobic and anaerobic process interactions in a sewer biofilm.

2.2.5.3.2 Nitrification

Oxidation of NH_4^+ to nitrite and nitrate is usually not assumed to be a significant process in sewer systems. However, nitrification may be observed in systems with a low concentration of organic compounds. The nitrifying bacteria are not important in sewer systems due to their low growth rate, and in the biofilm they are not able to compete effectively for oxygen with the aerobic heterotrophic bacteria, which have a much faster growth rate.

2.2.5.3.3 Denitrification

Nitrate is used for oxidation of organic matter in the absence of oxygen. This may take place in anoxic conditions in gravity sewers and pressure mains or in the deeper parts of biofilms in otherwise aerobic gravity sewers. The nitrate concentration is usually low in sewer systems, however, the oxidation capacity is much higher for nitrate than for oxygen (2.86 times more COD per g of N than per g of O_2), and it is therefore important for the consumption of any organic matter present. The yield coefficient is approximately the same as for oxygen. Very few nitrate consumption rates have been published from studies of full-scale sewer systems. A typical range is expected to be 0.1–0.2 g NO_3-N/m^2h at 20 °C, based on pressure main studies where nitrate has been added to control sulphide formation.

2.2.5.3.4 Sulphate reduction

Sulphate reduction takes place in sewers in the absence of oxygen and nitrate. In aerobic gravity sewers it occurs deep within biofilms and in sewer deposits, and may be responsible for a significant part of the transformation of organic matter and, indirectly, a part of the oxygen consumption via sulphide production (Norsker *et al.*, 1995) (Figure 2.2.22). The yield coefficient for sulphate reduction is low: 0.05–0.1 g biomass produced/g substrate consumed (in COD units), about 0.2 g biomass produced/g SO_4^{2-}-S reduced. The process is therefore insignificant for the production of biomass in gravity sewers compared with growth under aerobic and anoxic conditions. In pressure mains, however, it is the most important process for biomass production in addition to fermentation. Typical maximum sulphate reduction rates or sulphide production rates in pressure mains are 50–150 mg S/m^2h at 20 °C (Nielsen *et al.*, 1992). Hydrogen sulphide formation, effects and control in sewers have been dealt with in various studies. These phenomena are described in a number of publications (e.g. Hvitved-Jacobsen, 2002). However, limited discussion is given in Section 3.3.

2.2.5.4 Detachment

The biomass produced is detached from the biofilm to the bulk water by erosion and by sloughing. Single bacteria and small parts of the biofilm erode continuously from the surface, whilst bigger parts of the biofilm may slough intermittently. Sloughing may take place when large changes in shear stress or substrate conditions occur, but the mechanisms are not well understood. In a steady state situation or during longer time periods, the total detached biomass will be equal to biomass production. If oxygen is the primary electron acceptor, a biomass production of 0.5–3 g COD/(m^2h) at 20 °C can be expected to detach continuously. The corresponding increase in biofilm thickness (before detachment) has been studied in biofilm reactors using domestic wastewater, and a net rate of 60–120 µm/day was observed, depending on the DO concentration (Norsker *et al.*, 1995). As for other biological processes (Section 2.2) the growth of sewer slime may be considered subdivided into three phases:

- slow lag phase;
- fast exponential phase; and
- equilibrium or plateau phase with nearly constant biofilm thickness.

Dauber and Novak (1982) determined a growth rate of 7–30 (mean 14) g dry matter/(m^2 d) for erodible sewer slime in a sewerage system in Zürich-Dübendorf, Switzerland. The slime remaining on the sewer wall had a mass of 30 g dry matter/m^2. Gutekunst (1988) found growth rates between 1.7 and 2.2 g dry matter/(m^2 d) in three different sewers in Germany. The total mass of sewer slime there ranged from 25–39 g dry matter/m^2, the mass of inerodible slime comprised 10–15 g dry matter/m^2. Brombach *et al.* (1990) determined a growth rate of 15 g dry matter/(m^2 d) after disturbance of slime layer following a runoff event. The growth of sewer slime is influenced mainly by hydraulic conditions. Reiff (1992) found a maximum growth at a shear stress of approximately 1.5 N/m^2, whereas above and below this value the growth tended to zero. During the experiments in a prototype test rig there was only little growth at a shear stress higher than 5 N/m^2. Duddridge *et al.* (1982) investigated the growth of specific biofilm bacteria and determined an optimum shear stress of 6 N/m^2 as well as decreasing growth below a shear stress threshold of 4 N/m^2.

The maximum growth of sewer slime is independent of the oxygen concentration in the sewage and, within the usual variation range of COD in municipal sewage of 250 to 800 mg/l, also independent of the COD concentration. However, the growth rate of slime and the period of time until the maximum thickness is reached does depend significantly on the oxygen and COD concentration. Any dependency of sewer slime growth on sewer wall material has not been found.

2.2.5.5 Adsorption of particles

Particles may be trapped on the biofilm surface or voids in the biofilm, where any organics may be hydrolysed by exoenzymatic activity and further degraded. A number of factors influence the adsorption/desorption of particles, such as particle size, surface charge, pH etc., as well as biofilm surface properties and bulk water flow pattern. Studies of model biofilms have shown that water flows into the biofilm in small channels, making the prediction of transport of particles as well as soluble compounds very complex (Norsker *et al.*, 1995).

An exchange of heavy metals between the bulk water phase and the biofilm takes place (Gutekunst, 1990). The concentration of heavy metals in the biofilm may be considered as an indicator of the preceding wastewater transport of heavy metals. Heavy metals originating from short-term increases in concentration in the bulk sewage may be trapped in the biofilms and slowly released.

2.2.6. Sediment analysis

The sampling of sewage and sewer sediments for analytical purposes is outlined in the Appendix. Generally, methods recommended for analysis of solids and suspended solids may also be applied to sewer solid samples, e.g. APHA, AWWA and WPCF Standard Methods (1995). In the following only selected methods for analysis of sewer solids will be described.

2.2.6.1 Physical characterisation – sieving and settling

Typically, the simplest technique used to assess the properties of sewer solids is sieving. In spite of the practicability of sieving, the analysis and interpretation of the results needs to be done very carefully. As an example, Figure 2.2.23 shows the distribution of grain size of sediments from a combined sewer with diameter 1100 mm, gradient 0.21 % and about 14 l/s of dry weather flow (Michelbach and Wöhrle, 1994). More than 65 % of the material has a grain size between 1 and 20 mm. This is gravel, which has been used on the roads during winter time instead of salt. Figure 2.2.23 also shows the loss on ignition of these sediments (the organic fraction). Above 0.2 mm most of the particles are mineral, and the smaller fractions have a higher organic content. To understand this figure one has to know what will happen whilst sieving sewer sediments, as all of the soft organic grains will be damaged by squeezing through the sieve, resulting in smaller grain sizes. Any apparent information about the original organic particles is therefore incorrect. Artières and Stotz (1989) described this organic material within the sediment as a paste of fat, soft organic material and microorganisms. During storms and high shear stresses, this paste will be eroded, together with mineral particles, and it is not known which kind of transformation with respect to grain size and transportation of solids, etc., occurs. As an example, particles having the same settling velocity of 2.8 mm/s may correspond to very different types of particles, cf. Figures 2.2.23 and 2.2.24. The specific gravities range between 2.65 and 1.05 or less. Hence, grain size is presumed to vary between 60 μm and 330 μm. Such differences in size cannot be measured by sieving. An alternative, using laser techniques to assess size, may give information about the transportation and settling behaviour of the solids, but no information about the nature of the particle.

The best estimate of settling velocities for sewer solid particles is obtained by measurement and not by calculation, based on the physical parameters for the suspended solids. However, at present there is no universally accepted apparatus for determining sewage settling velocity. Indeed, even for a given apparatus it is not uncommon for different users to either modify the equipment or the experimental procedure (or do both) to meet their own needs. The following two examples illustrate this point (see also Appendix).

1. In Germany, prior to the use of the UFT settling velocity device (Michelbach and Wöhrle, 1992) the sample is pre-settled in an Imhoff-cone, and only those particles which settle within a 2 hour period are introduced into the UFT settling device for analysis. The argument for the adoption of this process is that particles collected under these conditions will have a minimum settling velocity of 0.5 mm/s. If conventional German (ATV) design processes for stormwater treatment are used, then such low settling velocities or surface loadings will lead to excessive storage volumes, e.g. of more than 1000 m^3/ha of contributing area, which will then have the capacity to retain rainfall of more than 100 mm.

2. By comparison, in the UK the Aston Column (Tyack *et al.*, 1996) has been designed to enable the floating fraction to be separated from that which sinks, and the full spectrum of settling velocities is determined. There are two reasons for this procedure. First, by considering the whole sample no information is lost, and once a full picture has been obtained, any specific portion of the resulting curve can be considered in detail. Second, much of the pollution load is associated with the fine fraction of particles that possess very low settling velocities. If advances are to be made in both the design of treatment processes and in computer modelling of sewer solids, transport decisions need to be made based upon a knowledge of the full spectrum of settling velocities in a given sewage sample.

When the above are taken into consideration, it is clearly essential that anyone wishing to utilise published settling velocity results ascertain the exact procedure used in their determination. Details of the main settling velocity devices currently in use or under development, together with recommended standard terminology and methods of results presentation, are given in the Appendix.

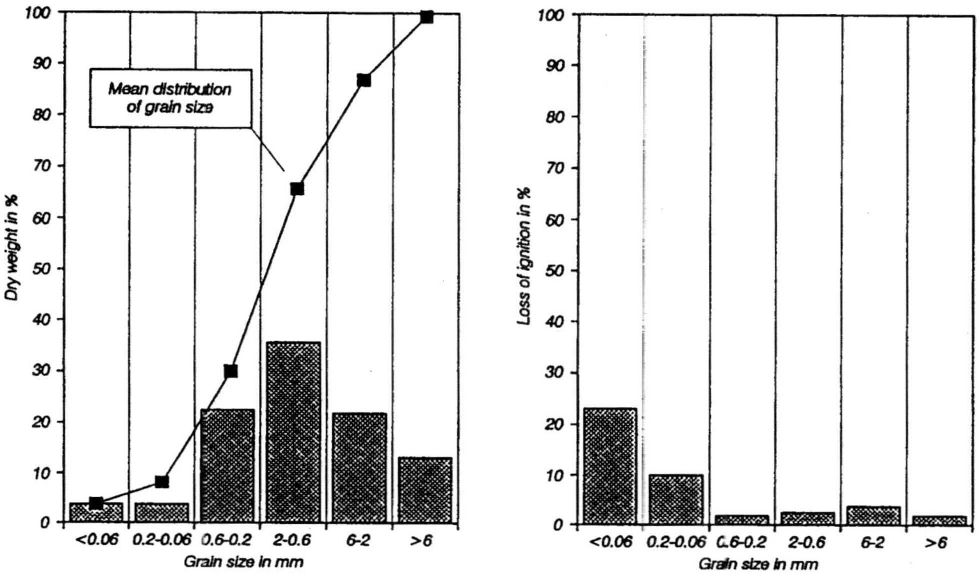

Figure 2.2.23. Grain size of sewer sediments in a combined sewer and loss on ignition (Michelbach, 1995).

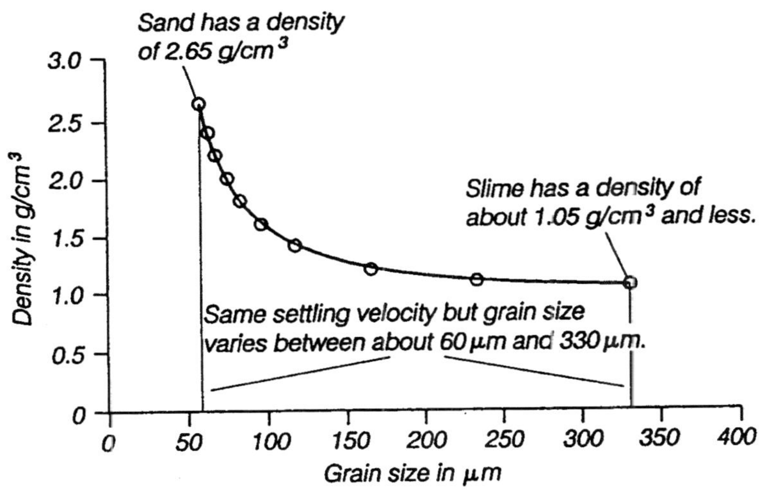

Figure 2.2.24. Density and grain size of particles with a settling velocity w_s = 2.8 mm/s for a surface loading of 10 m/h or 10 $m^3/m^2 \cdot h$.

2.2.6.2 Chemical characterisation

Only limited detailed analyses of sewer sediments have been reported. Most analyses utilise standard aqueous based tests to determine bulk properties (APHA, AWWA and WPCF Standard Methods, 1995). Sample preparation may require dilution and pre-mixing to release the chemical/biological pollutants from adherence to or incorporation in the sediments, and special preparation protocols are required (e.g. Crabtree and Forster, 1989). Aggressive blending of samples with distilled water is presumed to 'release' the pollutants into the aqueous phase, although bacteriological studies (McGregor *et al.*, 1996) suggest that this may affect the viability of some micro-organisms.

Several studies have reported data in which chemical pollutants are ascribed to different fractions or phases of the sediment or to the interstitial liquid. This approach accords with the basic modelling strategies used, for example, in MOUSETRAP, in associating 'potency' factors for the pollutants with the sediments or dissolved and suspended particulates. In the UK, the classification of sediments into characteristic types (Table 2.2.6), based principally on visual observations of their location and nature, has also provided a system by which chemical pollutant concentrations may be ascribed in terms of their occurrence and disposition within combined sewer systems. Table 2.2.14 outlines the sample preparation techniques currently in use (McGregor *et al.*, 1993).

Table 2.2.14. Sediment sample pre-preparation prior to carrying out standard chemical tests

Preparation method	Protocol
1. Blending (vigorous industrial machine)	100 ml of sediment blended, using an industrial blender, in 200 ml of distilled water and diluted to 1 litre
2. Mixing/inversion (in a glass tube by turning end-over-end)	Repeated inversion of 100 ml of sediment in 900 ml of distilled water in a measuring cylinder
3. Stirring	Stirring 100 ml of sediment with 900 ml of distilled water in a cylindrical container at variable rates for different time periods

The sediment samples analysed using these comparative procedures provide several distinct polluting factions. These fractions are summarised in Table 2.2.15.

Table 2.2.15. Pollutant fractions obtained from different sample preparation and test protocols

Pollutant fraction	Obtained from
1. Total sediment load (TSL)	Blending
2. Readily erodible (RE)	Inverting or stirring
3. Dissolved and fine (DF)	Settlement of RE Fraction for 15 minutes
4. Associated with sediment fraction (AS)	TSL Fraction – RE Fraction
5. Readily settleable fraction (RS)	RE Fraction – DF Fraction

The significance of these 'fractions' is their propensity to remain in (or associated with) the bed, or to be released into the flow during the application of an applied shear stress, possibly leading to foul flushes. They are also useful as a means of characterising sediments and pollutants for sewer flow quality modelling.

The release of pollutants from eroding beds is difficult to quantify and in an attempt to determine what may be released in a sewer, near bed sediments (Type C), collected in traps were subject to the inversion preparation and the loads released into the overlying liquid were: COD (52 %), BOD (53 %), TS (56 %), and VS (53 %), with ammonia-N (71 %). Of the released pollutants, the major fraction remained in the fine/dissolved phase: COD (76 %), BOD (70 %), ammonia (79 %), TS (60 %) and VS (57 %).

Type A (granular bed deposit) sediments collected in Dundee were subjected to a stirring preparation protocol at varying bed shear rates from 0.06 N/m^2 up to 0.56 N/m^2 and for differing time durations. This comprised sample re-constitution in a laboratory beaker with overlying distilled water and stirring at controlled rates using a paddle. Little surface disturbance of the sediment occurred for the rates of shear used, and the erosion and initiation of pollutant release corresponded with the highest shear stress applied. Table 2.2.16 shows the results and illustrates the differing proportions of pollutants released as a consequence of the varying applied shear rates. In actual sewers the bed shear stresses are very much greater than those utilised in the laboratory, with typical ranges up to more than 200 N/m^2. However, evidence suggests that erosion in fact often occurs at quite low shear stresses (circa 1 N/m^2) due to a 'fluidisation' of the bed surface (Ristenpart *et al.*, 1995), and the laboratory results described here are thus believed to be representative.

Table 2.2.16. Distribution of pollutants attached to and released from different sewer sediment types

	Near bed (Type C) sediment			Type A sediment		
	Inverted			Stirred (0.06–0.56 N/m^2)		
	% fine *	% coarse **	% AS #	% fine *	% coarse **	% AS #
COD	76	24	48	57–28	43–72	91–75
BOD	70	30	47	83–46	17–54	93–69
amm-N	79	21	29	100	0	79–58
TS	60	40	44	100–27	0–73	97–89
VS	57	43	47	88–29	12–71	92–72

* Fine fraction: < 0.5 mm; ** Coarse fraction: > 0.5 mm; # Associated with sediment

It can be seen that at the highest shear of 0.56 N/m^2, some 25 % of the total COD load was released, compared with 9 % for the lower shear of 0.06 N/m^2. Of the 25 % released from the sediment, some 28 % was associated with the fine fraction and 72 % with the coarse fraction. The distribution of COD within the fine and coarse sediment phases at the lowest shear was 57 % and 43 % respectively at 0.06N/m^2. This suggests that as the shear stress decreases, the relative proportion of COD associated with the fine fraction also increases. This trend may be seen for BOD, TS and VS, albeit to differing degrees depending upon the nature of the pollutant. All of the ammonia released from the sediment for all shear stresses was found to remain in the fine/dissolved phase (i.e. remains suspended or in solution).

At higher shear stresses the sediment bed will be more disturbed, subsequently releasing coarser material and larger quantities of interstitial liquid into the erodible fraction. At these shear stresses there will be sufficient energy to sustain the larger particles in suspension, accounting for the greater percentage found in the coarser phase of the readily erodible fraction. From these data it can be seen that pollutant release from sewer sediments may be quantitatively apportioned to either a settleable phase or a very fine/dissolved phase of the erodible fraction. It is concluded that the degree of association to either fraction will be dependent upon the level of applied shear, the characteristics of the sediment type and also the nature of each pollutant.

Aggressive sample preparation using an industrial blender is believed to result in the 'release' into a measurable state, of as much of the pollutants associated with the sediment as could conceivably be released by storm flow erosion. This approach has also been shown to give results which are more repeatable than those obtained using other preparation methods. There are problems, however, associated with the blending preparation, particularly regarding the determination of ammoniacal nitrogen and bacteria testing. The problems are apparently caused by an excessive 'release' of ammonia from the particulate to the dissolved phase caused by the vigour of the blending process. Good correlations have been obtained between the results of the tests using the blending sample preparation protocol and the other preparations (Table 2.2.15). Table 2.2.17 (McGregor, 1993) shows the pollutant loads associated with each of the fractions for Class A and C sediments, given in terms of percentages of the results obtained by carrying out blending tests. The Class A results are from the highest shear stresses used (0.56 N/m^2) and Class C from the mixing protocol tests.

Table 2.2.17. Pollutants/solids in different fractions as percentages of results obtained from blending preparation techniques for samples of sewer sediments

Determinand	Readily erodible		Dissolved/fine		Associated with sediment		Readily suspendable	
Sediment type	A	C	A	C	A	C	A	C
COD	25	57	7	42	75	50	18	15
BOD_5	35	58	16	39	69	50	19	19
Amm-N	42	82	42	67	58	25	2	15
TS	11	67	3	33	89	44	8	34
VS	28	70	8	33	72	49	20	37

There is some evidence that there are differences between summer and winter chemical characteristics of sewer sediments which may be related to temperature and de-icing activities. Attempts to relate chemical characteristics of sewer sediments to physical properties have shown the variability. Some correlations have been demonstrated between COD and volatile solids, but generally there are no definite relationships.

Other pollutants (except for sulphides and possibly methane) are transferred into sediment beds via attachment to depositing particles or in the liquid phase. Once a bed is established, additional pollutants may become transferred from overlying sewage by sorption and diffusion processes. Concurrently, in most sewers with bed deposits, daily erosion of the bed will occur at the dry weather peak, and evidence from French and Belgian studies, suggests that the finer solids become elutriated from the bed, so that even in a sewer with a 'stable' bed the mechanisms act together to:

- daily erode some of the older sediment and allow it to be replaced by 'fresher' material;
- facilitate mixing with the flow and allow biofilm bacteria exposure to substrate;
- gradually increase the overall particle size in the deposit (by removing the fines);
- encourage oxygen transfer in the upper layers of the bed;
- allow diffusion of 'fresh' pollutants into the bed and release of pollutants from biofilms into the liquid phase.

With thicker, stable bed deposits, conditions will become anoxic at some point in the vertical profile (Section 2.2.3).

2.2.6.3 Microbiological characterisation

The importance of the biological nature and consequent growth of biofilms, together with associated biochemical processes, within sewer solid deposits and attached to sewer walls, has not as yet been fully explored. The problems of sampling and analysis to determine with a reasonable degree of precision the populations and nature of bacteria present in sewerage systems are well known. These problems are further compounded when dealing with sediments, because of the necessity to sever the bacteria attachments (in the immobilised film phase) to the particulates in order to carry out the standard tests. The study of bacteria in sewer sediments is in its infancy and very few data have been published. Preliminary studies in the UK co-ordinated by WRc have been carried out in Birmingham and Dundee, and a notable study of storm drain sediment in the USA (Marino and Gannon, 1991). Most recent publications report bacteria concentrations in sewage flows as these are primarily concerned with the control of CSO discharges, however, there are clearly interactions between sewage flows and sediments (Ashley and Dabrowski, 1995).

The association with solids, providing the opportunity for bacterial attachment and also a reduction in penetration of solar radiation, by turbidity, thus reducing the die-off rate of enteric bacteria (Irving, 1977) is clearly very important for environmental systems.

The examination of bacteria populations in sewer sediments has so far concentrated largely on populations of Faecal Colifoms (FC), Total Coliforms (TC), and Faecal Streptococci (FS) as indicators of potential presence of pathogens and viruses. Whilst a number of studies have examined sewage quality, few have considered sediments. The studies undertaken in Dundee to assist with the development of guidelines for the control of sewage discharges into coastal areas and the inclusion of bacteria in the MOUSETRAP model have also been principally concerned with sewage, but have looked for evidence of sediment erosion adding to bacteria numbers. The sediment data for bed deposits are summarised (Ashley and Dabrowski, 1995) and compared with sewage data in Table 2.2.18.

Table 2.2.18. Bacteria numbers in sediment and sewage in Dundee

	Bacteria numbers (million/100 ml)				
		DWF		Storms	
	Sediment*	Summer	Winter	Summer	Winter
TC	0.9–2631	20.5	6.3	3.8–113	2.2–59
FC	0.01–314	2.6	-	0.3–27	0.1–8.5
FS	< 0.1–39	0.14	0.28	0.03–1.25	0.04–2.0

* Lower counts were observed for winter samples.

The results show clearly that sewer sediments are bacteria accumulators and nurseries. Later results from Dundee (McGregor *et al.*, 1996) investigated the sensitivity of the indicator species to test procedure protocols. The results presented in Table 2.2.18 were obtained using the standard blending protocol preparation for samples, whereas the later tests used each of the protocol options as given in Table 2.2.15, in order to apportion bacteria to the relative solid-liquid phases. The results showed that the blending protocol, assumed to provide a test sample which will yield extreme values for physical and chemical testing, may well be compromising the viability of the microorganisms. Thus the earlier results (Table 2.2.18) may give an underestimate of sediment bacteria numbers.

The inter-relationship between microorganisms and sewer solids is important both in terms of in-sewer processes (Hvitved-Jacobsen *et al.*, 1995a) and for the effects once sewage is discharged into watercourses. Current studies are investigating biodegradation and associated oxygen consumption of sewer sediments in suspension under aerobic conditions following, for example, erosion during a storm event (Vollertsen and Hvitved-Jacobsen, 1997). The oxygen utilisation rate (OUR) for sewer sediments in suspension is a major characteristic in this respect.

2.2.7 Modelling of solids and pollutant transformation in sewers

Knowledge concerning the integrated physical, chemical and microbial processes in sewers is now at a level where the transformations between these phases can be included in a model for wastewater transport. For example, MOUSETRAP is a commercial model for sewer solids and pollutant transport, which includes sewer processes (Garscal *et al.*, 1995).

The concept described below for modelling wastewater and sewer solids transport and transformation in sewers can be considered as the basis for further development (Hvitved-Jacobsen *et al.,* 1998). Process studies and modelling are being undertaken to investigate both solids and solute transport, and microbial transformation of wastewater organic carbon, including the associated electron acceptor system: dissolved oxygen, nitrate and sulphate. The aerobic processes taking place in gravity sewers are being emphasised. Processes taking place under anoxic or anaerobic conditions are of interest as these conditions may exist in gravity sewers as well as in pressure mains. Processes within the bulk water phase, the sewer biofilm and the sewer sediment and mass transfer between these subsystems are important.

2.2.7.1 A model concept for microbial wastewater transformations

The main objectives of a simulation model for wastewater transformations and transport in sewer systems are (Figure 2.2.25):

- simulation of dry weather flow processes which result in changes of the solids, the organic substrate and the biomass in the sewer. These transformations may affect the functioning of the sewer itself and the subsequent wastewater treatment;
- simulation of wet weather processes, especially concerning transport and transformation of sewer deposits and detached biofilms related to combined sewer overflow effects on receiving water bodies;
- advancement of understanding, especially for the evaluation of theories.

The modelling system may be based on a conventional hydrodynamic, transport-dispersion model concerning the bulk water phase. However, recent results (Portielje *et al.*, 1996) show that incorporation of stochastic properties in these deterministic models may be of particular interest in order to simulate varying conditions in the sewers.

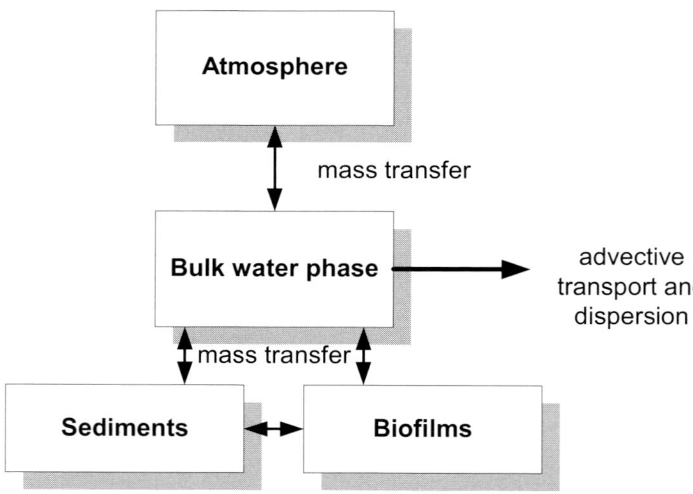

Figure 2.2.25. Structure of the modelling system for wastewater and solids transport and transformations in sewers.

Processes under aerobic conditions in the water phase have been investigated by Raunkjær (1993), Raunkjær *et al.* (1994, 1995, 1997) and Bjerre *et al.* (1995, 1998a, 1998b). A combined aerobic/anaerobic process model concept is shown in Table 2.2.19, based on the systems outlined in Figures 2.2.9 and 2.2.10. Description of the sulphate respiration process and methane production has not yet been determined and therefore not included in the model. The process model is the basis for further inclusion in a transport-dispersion model for wastewater transportation in sewers. The model outlined in Table 2.2.19 is based on investigations and considerations published in Bjerre *et al.* (1998b), Vollertsen and Hvitved-Jacobsen (1997), Tanaka and Hvitved-Jacobsen (1998); Hvitved-Jacobsen *et al.* (1998a, 1998b).

The model includes aerobic growth and maintenance energy requirements in the bulk water, aerobic growth in the biofilm, aerobic and anaerobic hydrolysis, fermentation and reaeration. Interaction with the sediments is ignored. The sulphur cycle is not included in the example shown in Table 2.2.19.

Transport of eroded sewer deposits and biofilm in the bulk water phase takes place mostly during storm flow conditions. Organic rich deposits may be weakly resistant to erosion, with low yield strength, hence the contribution of sewer sediments to the solids transport under such conditions is important to include in any model. Physical and microbial characteristics and processes of the sediments must therefore be integrated in the model (Vollertsen and Hvitved-Jacobsen, 1997).

Although the processes of importance in the biofilms are generally identical with the processes in the bulk water phase, the combined transport and transformation of substrate and electron acceptors within the biofilm must be taken into account when modelling the processes at a detailed level (Gujer and Wanner, 1990). The presence of substrate and electron acceptors at a specific position in the biofilm is also important. An improved understanding of the biomass as described for the bulk water phase may be incorporated in the modelling of the biofilm processes.

As yet, only a relatively simple modelling approach concerning the sewer biofilm has been developed, i.e. a simple surface flux model based on ½' order kinetics (Table 2.2.18). Justification for this type of model is partly based on the experimental results produced by Norsker *et al.* (1995), Raunkjær *et al.* (1997) and Bjerre *et al.* (1998a).

Under aerobic conditions in the bulk water phase, sewer sediment is considered as a biofilm surface as only the upper few millimetres (or less) are aerobic. However, anaerobic conditions in the underlying sediment may affect production of biodegradable substrates and reduced compounds (e.g. volatile fatty acids (VFA) and sulphides) which are important for the oxygen uptake rate at the surface. Theoretically, a model similar to that for the sewer biofilm is possible.

The following mass transfer processes are particularly relevant:

(a) Air-water mass transfer

- Oxygen mass transfer, i.e. reaeration. A relevant model description is based on the findings of Jensen (1995).
- Hydrogen sulphide mass transfer.

(b) Mass transfer between the biofilms and the water phase

- Transport of soluble, and to some extent colloidal, species of substrate and electron acceptors into and out of the biofilm by diffusion.
- Adsorption/desorption of particulate fractions of organic matter.
- Detachment of biofilms.

(c) Mass transfer between the sediments and the water phase

- Transport of soluble species of substrate and electron acceptors into and out of the sediments by diffusion.
- Adsorption/desorption of particulate fractions of organic matter.
- Sedimentation.
- Erosion (resuspension).

An example illustrating the application of this approach to a 20 km sewer, 1.20 m in diameter, is given in Hvitved-Jacobsen *et al.* (1998b).

Table 2.2.19. Integrated processes for transformation of wastewater organic matter in a sewer where aerobic as well as anaerobic processes may proceed

(Symbols are defined in Tables 2.2.1, 2.2.20 and 2.2.21.)

	S_F	S_A	X_{Bw}	X_{S1}	X_{S2}	$-S_O$	Process rate Equation
Aerobic growth in bulk water		$-1/Y_{Hw}$	1			$(1-Y_{Hw})/Y_{Hw}$	a
Aerobic growth in biofilm		$-1/Y_{Hf}$	1			$(1-Y_{Hf})/Y_{Hf}$	b
Maintenance energy requirement	-1					1	c
Aerobic hydrolysis, fast	1			-1			d, n=1
Aerobic hydrolysis, slow	1				-1		d, n=2
Anaerobic hydrolysis, fast	1			-1			e, n=1
Anaerobic hydrolysis, slow	1				-1		e, n=2
Fermentation in bulk water and biofilm	-1	1					f
Reaeration						-1	g

a: $\mu_H (S_F + S_A)/(K_S + (S_F + S_A))\ S_O/(K_O + S_O)\ X_{Bw}\ \alpha_w^{(T-20)}$

b: $k_{1/2}\ S_O^{0.5}\ Y_{Hf}/(1-Y_{Hf})\ A/V\ (S_F + S_A)/(K_{Sf} + (S_F + S_A))\ \alpha_f^{(T-20)}$

c: $q_m\ S_O/(K_O + S_O)\ X_{Bw}\ \alpha_w^{(T-20)}$

d: $k_{hn}\ (X_{Sn}/X_{Bw})/(K_{Xn} + X_{Sn}/X_{Bw})\ S_O/(K_O + S_O)\ (X_{Bw} + \varepsilon\ X_{Bf}\ A/V)\ \alpha_w^{(T-20)}$

e: $\eta_{fe}\ k_{hn}\ (X_{Sn}/X_{Bw})/(K_{Xn} + X_{Sn}/X_{Bw})\ K_O/(S_O + K_O)\ (X_{Bw} + \varepsilon\ X_{Bf}\ A/V)\ \alpha_w^{(T-20)}$

f: $q_{fe}\ S_F/(K_{fe} + S_F)\ K_O/(S_O + K_O)\ (X_{Bw} + \varepsilon\ X_{Bf}\ A/V)\ \alpha_w^{(T-20)}$

g: $K_L a\ (S_{OS}-S_O)$ where $K_L a = 0.86\ (1 + 0.20\ F^2)\ (s\ u)^{3/8}\ d_m^{-1}\ \alpha_r^{(T-20)}$

Table 2.2.20. Example of model parameters used in the sewer process model outlined in Table 2.2.19

Symbol	Definition	Value	Unit
μ_H	maximum specific growth rate for heterotrophic biomass	7	d^{-1}
Y_{Hw}	suspended biomass yield constant for heterotrophics	0.55	gCOD/gCOD
K_S	saturation constant for readily biodegradable substrate	1.0	$gCOD/m^3$
K_O	saturation constant for DO	0.5	gO_2/m^3
α_w	temperature coefficient in the water phase	1.07	-
Q_m	maintenance energy requirement rate constant	1.0	d^{-1}
$K_{1/2}$	1/2 order rate constant	2.5	$gO_2^{0.5}\ m^{-0.5}\ d^{-1}$
Y_{Hf}	biofilm yield constant for heterotrophic biomass	0.55	gCOD/gCOD
K_{Sf}	saturation constant for readily biodegradable substrate	1.0	$gCOD/m^3$
E	efficiency constant for the biofilm biomass		-
α_f	temperature coefficient in the biofilm	1.03	-
K_{h1}	hydrolysis rate constant, fraction 1 (fast)	4.0	d^{-1}
K_{h2}	hydrolysis rate constant, fraction 2 (slow)	1.0	d^{-1}
K_{X1}	saturation constant for hydrolysis, fraction 1	0.5	gCOD/gCOD
K_{X2}	saturation constant for hydrolysis, fraction 2	0.2	gCOD/gCOD
η_{fe}	anaerobic hydrolysis reduction factor	0.14	-
Q_{fe}	maximum rate for fermentation	3	d^{-1}
K_{fe}	saturation constant for fermentation	20	gCOD/gCOD

w: water phase; f: biofilm

Table 2.2.21. Reaeration and flow characteristics used in the sewer process model outlined in Table 2.2.18

Symbol	Definition	Value	Unit
K_La	oxygen transfer coefficient		d^{-1}
T	temperature		^oC
S_{OS}	dissolved oxygen saturation concentration at T ^oC		$GO2/m^3$
F	Froude number = $u/(g\ dm)^{0.5}$		-
U	mean flow velocity		M/s
G	gravity acceleration	9.81	M/s^2
S	slope		M/m
D_m	hydraulic mean depth		M
α_T	temperature coefficient for reaeration	1.024	-

2.2.8 Concluding remarks

The sewer is an integral part of the urban wastewater system: the sewer, the wastewater treatment plant and the local receiving waters. However, when compared with wastewater treatment plants and receiving waters, the sewer is often neglected as a reactor in terms of the intrinsic physical, chemical and biological processes. As advanced wastewater treatment is becoming more widespread, and combined sewer overflow impacts on receiving waters less acceptable, the relative importance of sewer processes in the design and function of sewer systems is increasing. The present level of knowledge of in-sewer processes needs to be improved in order to meet future aspirations to treat all parts of the system in an integrated way.

Wide variability of sewer solids properties in time and space have been observed. This variability is often not possible to correlate with other parameters. The behaviour of the sewer as a reactor, with dominant dry as well as wet weather processes, has clearly been demonstrated in a range of studies. However, the various processes and links are often difficult to predict.

The following are acknowledged as making significant contributions to the interactions occurring in the sewer involving the transformations of sewer solids, i.e. suspended particles, biofilm and sediment, and therefore require further investigation:

- microbial transformation processes under aerobic as well as anaerobic conditions, including reaeration and the resulting dissolved oxygen balance;
- accumulation and solidification processes of solids in the sewer sediment;
- sedimentation and erosion of sewer sediments and detachment of biofilms.

2.2.9 References

APHA, AWWA and WPCF (1995) *Standard Methods for the Examination of Water and Wastewater*. APHA, AWWA and WPCF.

Artières, O. and Stotz, G. (1989) *Caractéristiques des dépôts en réseaux d'assainissement unitaires. Conséquences sur leur transport.* La Houille Blanche, 3/4, 243–246.

Ashley, R.M., Coghlan, B.P. and Jefferies, C. (1989) The quality of sewage flows and sediment in Dundee. *Wat. Sci. Tech.*, 24(6), 39–46.

Ashley, R.M. and Crabtree, R.W. (1992) Sediment origins, deposition and build-up in combined sewer systems. *Wat. Sci. Tech.*, 25(8), 1–12.

Ashley, R.M. and Dabrowski, W. (1995) Dry and storm weather transport of coliforms and faecal streptococci in combined sewage. *Wat. Sci. Tech.*, 31(7), 311–320.

Ashley, R.M., Longair, I.M., Wotherspoon, D.J.J. and Kirby, K. (1993) Flow and sediment movement monitoring in large sewers. *Wat. Sci. Tech.*, 28(11–12), 55–62.

Ashley, R.M. and Verbanck, M. (1996) Mechanics of sewer sediment erosion and transport. *J. Hydr. Res.*, 34(6), 753–769.

Ashley, R.M., Wotherspoon, D.J.J., Coghlan, B.P. and McGregor, I. (1992a) The erosion and movement of sediments and associated pollutants in combined sewers. *Wat. Sci. Tech.*, 25(8), 101–114.

Ashley, R.M., Wotherspoon, D.J.J., Goodison, M.J., McGregor, I. and Coghlan, B.P. (1992b) The deposition and erosion of sediments in sewers. *Wat. Sci. Tech.*, 26(5–6), 1283–1293.

ATV-Arbeitsblatt A 128 neu (1992) *Richtlinien für die Bemessung und Gestaltung von Regenentlastungsanlagen in Mischwasserkanälen.* Gesellschaft zur Förderung der Abwassertechnik, St. Augustin, 1992.

Bachoc, A. (1992) Location and general characteristics of sediment deposits into man-entry combined sewers. *Wat. Sci. Tech.*, 25(8), 47–55.

Bjerre, H.L., Hvitved-Jacobsen, T., Schlegel, S. and Teichgräber, B. (1998a) Biological activity of biofilm and sediment in the Emscher river, Germany. *Wat. Sci. Tech.*, 37(1), 9–16.

Bjerre, H.L., Hvitved-Jacobsen, T., Teichgräber, B. and te Heesen, D. (1995) Experimental procedures characterizing transformations of wastewater organic matter in the Emscher river, Germany. *Wat. Sci. Tech.*, 31(7), 201–212.

Bjerre, H.L., Hvitved-Jacobsen, T., Teichgräber, B. and Schlegel, S. (1998b) Modelling of aerobic wastewater transformations under sewer conditions in the Emscher river, Germany. *Wat. Env. Res.*, 70(6), 1151–1160.

Boon, A.G. and Lister, A.R. (1975) Formation of sulphide in rising main sewers and its prevention by injection of oxygen. *Prog. Wat. Tech.*, 7(2), 289–300.

Brombach, H., Michelbach, S. and Wöhrle, C. (1990*) Sedimentations- und Remobilisierungsvorgänge im Abwasserkanal.* Annual Report 1990, working group 'Niederschlagsbedingte Schmutzbelastung der Gewässer', joint project 'Niederschlag' of German Ministry of Research and Technology.

Brombach, H., Michelbach, S. and Wöhrle, C. (1992*) Sedimentations- und Remobilisierungsvorgänge im Abwasserkanal*. Schlussbericht zum BMFT-Verbundprojekt Niederschlag, Teilprojekt 3, Karlsruhe.

Characklis, W.G. and Marshall, K.C. (eds) (1990) *Biofilms*. John Wiley, New York, USA.

Chebbo, G. (1992) *Solides des rejets urbains par temps de pluie: Caractérisation et traitabilité.* Thèse de Doctorat de l'Ecole Nationale des Ponts et Chaussées, 410 pp. + annexes.

Chebbo, G. and Bachoc, A. (1992) Characterisation of suspended solids in urban wet weather discharges. *Wat. Sci. Tech.*, 25(8), 171–179.

Chebbo, G., Bachoc, A., Laplace, D. and Le Guennec, B. (1995) The transfer of solids in combined sewer networks. *Wat. Sci. Tech.*, 37(1), 95–105.

Chebbo, G., Musquère, P. and Bachoc, A. (1990) Solids transferred into sewers. *Proc. 5th Int. Conf. Urban Storm Drainage*, Osaka (Japan), 885–890.

Crabtree, R.W. (1989) Sediment in sewers. *Journal IWEM*, 3(6), 569–578.

Crabtree, R.W, Ashley, R.M, and Saul, A.J. (1991) *Review of Research into Sediments in Sewers and Ancillary Structures.* Foundation for Water Research, Report No. FR 0205.

Crabtree, R.W. and Foster, C. (1989) *Preparation Protocols for the Analysis of Combined Sewer Sediment Samples*. WRc Report ER354E.

Crabtree, R.W., Foster, C.F., Nalluri, C. and Williams, D. (1989) *Recommendations for Further Research into Combined Sewer Sediment Deposits*. Foundation for Water Research, Report No. FR 0040.

Dauber, L. and Novak, B. (1982) *Quellen und Mengen der Schmutzstoffe in Regenabflüssen einer städtischen Mischkanalisation.* Separatum Nr. 927 der Eidgenössischen Anstalt für Wasser-versorgung, Abwasserreinigung und Gewässerschutz, Zürich.

Dreywood, R. (1946) Qualitative test for carbohydrate material. *Industrial and Engineering Chemistry*, 18, 499.

Duddridge, J., Kent, C. and Laws, J. (1982) Effect of surface shear stress on the attachment of pseudomonas fluorescens to stainless steel under defined flow conditions. *Biotechnology and Bioengineering*, Vol. 24, 153–164.

Frølund, B., Palmgren, R., Keiding, K. and Nielsen, P.H. (1996) Strategies for extraction and characterization of activated sludge exopolymers. Extraction of polymers from activated sludge using a cation exchange resin. *Water Res.* 30(8), 1749–1758.

Garsdal, H., Mark, O., Dørge, J. and Jepsen, S.-E. (1995) MOUSETRAP: Modelling of water quality processes and the interaction of sediments and pollutants in sewers. *Wat. Sci. Tech.*, 31(7), 33–41.

Gent, R., Crabtree, R.W. and Ashley, R.M. (1996) A review of model development based on sewer sediments research in the UK. *Wat. Sci. Tech.*, 33(9), 1–7.

Gujer, W. and Wanner, O. (1990) Modeling mixed population biofilms. *Biofilms*, edited by W.G. Characklis and K.C. Marshall, J. Wiley, New York, 397–443.

Gutekunst, B. (1988) *Sielhautuntersuchungen zur Einkreisung schwermetallhaltiger Einleitungen.* Veröffentlichungen des Instituts für Siedlungswasserwirtschaft der Universität Karlsruhe, Heft 49.

Henze, M., Grady Jr., C.P.L., Gujer, W., Marais, G.v.R. and Matsuo, T. (1987) *Activated Sludge Model No. 1.* Scientific and Technical Report No. 1. International Association on Water Pollution Research and Control, 33 pp.

Henze, M., Gujer, W., Mino, T., Matsuo, T., Wentzel, M.C. and Marais, G.v.R. (1995) *Activated Sludge Model No. 2*. Scientific and Technical report No. 3. International Association on Water Quality, 32 pp.

Hermann, R. and Kari, G. (1990) Grain size dependent transport of non-polar organic trace pollutants (PAH, PCB) by suspended sediments during urban storm runoff. *Proceedings of 5th ICUSD*, Suita, Osaka, 23–27 July, 499–503.

Heukelekian, H. and Balmat, J.L. (1959) Chemical composition of the particulate fractions of domestic sewage. *Sewage and Industrial Wastes*, 31, 413.

Hunter, J.V. and Heukelekian, H. (1965) The composition of domestic sewage fractions. *J. WPCF*, 37, 1142–1163.

Hvitved-Jacobsen, T. (1982) The impact of combined sewer overflows on the dissolved oxygen concentration in a river. *Wat. Res.*, 16, 1099–1105.

Hvitved-Jacobsen, T. (2002) *Sewer Processes.* CRC Press. ISBN 1-56676-926-4.

Hvitved-Jacobsen, T., Jütte, B., Nielsen, P.H. and Jensen, N.Aa. (1988) Hydrogen sulphide control in municipal sewers. *Pretreatment in Chemical Water and Wastewater Treatment*, proceedings of 3rd International Gothenburg Symposium, edited by H.H. Hahn and R. Klute, Gothenburg, Sweden, June 1–3, 1988, Springer-Verlag, 239–247.

Hvitved-Jacobsen, T., Nielsen, P.H., Larsen, T. and Jensen, N.Aa. (eds) (1995b) The sewer as a physical, chemical and biological reactor. *Wat. Sci. Tech.*, 31(7), 330 pp.

Hvitved-Jacobsen, T., Raunkjær, K. and Nielsen, P.H. (1995a) Volatile fatty acids and sulfide in pressure mains. *Wat. Sci. Tech.,* 31(7), 169–179.

Hvitved-Jacobsen, T., Vollertsen, J. and Tanaka N. (1998a) Wastewater quality changes during transport in sewers – an integrated aerobic and anaerobic model concept for carbon and sulfur microbial transformations. 19th Biennial Conf. of IWA, Vancouver, Canada, 21–26 June. *Wat.Sci.Tech.* 38(10), 257–264 (see text 249–256). Errata: *Wat. Sci. Tech.*, 39(2), 242–249.

Hvitved-Jacobsen, T., Vollertsen, J. and Nielsen, P.H. (1998b) A process and model concept for microbial wastewater transformations in gravity sewers. *Wat. Sci. Tech.*, 37(1), 233–241.

Irving, T.E. (1977) *Preliminary Investigations on the Effects of Particulate Matter on the Fate of Sewage Bacteria in the Sea.* Water Research Centre, Technical Report TR55, September.

Jahn, A. and Nielsen, P.H. (1996) Extraction of extracellular polymeric substances (EPS) from biofilms using a cation exchange resin. *Wat. Sci. Tech.*, 32(8), 157–164.

Jahn, A. and Nielsen, P.H. (1997) Cell biomass and exopolymer composition in sewer biofilms. *Wat. Sci. Tech.*, 37(1), 17–24.

Jensen, N.Aa. (1995) Empirical modelling of air-to-water oxygen transfer in gravity sewers. *Wat. Env. Res.*, 67(6), 979–991.

Jensen, N.Aa. and Hvitved-Jacobsen, T. (1991) Method for measurement of reaeration in gravity sewers using radio tracers. *Research Journal of the Water Pollution Control Federation*, 63(5), 758–767.

Laplace, D., Sanchez, Y. and Dartus, D. (1990) Trunk sewers clogging development observation and first interpretations. *Wat. Sci. Tech.*, 22(10/11), 259–266.

Larson, M., Berndtsson, R., Hogland, W., Spanberg, A. and Bennerstedt, K. (1990) Field measurements and mathematical modelling of pollution build-up and pipe deposit wash-out in combined sewers. *Proc. 5th Int. Conf. Urban Storm Drainage*, Osaka, Japan, 325–332.

Levine, A.D., Tchobanoglous, G. and Asano, T. (1985) Characterization of size distribution of contaminants in wastewater: treatment and reuse implications. *J. WPCF*, 57, 805–816.

Levine, A.D., Tchobanoglous, G. and Asano, T. (1991) Size distributions of particulate contaminants in wastewater and their impact on treatability. *Wat. Res.,* 25, 911–922.

Logan, B.E. and Jiang, Q. (1990) Molecular size distributions of dissolved organic matter. *J. Env. Eng.*, 116, 1046.

Lowry, O.H., Rosebrough, N.J., Farr, A.L. and Randall, R.J. (1951) Protein measurement with the folin phenol reagent. *J. Biol. Chem.,* 193, 265.

Lucas-Aiguier E., Chebbo, G., Bertrand-Krajewski, J., Gagné, B. and Hedges, P. (1998) Analysis of the methods for determining the settling characteristics of sewage and stormwater solids. *Wat. Sci. Tech.*, 37(1), 53–60.

Marino, R.P. and Gannon, J.J. (1991) Survival of faecal coliforms and faecal streptococci in storm drain sediment. *Wat. Res.*, 25(9), 1089–1098.

Marsalek, J. (1985) Toxic contaminants in urban runoff: a case study. Urban runoff pollution: NATO ASI series, editors: H.C. Torno, J. Marsalek and M. Desbordes, 39–57.

McGregor, I. (1993) *Release of Characteristic Pollutants from Combined Sewer Sediments.* MPhil thesis, Dundee Institute of Technology.

McGregor, I., Ashley, R.M. and Oduyemi, K.O.K. (1993) Pollutant release in sewer systems and their release into receiving waters. *Wat. Sci. Tech.*, 28(8–9), 161–169.

McGregor, I., Souter, N. and Ashley, R.M. (1996) Sewer sediments: Preparation methods for enumerating bacteria. *Wat. Sci. Tech.*, 33(9), 179–186.

Michelbach, S. (1995) Origin, resuspension and settling characteristics of solids transported in combined sewage. *Wat.Sci.Tech.*, 31(7), 69–76

Michelbach, S. and Wöhrle, C. (1992) Settleable solids in a combined sewer system: measurement, quantity, characteristics. *Wat, Sci. Tech.*, 25(8), 181–188.

Michelbach, S. and Wöhrle, C. (1994) Settleable solids out of a combined sewer system: settling, behaviour, stormwater treatment and sedimentation rates in rivers. *Wat. Sci. Tech.*, 29(1–2), 95-102.

Nalluri, C. and Alvarez, E.M. (1992) The influence of cohesion on sediment behaviour. *Wat. Sci. Tech.*, 25(8), 151–164.

Nielsen, P.H., Raunkjær, K., Norsker, N.H., Jensen, N.Aa. and Hvitved-Jacobsen, T. (1992) Transformation of wastewater in sewer systems: a review. *Wat. Sci. Tech.*, 25(6), 17–31.

Norsker, N.H., Nielsen, P.H. and Hvitved-Jacobsen, T. (1995) Influence of oxygen on biofilm growth and potential sulphate reduction in gravity sewer biofilm. *Wat. Sci Tech.*, 31(7), 159–167.

Ødegaard, H. (1987) Particle separation in wastewater treatment. *7th European Sewage and Refuse Symposium*, Munich. EWPCA, Bonn, F.R.G.

Painter, H.A. and Viney, M. (1959) Composition of a domestic sewage. *J. Biochem. Microbiol. Technol. Eng.*, 1, 143.

Pisano, W.C. (1996) Summary: United States 'sewer solids' settling characterization methods, results, uses and perspective. *Wat. Sci. Tech.*, 33(9), 109–115.

Pisano, W.C. and Queiroz, C. (1977) *Procedures for Estimating Dry Weather Deposition in Sewerage Systems.* US EPA Report EPA-600/2-77-1. 20 July.

Pisano, W.C., Thibaults, N. and Forbes, G. (1990) The Vortex solids separator. *Wat. Env. Tech.*, May, 64–71.

Pomeroy, R.D. and J.D. Parkhurst (1977) The forecasting of sulfide buildup rates in sewers. *Prog. Wat. Techn.*, 9(3), 621–628.

Portielje, R., Hvitved-Jacobsen, T. and Schaarup-Jensen, K. (1996) Stochastic reliability methods in modelling effects of urban pollution runoff. *Proceedings of 7th International Conference on Urban Storm Drainage*, Vol. III, Hannover, Germany, 9–13 September 1996, 1491–1496.

Raunkjær, K. (1993) *Characterization and Transformation of Wastewater Organic Matter in Sewer Systems*, PhD dissertation, Environmental Engineering Laboratory, Aalborg University.

Raunkjær, K., Hvitved-Jacobsen, T. and Nielsen, P.H. (1994) Measurement of pools of protein, carbohydrate and lipid in domestic wastewater. *Wat. Res.*, 28, 251–262.

Raunkjær, K., Hvitved-Jacobsen, T. and Nielsen, P.H. (1995) Transformation of organic matter in a gravity sewer. *Wat. Environ. Res.*, 67, 181–188.

Raunkjær, K., Nielsen, P.H. and Hvitved-Jacobsen, T. (1997) Acetate removal in sewer biofilms under aerobic conditions. *Wat. Res.*, 31(11), 2727–2736.

Reiff, H. (1992) *Wachstum und Abtrag der Sielhaut in Mischwasserkanälen. Wasser – Abwasser – Abfall, Heft 9. Schriftenreihe des Fachgebietes Siedlungswasserwirtschaft*, Universität-Gesamthochschule Kassel.

Rickert, D.A. and Hunter, J.V. (1971) General nature of soluble and particulate organics in sewage and secondary effluents. *Wat. Res.*, 5, 421.

Ristenpart, E. (1995a) *Feststoffe in der Mischwasserkanalisation Vorkommen, Bewegung und Verschmutzungpotential*, PhD thesis, University of Hannover, Germany.

Ristenpart, E. (1995b) Sediment properties and their changes in a sewer. *Wat. Sci. Tech.*, 31(7), 77–83.

Ristenpart, E., Ashley, R.M. and Uhl, M. (1995) Organic near-bed fluid and particulate transport in combined sewers. *Wat.Sci.Tech.*, 31(7), 61–68.

Schmitt, F. and Seyfried, C.F. (1992) Sulphate reduction in sewer sediments. *Wat. Sci. Tech.*, 25(8), 83–90.

Skipworth, P.J., Tait, S. and Saul, A.J. (1996) Laboratory investigations into cohesive sediment transport in pipes. *Wat. Sci. Tech.*, 33(9), 187–193.

Sollfrank, U. and Gujer, W. (1991) Characterization of domestic wastewater for mathematical modeling of the activated sludge process. *Wat. Sci. Tech.*, 23(4/6), 1057–1066.

Stotz, G. and Krauth, K.H. (1986) Depositions in combined sewers and their flushing behaviour. *Proc. Int. Symp. on Comparison of Urban Drainage Models with Real Catchment Data.* UDM 86, Dubrovnik, edited by C. Maksimovic, and M. Radojkovic.

Tanaka, N. and Hvitved-Jacobsen, T. (1998) Transformations of wastewater organic matter in sewers under changing aerobic/anaerobic conditions. *Wat. Sci. Tech.*, 37(1), 105–113.

Thistlethwayte, D.K.B. (ed.) (1972) *The Control of Sulfides in Sewerage Systems*. Butterworth, Sidney, Australia.

Torfs, H. (1995) *Erosion of mud/sand mixtures*. PhD thesis, Katholieke Universiteit Leuven.

Tyack J N., Hedges P.D. and Smisson, R.P.M. (1996) The relationship between settling velocity grading and the characteristics of the contributing catchment. *Wat.Sci. Tech.*, 33(9), 135–142.

USEPA (1985) *Design Manual: Odor and Corrosion Control in Sanitary Sewerage Systems and Treatment Plants*. EPA/625/185/018, Cincinnati, Ohio, USA.

Verbanck, M. (1990) Sewer sediment and its relation with the quality characteristics of combined sediment deposits. *Wat. Sci. Tech.*, 22(10–11), 247–258.

Verbanck, M. (ed.) (1992) Workshop on origin, occurrence and behaviour of sediments in sewer systems. *Wat. Sci. Tech.*, 25(8), 234 pp.

Verbanck, M. (1992a) Field investigation on sediment occurrence and behaviour in Brussels combined sewers. *Wat.Sci.Tech.*, 25(8) 71–82.

Verbanck, M., Ashley, R.M. and Bachoc, A. (1994) International workshop on origin occurrence and behaviour of sediments in sewer systems: summary of conclusions. *Water Research*, 28(1), 187-194.

Vollertsen, J. and Hvitved-Jacobsen, T. (1997) Aerobic microbial transformations of resuspended sediments in combined sewers – a conceptual model. *Wat. Sci. Tech.*, 37(1), 69–76.

Vollertsen, J., Hvitved-Jacobsen, T., McGregor, I. and Ashley, R.M. (1999) Aerobic microbial transformations of pipe and silt trap sediments from combined sewers. *Wat. Sci. Tech.*, 39(2), 231-241.

Williams, D.J.A., Williams, P.R. and Crabtree, R.W. (1989) *Preliminary Investigation into the Rheological Properties of Sewer Sediment Deposits and the Development of a Synthetic Sewer Sediment Material for Laboratory Studies.* Foundation for Water Research Report FR 0016.

Wotherspoon, D.J.J. (1994) *The Movement of Cohesive Sediment in a Large Combined Sewer.* PhD thesis, University of Abertay Dundee.

Wotherspoon, D.J.J. and Ashley, R.M. (1992) Rheological measurements of the yield strength of combined sediment deposits. *Wat. Sci. Tech.*, 25(8), 165–169.

Xanthopoulos, C. and Augustin, A. (1992) Input and characterisation of sediments in urban sewer systems. *Wat. Sci. Tech.*, 25(8), 21–28.

2.3 TRANSPORT, DEPOSITION AND EROSION

2.3.1 Introduction

The transport of solids within a fluid is sustained by some of the energy of the flow. When the sewage flow, as a transport medium, contains sufficient energy, it promotes the movement of solid particles. The flow is itself influenced by the solids being transported, e.g. fluid characteristics noticeably change at high solids concentrations, typically above 500 mg/l, and the boundary resistance also varies due to the presence of a deposited sediment bed. The interactions between the sewage flow and solids movement, together with the complex interrelationships between the different solids conveyed and the general heterogeneity of these solids, lead to difficulties for the precise description of solid transport processes.

In view of these difficulties, the first two parts of this section deal with the hydraulics of flow in pipes (Section 2.3.2) and storage chambers (Section 2.3.3). They only describe briefly the detailed fluid dynamic principles responsible for solids transport and movement; more details can be found in other publications. The next section is primarily concerned with the transport of solids in sewers (Section 2.3.4), specifically: suspended solids, gross solids, near bed solids, and granular (traditional) bed load. Section 2.3.5 deals with deposition in sewers and 2.36 with the occurrence of biofilms. An introduction to modelling is given in Section 2.3.7.Solids movement is considered in ancillaries and special structures in drainage systems, such as tanks, basins and CSO chambers (Section 2.3.8).

2.3.2 Sewer flow hydraulics

Sewers are generally circular (or near circular) in cross section and are designed to convey the flow with or without solids under partially filled or full flow conditions. If the velocity of flow or the boundary shear is large enough, in relation to the incoming sediment load, a 'clean' boundary with no deposition of solids (self-cleansing sewer) can be maintained. Lengths of sewer with mild slopes or local depressions, and with sufficiently high sediment input from upstream, are subject to a continuous sedimentation-erosion process due to the time varying nature of flows in sewers. Virtually all systems, separate or combined, experience some sedimentation (Section 2.2.4). Deposited sediment affects the velocity field because the cross section is transformed by sedimentation into a different (reduced area) shape, with an altered wetted perimeter. In addition, the sediment bed has a surface roughness which is different from that of the sewer side walls, and the overall effect is to cause a greater resistance to flow. Hence overall there is a decrease in flow capacity. Since sediment deposits are often found in systems where the available (flow) energy is low, the decrease in flow capacity caused by the accumulation of sediment can lead to serious problems (Chapter 3).

2.3.2.1 Mean velocity, shear stress and turbulence distribution over the flow cross section

A knowledge of the shear stress distribution over the boundary is required in order to determine the flow resistance due to sediment deposits. Assuming steady, two dimensional (2D) turbulent flow, shear stress can be estimated by assuming that the velocity distribution follows a logarithmic law. It has been a common practice to use the 'law of the wall' as the equation to determine the average velocity u_m in a vertical cross-section profile for a rough turbulent regime:

$$\frac{u_m}{u_*} = \frac{1}{\kappa}\ln\left[\frac{12R}{k_s}\right]_a \tag{2.3.1}$$

where: u_* is the shear velocity (m/s)
κ is the von Karman constant (≈ 0.4)
R is, hydraulic radius (m)
k_s is sand roughness (m)

This equation is suitable for wide, open channels but it may also be applied to part-full sewers, which can be considered analogous to open channels with a composite roughness (Perrusquía *et al.*, 1987; Perrusquía, 1990).

Both velocity and shear stress distributions are produced principally by the turbulence of the sewage flow. The turbulence levels (all three components) are dependent on the proportionate depth of flow (Nalluri and Novak, 1973, 1975). At the highest depths, the crowning effect of the channel cross section (if curved) causes an appreciable increase of turbulence levels at the free surface. The turbulence micro and macro scales of the flow field in the main flow direction approach maxima at about mid-depth. Thus the relative flow depth also controls the type of flow, being approximately 2D in a wide channel (at low depths), three-dimensional (at higher depths) for part-full flow, with 2D flow prevailing at the higher depths and when there is pipe-full flow.

Often sewers contain a semi-permanent bed, which is responsible for variable distributions of the parameters controlling the flow field (Perrusquía *et al.*, 1995). Substantial research work has been carried out investigating the effects of deposited beds in part-full pipe channels at both laboratory and field scales (e.g. Alvarez, 1990; Perrusquía, 1991; Kleijwegt, 1992; Ristenpart, 1995).

The boundary shear distribution is notably different between the bed (rougher texture) and side walls (relatively smoother), as illustrated in Figure 2.3.1. The effective shear responsible for erosion of the bed is larger than the shear on the walls. In all conditions, the thickness of the bed and relative flow depths over the bed are important parameters controlling the velocity, shear and turbulence distributions.

The relationship between the bed shear stress and the average shear stress in a sewer with a deposited bed is illustrated by Ackers *et al.* (1996) (Figure 2.3.2). A 'bed shear ratio' was defined there as the bed shear stress τ_b divided by the average shear stress τ_0 (full-pipe flow conditions, the latter without deposits).

Figure 2.3.2 shows a reduction of the bed shear ratio at increased sediment bed depths as well as for decreasing bed roughness. The ratio is virtually independent of pipe diameter as well as gradient and is affected to only a minor extent by the pipe wall roughness value k_0.

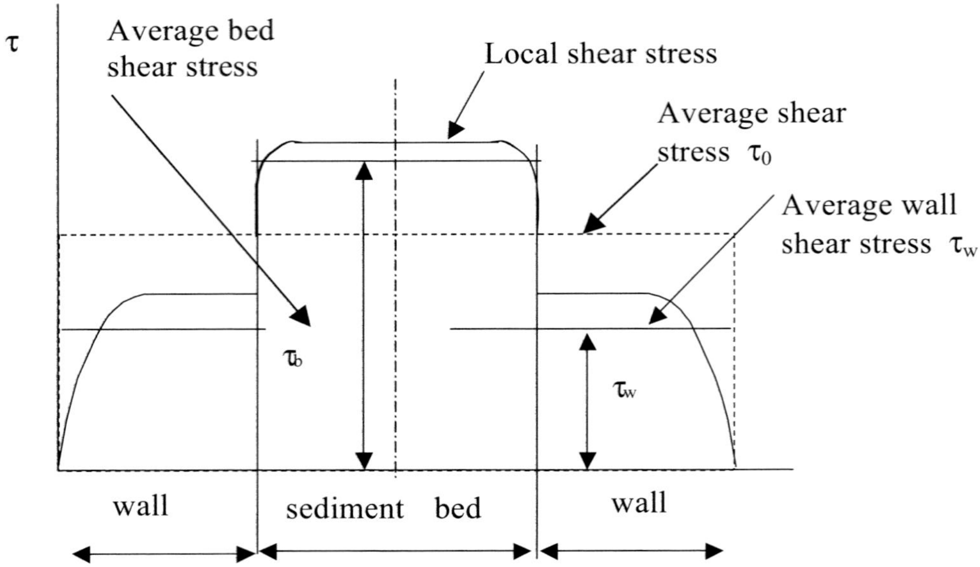

Figure 2.3.1. Boundary shear stress distribution of part-full pipes with composite roughness (adapted from Perrusquía *et al.*, 1995).

2.3.2.2 Composite roughness and sidewall elimination procedures

Several methods for the estimation of the composite roughness in open channels have been published. However, these have usually been applied to flow in rectangular channels where the influence of the side-wall roughness is not as significant as for sewers with sediment deposits. Einstein (1942) proposed a method for the calculation of flow resistance caused by channel beds when the flow resistance of the side walls is known. This method is often referred to as the 'side-wall elimination procedure', and is used to separate the effects of the smoother side walls and the rougher beds in channels. In practical engineering work the problem is to compute the composite roughness based on the known roughness of both bed and side walls. In sewers, the cross section usually cannot be regarded as wide, as is often the case for open channels. Therefore, the properties of the sediment bed need to be computed separately from those of the side walls. Procedures commonly used to determine the hydraulic characteristics of the sediment bed are those proposed by Einstein (1942) and Vanoni and Brooks (1957). The former uses Manning's equation whilst the latter uses the Darcy-Weisbach equation to estimate the hydraulic roughness associated with the smooth walls and rough beds. Detailed descriptions of application of the procedure are given in the literature (e.g. Kleijwegt, 1992; Perrusquía *et al.*, 1995). The bed shear calculations using the Einstein and the Vanoni Brooks separation techniques appears to be applicable in sewers, particularly when there are shallow depths (Alvarez, 1990; Wotherspoon, 1994; Skipworth, 1996).

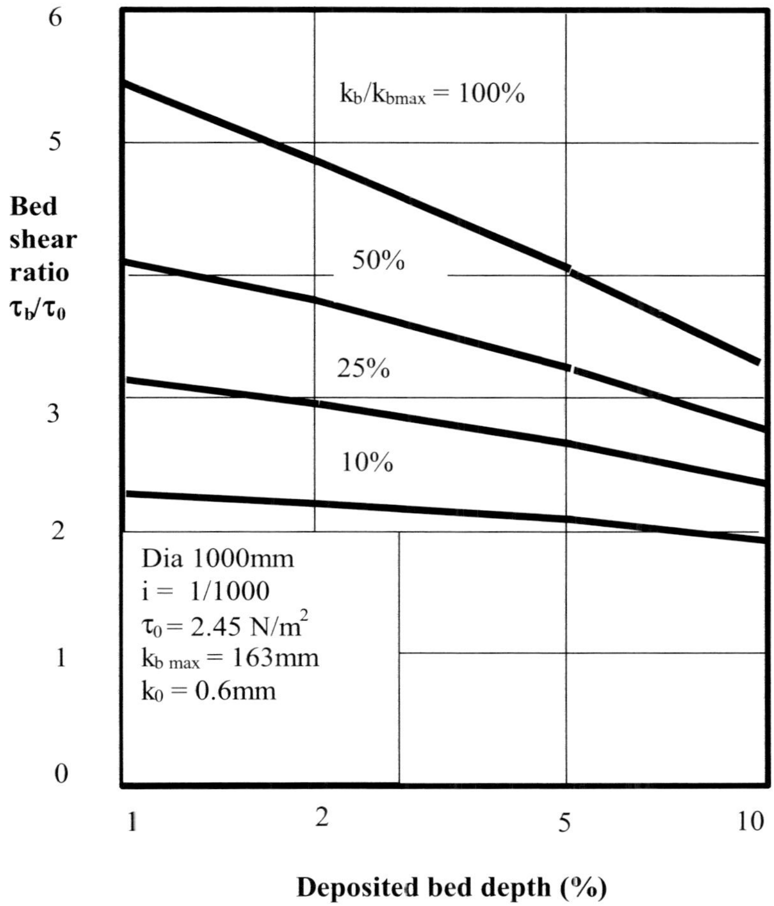

where k is roughness coefficient (Colebrook-White equation)
τ is the average shear stress exerted by fluid on boundary
suffixes b and 0 refer to the bed and clean pipe wall components respectively
i is hydraulic gradient

Figure 2.3.2. Variation in bed shear stress with bed depth and roughness for full pipes (adapted from Ackers *et al.*, 1996).

2.3.2.3 Flow resistance with sediment deposits

The hydraulic section of a part-full circular sewer with sediment deposits is that of a modified or truncated circular open channel, which will produce consequential shape effects on the flow. The effect of the shape of the cross section (which varies with flow depth) on flow resistance in part-full (open channel) pipe flow cannot be properly accounted for using the concept of mean hydraulic radius. This is because the hydraulic radius is an arbitrary linear parameter and therefore a wide number of conduit forms may be characterised by the same value. Kazemipour and Apelt (1979) as well as Myers (1982) showed that open channel friction factors are significantly different from those obtained using the Prandtl-Karman equations established for full-pipe flow conditions.

The friction coefficient governing the resistance law in sediment-laden flow is some 10-15% larger than that of the corresponding clear water flow and is effectively found to be a function of the limit of deposition concentration, pipe diameter, flow depth and sediment size (Nalluri *et al.*, 1994).

Attempts have been made to establish semi-empirical limit of deposition transport models which can be used to determine flow characteristics either with or without deposited beds in circular pipe channels. Corresponding frictional resistance coefficients have been determined for flows with sediment in transport. For example, in an experimental study to investigate the flow in pipe channels over deposited sediment beds, Perrusquía *et al.* (1987) evaluated the effects on flow capacity caused by different levels of sediment accumulation. It was shown that in some cases sediment deposition equivalent to 10% of the pipe diameter produced a reduction of 25 % in the flow conveyance, whilst a deposit of 20 % reduced the conveyance to half of the corresponding clear water flow, as illustrated in Figure 2.3.3.

Perrusquía (1988) computed the flow resistance of movable plane beds. The results suggested that the mean equivalent bed roughness is in the range of approximately 2 to 4 times the d_{50}. If there are bedforms in the bed, flow resistance prediction becomes more complicated. The flow resistance is assumed to consist of two components: one depending on grain friction and the other on form friction. Perrusquía (1991) found that the flow resistance could be predicted, to a certain extent, using the available methods for alluvial channels.

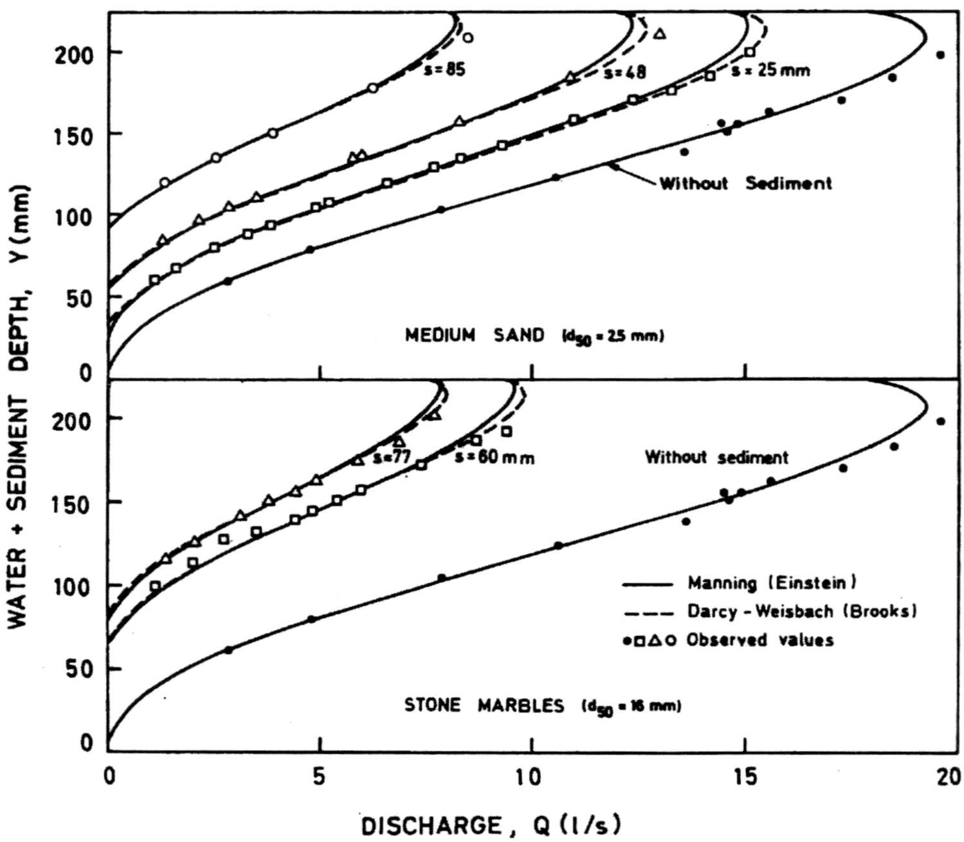

Figure 2.3.3. Rating curves for flow in circular channels with deposited beds (adapted from Perrusquía *et al.*, 1987).

2.3.3 Tank, basin and CSO hydraulics

Detailed information about tanks, basins and CSOs is given in hydraulic texts and reports (e.g. Bos, 1989; Hager, 1994; Read, 1997; Water Research Centre 1994). Storage mass balance equations are included in Chapter 3, and are not discussed further in this section. New methods such as Computational Fluid Dynamics (CFD) for the simulation of the flow field and the trajectories of discrete particles in hydraulic structures have emerged recently (Stovin, 1996; Harwood and Saul, 1996). These are now being applied for the estimation of efficiency of storage chamber performance (hydraulics and sediment loads) as well as the prediction of sediment deposition.

Flow patterns in storage chambers (as these affect sediment behaviour) are described below (Section 2.3.5). During a storm event the following sequence of overall flow conditions is typically observed in a storage chamber (Ellis, 1992). Initially the dry weather flow (DWF) is contained within the DWF channel, but as the inflow into the structure begins to rise, the flow in the chamber overtops the DWF channel and spreads out over the storage chamber benching. A substantial amount of turbulence may be observed just as the flow overtops the DWF channel. As the inflow increases further, the depth of flow in the chamber rises and the stored volume increases. The actual flow rate at which this transition occurs, and the rate of filling, are functions of the continuation flow control head-discharge characteristic and the available volume of storage. Under these partially filled conditions the flow is generally tranquil. Depending upon the internal configuration of the chamber, the flow pattern may or may not be symmetrical, and some dead zones in the stored volume may develop. This is illustrated by an exemplary sketch of a flow field observed in a model storage chamber (Figure 2.3.4).

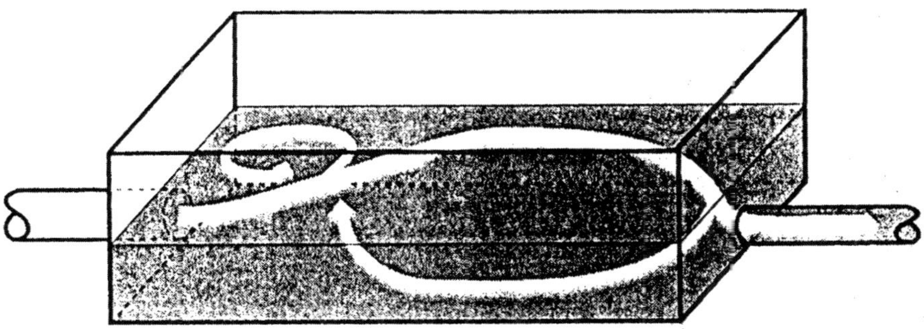

Figure 2.3.4. Sketch of a flow field observed in a model storage chamber (Stovin and Saul, 1996).

If the chamber fills to the point of overflowing then there will be a tendency for the flow to short-circuit the stored volume and pass directly from inflow to overflow. The main body of the stored volume remains tranquil. In situations where the overflow weir is positioned on a side wall of the chamber, this tends to result in the production of asymmetric flow circulations. As the chamber drains, the flow generally remains tranquil until the benching is reached. During the last stages of draining, the flow may become more vigorous.

The flow patterns which develop during the comparatively tranquil spilling stage are particularly important in determining the locations where sediments deposit on the chamber floor. The flow patterns depend on the internal geometry of the chamber as well as the hydraulic characteristics of the inflow and the chamber outlets. These issues are discussed further in Chapter 3.

2.3.4 Solids movement in sewers

Sewer flows contain both solid particles and gases. The solid particles comprise an inorganic component of which sand is the main fraction, as well as fibrous and organic material (Section 2.2). In the gas phase, bubbles enter the sewage flow by turbulent mixing, surface aeration and due to biological processes in the water column and bed deposits.

An extensive literature about solids transport is available for flows in alluvial channels. However the conditions related to solid transport in sewers are significantly different The main differences were outlined by Verbanck *et al.* (1994) and Berlamont and Torfs (1996).

- Sewer solids are complex mixtures of organic and inorganic materials, often found at the bed in stratified layers.
- Temporal and spatially variable effects are important over both long and short time and length scales. In sewers the effects of unsteady and non-uniform flows are more important; for example, classical suspended load theories presuppose that quasi-steady state conditions prevail; this is rarely the case in many sewers.
- The confined cross-sectional shapes of sewers introduces important turbulence and boundary effects: with changes in friction effects due to the relatively large differences in roughness contributions from lateral walls and sediment bed respectively.
- Empirically calibrated transport equations developed for fluvial conditions do not account for these effects or any sediment supply limitations in sewers.
- Amounts of sediment transported in sewage flows can be a fraction of the solid transport capacity, notably in lengths where there is an insufficient source of deposited material to be re-eroded to satisfy the transport capacity of the prevailing flow.

The movement of solids (and associated pollutants) in sewers can resemble that in rivers in that particles move in suspension, near, or in contact with the bed, or can remain within the bed, under different hydraulic conditions. When considering sediment motion from the point of view of general physics, a major distinction can be drawn between suspended and other modes of transport, the latter modes comprising rolling, saltation and sliding. Particles not in suspended transport (the so-called bed load particles) receive no upward impulses other than those due to the successive contacts between the solid and the bed. Particles held in suspension by the flow forces are completely surrounded by the fluid.

In sewer flows, there is a continuous process of particles interchanging between suspension, the bed and the near bed region. All processes occur simultaneously as illustrated in Figure 2.3.5.

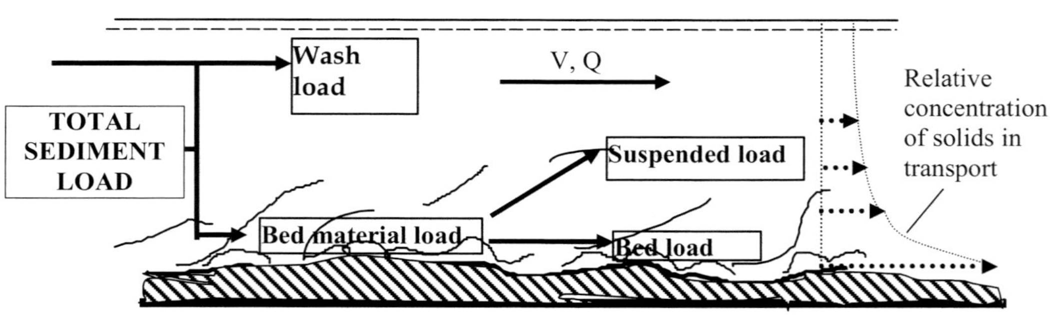

Figure 2.3.5. Sediment moving and deposited in a sewer.

Laboratory pipe flume studies have been used extensively to investigate transport in pipes with clean and deposited beds. Prolonged studies at Hydraulics Research in the UK have provided the foundation for the current methods of controlled sewer sediment design (Section 4.2). Figure 2.3.6, from these studies, illustrates diagrammatically the various transport regimes observed in circular pipe flows.

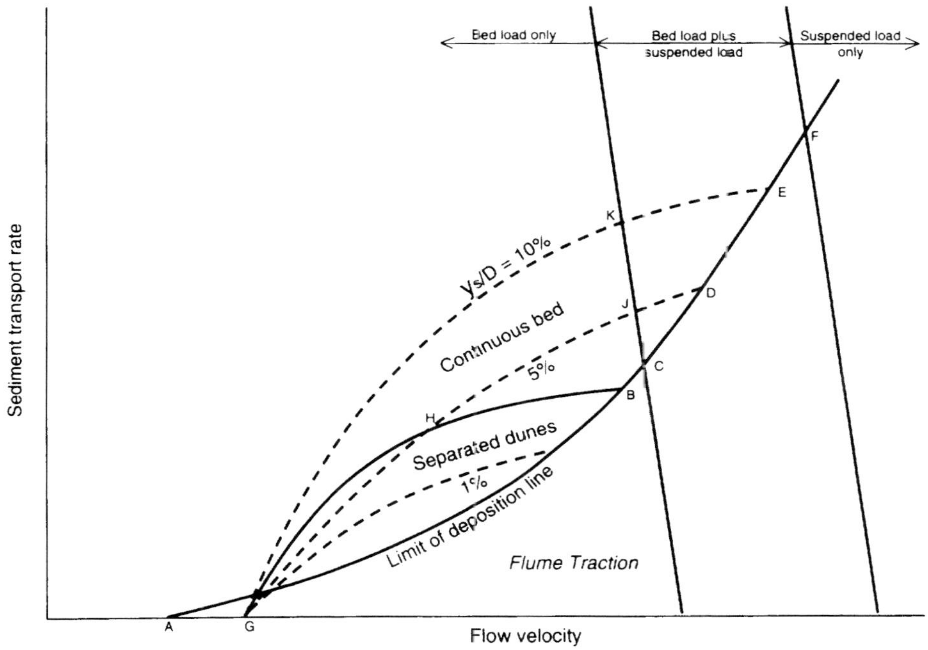

Figure 2.3.6. Diagrammatic representation of sediment transport regimes in pipes (adapted from May, 1994).

Ashley and Verbanck (1996), when discussing different sewer sediment transport modes, give an approximate classification system adapted from fluvial channels (Raudkivi, 1990), based on the relationship for the sedimentation parameter:

$$\eta = \frac{w_s}{\kappa u_*} \tag{2.3.2}$$

where η is the sedimentation parameter
w_s is the particle settling velocity (for homogeneous suspensions)
u_* is the fluid bed shear velocity and
κ is Von Karman's constant

The ranges for dominant transport mode defined by η are:

bed load $5 < \eta < 15$
saltation $5 < \eta < 3$
suspension $\eta \le 3$

Ashley *et al.* (1994) used field measurements of material collected in bed load traps in combined sewers, to show that the definition of equation (2.3.2) for suspended and bed loads appeared to correspond with the material sampled from suspension and in the traps.

2.3.4.1 Suspended solids movement in sewers

A mass of sediment particles is said to be in a state of suspension if it is kept above the bed by the eddies of fluid turbulence. In this case, there are vertical, upward impulses provided to the particles by the fluid, and such processes can only be maintained in a turbulent medium. Flume studies conducted by Grass (1974) and Sumer and Deigaard (1979) show that suspension is achieved through the advection of masses of fluid away from the near-bed region into the turbulent boundary layer. 'Burst' processes (Kline *et al.*, 1971) have been described as extracting significant masses of buffer layer fluid (together with the sediment it contains) and projecting it into the turbulent flow. Bagnold (1966) reasoned that suspension occurs when the upward directed components of the turbulent velocity fluctuations w'_{up} exceed the fall velocity of the particles w_s, both expressed in m/s. This can be expressed more generally using an empirical relationship between the root mean square component w'_{up} and the shear velocity u_*.

In general, for a full suspension $\eta \leq 3$ (Equation 2.3.2). Based on diffusion considerations, all conditions for which $\eta < 1$ correspond to flow situations where a given particle released in the flow will be unlikely to enter the constant-stress layer close to the bed (Batchelor, 1965). The threshold value for suspended load, expressed by Equation 2.3.2, is similar to that recommended in a revised UK sewer design method (Ackers *et al.*, 1996). It is to be noted that there are many difficulties in evaluating the appropriate shear stress to determine the shear velocity in sewer flows, and the accurate measurement of settling velocity of the material is open to question, as discussed in the Appendix.

The longitudinal velocity of suspended particles is comparable with u_m the average velocity of the fluid, whilst the bed load (granular) particles move slower at a longitudinal velocity of the order of the shear velocity u_*.

Thus the suspended load q_{SS} (g/s) is traditionally calculated as the integral over the cross-sectional area A (m^2) of the concentration C (mg/l) multiplied by the velocity u (m/s). The advection transport is given by:

$$q_{SS} = \int_A u(y,z) \cdot C(y,z)\, dA \qquad (2.3.3)$$

A transport mode that occurs but is not taken into account in Equation 2.3.3 is the dispersion phenomenon. Due to spatial sediment concentration variations and velocity gradients, mixing effects occur that tend to attenuate solids concentration peaks in the water column. However, this process applies only to the dissolved and fine suspended matter. A detailed description and mathematical formulation is given by Markofsky (1980), and some recent results are described by Guymer *et al.* (1998).

Most of the particles able to pass over the high level weirs of storm overflows (and the associated pollution) will derive from the suspended particle transport process. Hence, in terms of total mass fluxes of both solids and pollutants, suspended transport is the primary mode of conveyance of sewer solids. When integrated over the flow cross-section, the relatively large volume of water conveying (fairly dilute) suspensions of particles is such that the lower concentrations ($\cong$ 250 mg/l), compared with the material moving near the bed

(≅ 5000 mg/l), still lead to larger total mass fluxes. Typically, some 80–90 % of the total solids by mass are conveyed as the material moving in suspension (Ashley *et al.*, 1994).

The suspended load in sewer flows has been investigated extensively in many studies, with samples drawn from the flow at different depths simultaneously using small bore tubes (≈ 10 mm diameter) (Chebbo, 1992; Ashley and Crabtree, 1992; Ristenpart, 1995; Ristenpart *et al.*, 1995; Verbanck, 1993, 1995a). The predominant particles in suspension during dry and wet weather are about 40 µm in size, and attributed to 'sanitary' solids. Most of the suspended particle load in combined sewer flows is organic (≅ 80 %) and has high specific surface area, i.e. biochemically active particles with a significant capacity to adsorb micro-pollutants (Section 2.2). Due to the diversity of sources of suspended solids, there are distinct temporal variations in concentration. Figure 2.1.15 in Section 2.1.4.1 shows a typical depth-averaged DWF pattern of suspended solids concentration in a combined sewer.

In both dry and wet weather flows, the particles conveyed in suspension have low settling velocities. For example, in a review of US data, while warning of the limitations caused by sampling procedures and type of settlement test used (see Appendix), w_{s50} was given as 0.6 mm/s for dry-weather sewage, 0.2 mm/s for stormwater and 12 mm/s for sediment (Pisano, 1996).

2.3.4.1.1 Suspended sediment concentration profiles

The suspended sediment (SS) transport rate is often determined numerically as the integral through the whole wetted area of the suspended sediment flux, i.e. the product of local velocity and local concentration (Equation 2.3.3). This requires an adequate law describing the variation of SS concentration throughout depth. Based on an analogy with molecular diffusion, the well-known 'convection-diffusion' equation was developed by Schmidt (1925) in connection with studies of dust in the atmosphere and by O'Brien (1933) in studies of suspended sediments in streams. After partial integration, this gives:

$$\ln \frac{C_y}{C_{a*}} = - w_s \int_{a*}^{y} \frac{dy}{\varepsilon_s} \tag{2.3.4}$$

where C_y is volumetric solids concentration at height y above the boundary (mg/l)
C_{a*} is volumetric solids concentration at height a_* above the boundary (mg/l)
ε_s is sediment diffusivity of mass (m^2/s)
ws is fall velocity of suspended particles.

One solution of this equation has been extensively used in fluvial and other natural environments. It is based on the assumption that the diffusion coefficients respectively for suspended sediment (ε_s) and for fluid momentum (ε_m) vary similarly, displaying a parabolic distribution throughout the depth (Rouse, 1937). Studies in the UK, Denmark and Germany have explored the potential application of the Rouse solution to sewer flows at both laboratory and field scale (e.g. Mark, 1995; Ristenpart, 1995; Skipworth, 1996). An example is given in Figure 2.3.7.

Limitations of the approach were found due, firstly, to unexpected values of the so-called 'Rouse exponent' and, secondly, to the difficulty in fixing the reference level a^* (to correspond with the given reference concentration C_{a*}). As an alternative, it is possible to solve Equation 2.3.4 using the continuum theory of mixtures giving accurate expressions for ε_s (McTigue, 1981). A two-layer model results, with an outer suspension region in which ε_s varies according to total flow depth (excluding any proximity of the water surface), and an inner suspension region in which ε_s is proportional to the distance from the bed boundary.

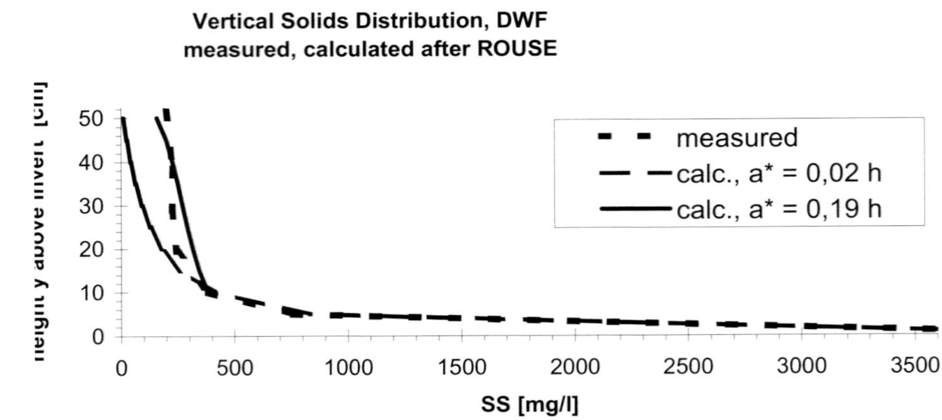

Figure 2.3.7. Suspended sediment profile in Hildesheim trunk sewer during dry-weather flow (adapted from Ristenpart, 1995).

Verbanck (2000) showed that the law valid in the outer flow may be given by:

$$\frac{C_y}{C_{a*}} = e^{\eta\,(1 - \frac{y}{a*})} \tag{2.3.5}$$

This was verified using data collected in dry weather flow in a large-sized sewer in the centre of Brussels, as shown in Figure 2.3.8. The second law, valid for the lower 25 % of the flow close to the bed (Coleman, 1982), is:

$$\frac{C_y}{C_{a*}} = \left(\frac{y}{a*}\right)^{-\eta} \tag{2.3.6}$$

This relationship still needs further experimental validation in the sewer environment, although Verbanck (2000) showed that this 'two-layer' model can be used to compute the respective contributions of the material moving close to the bed in the inner suspension region (Equation 2.3.6) and the outer suspension region (Equation 2.3.5), to the total SS mass flux in a sewer. This is further examined in Section 2.3.4.3.

These two laws expressed in terms of the sedimentation parameter η (Equation 2.3.2) have the merit, as pointed out by Coleman (1982), to utilise a^* as an intrinsic parameter characterising the turbulent flow. Earlier approaches linking C_{a*} with the bed load transport phase (e.g. Einstein, 1950) had serious shortcomings in view of the very different nature of the physical processes respectively involved in suspended and other modes of transport respectively.

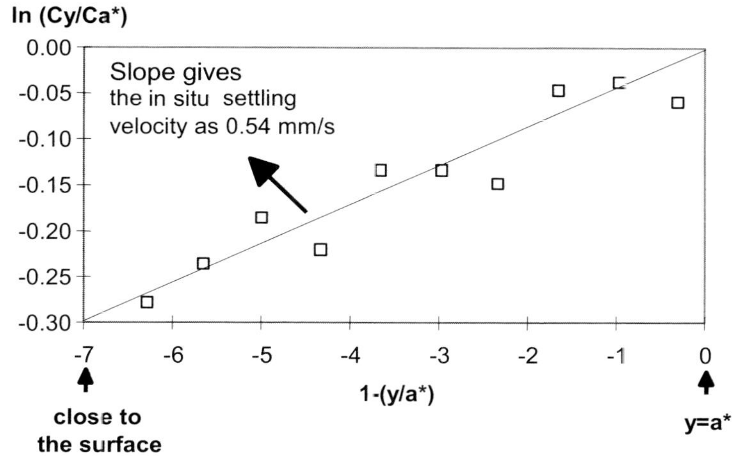

Figure 2.3.8. Illustrative suspended sediment profile in Brussels main trunk sewer during dry weather flow (adapted from Ashley and Verbanck, 1996).

2.3.4.1.1 Suspended sediment transport capacity in sewers

The estimation of sediment transport capacities in sewers is still uncertain with significant reliance on site-specific empirically calibrated relationships. There have been several attempts to re-calibrate available (fluvial) suspended load theories by applying data obtained in real sewers or in flume channels using pipe cross-sections or surrogate sewer sediments. However, most of the results have been disappointing. In terms of general applicability, Coghlan *et al.* (1993), Wotherspoon (1994) and Ackers *et al.* (1996) recently showed that reasonable results can be expected from the use of a computation procedure derived from that developed by Ackers and White (1973). For instance, Coghlan *et al.* (1993) compared solids concentrations observed in a sewer c_{obs} with estimated results c_{comp} and found that for nearly 70 % of the data the ratio c_{comp}/c_{obs} was greater than ½ and less than 2, criteria generally accepted as indicative of reasonable performance indicators for sediment transport formulae. In terms of suspended transport, simplification of the Ackers-White set of formulae (see Equations 2.3.16–2.3.18) is possible if w_s is utilised, instead of information limited only to sediment size (D_{gr}) (Verbanck 1995b). As different (turbulent, transition, laminar) flow regimes condition the settling behaviour of a suspended particle, a relationship utilising D_{gr} will necessarily be more complex than any approach using w_s directly.

Because of the various flow regimes experienced by a given particle, determination of the sediment settling velocity has often involved tedious trial-and-error calculations. Verbanck (1995b) has shown that notwithstanding the problems described in the Appendix, a single settling velocity formula, valid for the whole range of particle Reynolds number, was acceptable for the analysis of suspended load transport. This is derived from Rubey (1933):

$$w_i = \sqrt{\xi^2 + 1.09\ \Delta\ g\ d_i} - \xi \qquad (2.3.7)$$

for which $\xi = 13.95\ \nu/d_i$; $\Delta = (\rho_s-\rho)/\rho$ is the relative density of sediment in water

where ρ_s is particle density (kg/m^3); ρ is water density (kg/m^3)
d_i is particle size of the ith class (mm); ν is kinematic viscosity (m^2/s).

Bagnold (1966) showed, by application of streampower arguments, that heterogeneous suspensions may be described by an *effective* settling velocity in which all size classes i are taken into account according to a mass-weighted average:

$$w_s = \sum_i (p_i w_i) \tag{2.3.8}$$

where p_i is mass fraction of particle size fraction i
w_i is mean particle fall velocity in particle size fraction i.

The importance of this simple law when computing bulk suspended solids fluxes cannot be overemphasised, especially when the suspended material is heterogeneous, as is the case for suspended matter conveyed in sewer flows.

Relationships developed specifically for suspended sediment transport by Velikanov (1954) and Macke (1982) also utilise w_s, and have been found to be reasonably effective when applied to sewer flows (Ackers *et al.*, 1996; Bujon, 1988). When applying such relationships, however, the necessary field data for adequate description of the hydrodynamics of the sewer flows and of the sewer particles are rarely available, especially for the nature of the solids conveyed in suspension in actual sewers.

Unfortunately, tests conducted in laboratory flumes have largely ignored the suspended mode of transport, as this requires careful manipulation of the fine and lightweight particles. The finer particles generate cohesion and turbidity problems, and preclude the application of standard (optical) measurement procedures. Hence, as yet only limited sensitivity analysis has been undertaken on the various factors relevant to suspended transport formulae. An exception to this is recent work in the UK (Skipworth, 1996; Tait *et al.*, 1998), which estimated rates of erosion of mixed sediment beds in a pipe flume by measurements of suspended sediment concentration (in the absence of bed load transport) taken at the downstream end of the pipe at a number of heights above the bed surface and at regular time intervals, as it was concluded that the Rouse equation was not appropriate to predict the vertical suspended sediment distribution from a single measured reference concentration.

In most sewers, particularly when a sediment bed is present, the bulk hydraulic resistance corresponds to fully turbulent rough conditions. Inertia effects thus dominate over viscosity, facilitating the definition of a (dimensionless) Chezy coefficient in terms of k_s alone (i.e. independent of viscosity ν). This simplification has led to the development of a procedure to assess the average value of the shear velocity at the bed $u_{*,b}$, based on a knowledge only of the overall surface roughness k_s (mm), average velocity u_m (m/s) and total flow depth y_0 (m):

$$u_{*,b} = \frac{u_m}{\frac{1}{\kappa}\ln\left(\frac{12\ y_0}{k_s}\right)} \tag{2.3.9}$$

Equation (2.3.9), which is of the same form as Equation (2.3.1), was derived by Kleijwegt (1992) for circular sewers and subsequently by Verbanck (1995b) for rectangular-shaped conduits. Equation 2.3.9 is invaluable for the calculation of the average bed shear velocity $u_{*,b}$, as it does not rely on the evaluation of R_b (m), the part of the hydraulic radius associated with the bed flow zone. The traditional Einstein-Vanoni separation technique between bed and wall is of doubtful validity when very narrow (breadth/depth < 4) sections are considered (Kleijwegt, 1992). It appears that this limitation may explain some of the inaccuracies of applying classical fluvial sediment transport theories to flows having a very low breadth-to-depth ratio (Verbanck, 1995b).

Information about w'_{up}, the upward component of the turbulent velocity fluctuation in sewers is not available, hence the shear velocity at the bed $u_{*,b}$ is the key parameter for predicting suspended sediment transport rates. Based on Bagnold's (1966) dynamic theory of suspension, Verbanck (1995b) developed the following expression to assess the volumetric suspended sediment transport capacity in a sewer:

$$C_v = \frac{1}{5.16} \frac{u_{*,b}^3}{g \Delta w_s R_H} \tag{2.3.10}$$

which is applicable provided $\eta \leq 3$.

Equation 2.3.10 is based on the existence of residual Reynolds stresses in wall-bound shear flows, normal to the boundary, and produced by the asymmetric nature of shear turbulence (Irmay, 1960; Bagnold, 1966). The numerical constant 5.16, is also obtained theoretically, and should be replaced by 3.44 for a pure suspension, when there is no (granular) bed load transport to be maintained in the considered sewer section. The difference between the two numerical constants corresponds to the amount of energy necessary (and consumed) in maintaining the bed load transport in the flow area closest to the bed. Equation 2.3.10 has been applied successfully to a large-sized sewer in Brussels with a narrow lower section (Figure 2.3.9), and to large circular sewers in Dundee and Hildesheim (Verbanck, 1995b, 2000). More general validity of the approach, however, still needs to be demonstrated using more extensive data sets in sewers and laboratory channels.

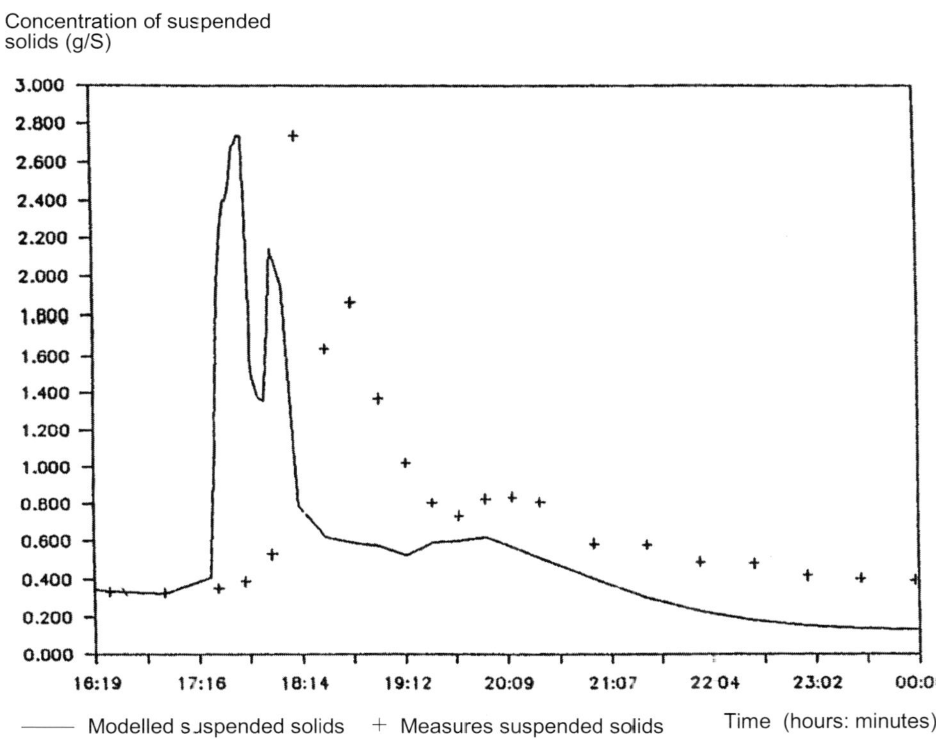

Figure 2.3.9. Suspended sediment transport (capacity and measured rate) in Brussels main trunk sewer (adapted from Verbanck, 1995b).

2.3.4.2 Near bed solids

There is considerable uncertainty about the nature of the material moving near the bed in combined sewer flows, and therefore whether the definitions used in fluvial hydraulics to define transport mode are appropriate (Section 2.3.3). Problems exist because of the nature of the particles (inorganic and organic, which are non-conservative and biochemically reactive). Problems include the definition and determination of particle size, density, shape and thus settling velocity (see Appendix). There is additional uncertainty about the interface between the solids conveyed in true suspension and those which comprise the bed, as evidence suggests an intermediate 'layer' of material at the base of the water column which may or may not be in motion, depending upon the hydraulic conditions, as illustrated in Figure 2.3.5. Observations using tubes extracting samples from the flow (Section 2.3.4.1) and near the bed by Wöhrle and Brombach (1991), Ashley *et al.* (1994), Verbanck (1995a) and Ristenpart *et al.* (1995), indicate that there may be a lower 'layer' of highly concentrated material close to the bed (equivalent to the Class C sediments described by Crabtree (1989), Tables 2.2.6 and 2.2.7, and Figure 2.2.13). There will be major differences both spatially within the same sewer and also temporally in response to changing inputs (particularly storm flows) and any other cause of unsteady flows. The studies by Lin *et al.* (1993) in Marseille, and Ashley *et al.* (1994) in Dundee, using traditional 'bed load' traps, have attempted to look at material moving in this mode. It is clear that there are important variations in observations related to this 'near-bed' transported material, especially regarding volatile solids. The organic proportion of this near bed transport has been termed 'dense undercurrents' and 'near bed (organic) layers'; as yet it is not clear whether the various reported studies are looking at similar material, or if transport occurs as a result of suspended (turbulent) processes, or in a bed-load mode (Ashley and Verbanck, 1996; Verbanck 2000) and if there are a range of types of this material. The knowledge on traditional granular bed load transport (analogous to alluvial processes) is summarised in Section 2.3.6. There is evidence to suggest that there may be more than one type of 'near bed solids', depending on ambient and historical flow conditions and where in the network the sewer is situated (Chebbo *et al.*, 2002).

For example, Ahyerre *et al.* (2000) studied a small sewer network in Paris, filming the interface between the coarse deposits and the sewage; the results showed a stratification of the deposits at the water-sediment interface, with a layer composed of organic matter and paper on the surface of coarse mineral deposits. This layer was immobile and varied in thickness from 1.5 cm to 7 cm according to the length of time it was observed. Above this organic layer, particles were in motion during dry weather, but in small quantities, and the boundary between the layer and the sewage was very well defined. Extensive research carried out to locate this layer and study its dynamics throughout the catchment showed that it was present in places where the shear stress was occasionally less than 0.1 N/m^2, that it accumulated during dry weather in a linear manner with a growth rate of 215 g/m^2/day and that it was eroded by low shear stresses (> 0.4 N/m^2). Subsequently, the term 'crème' has been applied to describe the immobile material (Chebbo *et al.*, 2002).

Verbanck (1995a) identified a highly concentrated 'layer' above the bed in a large sewer in Brussels (gradient 0.025%). Similar 'layers' were observed in Hildesheim by Ristenpart *et al.* (1995) and elsewhere in Germany by Wöhrle and Brombach (1991), as well as in Paris, more recently, by Ahyerre *et al.* (1998, 2000). Concentrations of solids were measured up to more than 3500 mg/l, compared with typical suspension concentrations of 200–500 mg/l. A three dimensional image of solids distribution measured in a circular sewer is given in Figure 2.3.10.

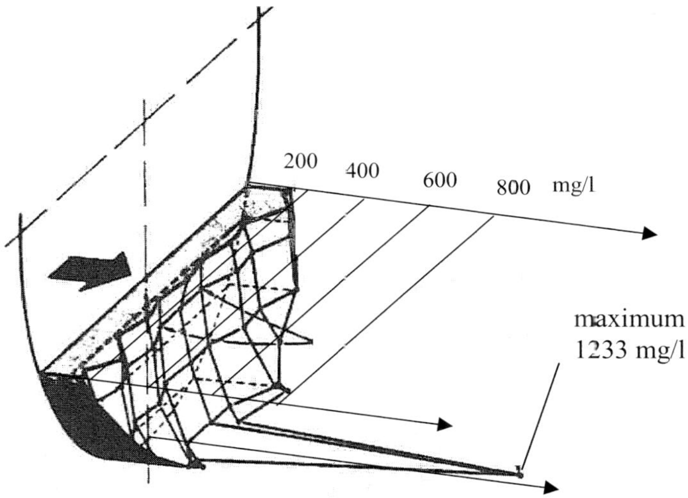

Figure 2.3.10. Measured solids distribution across a section of a 1.2 m diameter combined sewer (adapted from Wöhrle and Brombach, 1991).

As mentioned above, a variety of terms has been used to describe the near bed 'layer': heavy fluid layer; fluid mud; fluid sediment; dense undercurrent; organic near-bed fluid; near-bed solids.

Initially the term 'fluid sediment' was used because of the potential similarities between this material and 'fluid muds' found in estuarine studies. Fluid muds develop as a result of high-suspended sediment concentrations. In estuarine conditions a gradual increase in suspended sediment concentration in the fluid column towards the bed can reach a point where there is a rapid increase in concentration with zero or very little increase in depth (a *lutocline*). This increase can be to the mud layer that may comprise a mobile part overlying a stationary mud. Typical fluid mud bulk densities are 1,030–1,250 kg/m^3. At the base of the stationary mud, may be another sharp increase in concentration (bulk density) to the gel point which marks the boundary of an unconsolidated 'structure' and a consolidating bed which has developed some effective yield strength. It is not yet clear whether these concepts are applicable to sewer sediments as muds are largely homogeneous, whereas sewer sediments comprise a variety of particle sizes and characteristics and may be laid down in mixtures or stratified, in which case the density variation with depth will be more complex. For the time being, the terms 'dense undercurrents' and more recently, simply 'near bed solids' (Ashley and Verbanck, 1996, Arthur and Ashley, 1997; Ahyerre, 1999; Verbanck 2000; Chebbo *et al.*, 2002) have been adopted.

A possible illustration of the types of 'near bed solids' in sewers is given in Figure 2.3.11. The apparent nature of the near bed solids also may reflect the observational techniques in use. There are three typical approaches:

- sampling throughout the depth of the water column using (typically) small-bore tubes connected to vacuum pumps (Wöhrle and Brombach 1991; Coghlan 1995; Verbanck, 1995a, 1995b; Ristenpart 1995; Arthur 1996; Ahyerre 1999);
- using conventional bedload traps in sewer inverts (Lin, 1993; Coghlan 1995; Arthur 1996);
- direct in-situ visual observation using transparent sewers, boxes and an endoscope respectively (Arthur, 1996; Ahyerre 1999; Oms *et al.*, 2002).

Effective information about near bed solids will only be achievable once the variations introduced by these different approaches are properly investigated and quantified (Chebbo *et al.*, 2002).

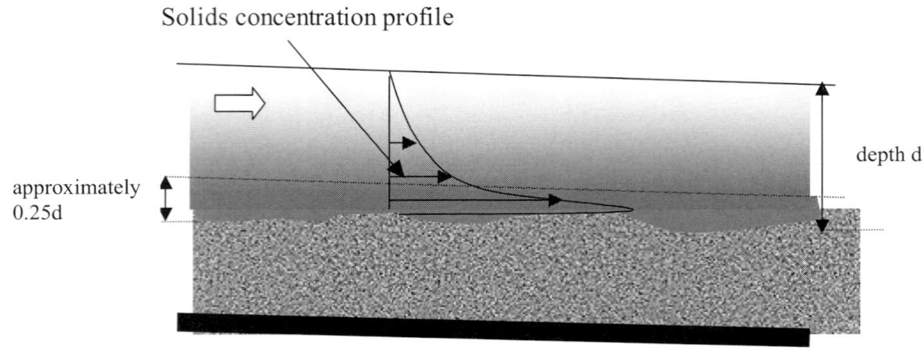

(a) Dense undercurrent or 'inner suspension' region

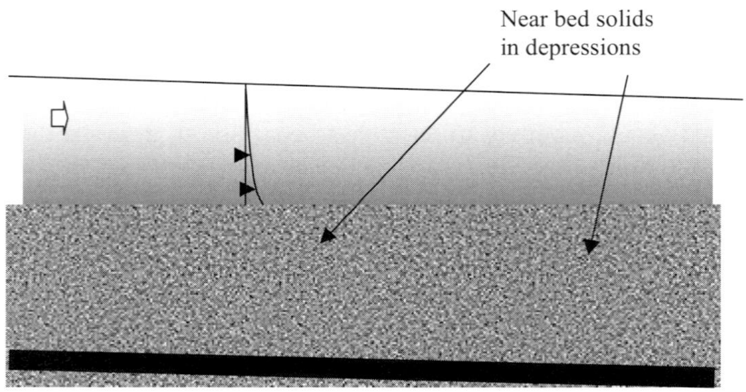

(b) Near bed solids being detained in depressions in bulk deposits during dry weather

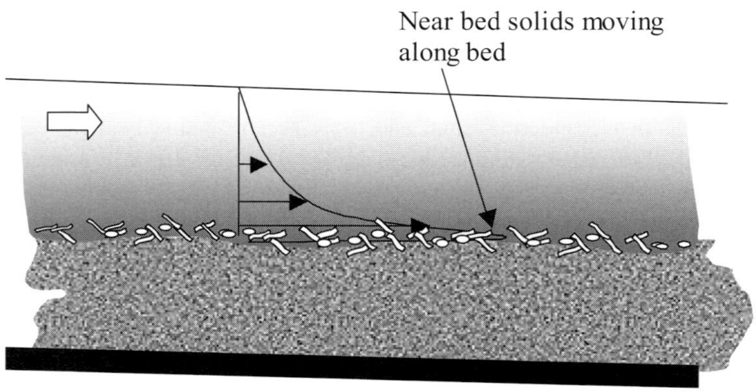

(c) Near bed solids moving along bed at higher flows

Figure 2.3.11. Possible types of near-bed solids in combined sewer systems.

2.3.4.2.1 Near bed (organic) solids

'Bed load' trap studies in Dundee and Marseille have shown the wide variability in materials collected in different sewers and at different times in a main network. Material moving near the bed in the large flat gradient sewer in Dundee varied considerably during dry and wet weather. The visual observations during dry weather indicated that the material moving along the bed, as 'bed load' transport over a deposited bed, had a typical size of 30 mm, and up to more than 50 mm, with an organic content up to more than 90 %. This maintained a transitory contact with the bed and moved somewhat slower than the average flow velocity. Additionally, gross (sanitary) solids (size > 50 mm) were seen to move at all depths in the flow column. During storms the granular fraction of the solids collected in the traps was found to increase as did the mean particle size (Table 2.3.1).

Table 2.3.1 Near bed solids collected in traps in Dundee combined sewer (Ashley *et al.*, 1994)

Condition	Volatile solids fraction (%)	Particle size d_{50} (mm)
Dry weather	2–19	2–5
Wet weather	2–29	2–> 20

There are interactions between the granular 'sewer sediments' and the much larger organic 'gross' solids, as found extensively in the bed-load traps, which are not sampled using conventional automatic (small-bore tube) sewage samplers. The fraction of material collected in the traps and transported in the 'near-bed' mode may be a significant proportion of the total, comprising up to some 12 % by dry weight in Dundee, but because of the low density, it in fact represents a much greater proportion of the total volumetric solids in motion. The coarsest particles appear to be transported during the highest flows and in the steepest sewers, and material moving near the bed may vary in the same sewer from traditional granular bed load (see Section 2.3.4.4), to a 'dense undercurrent' type suspension and/or (perhaps simultaneously) as the more discrete larger organic near-bed material. This is believed to depend on the upstream conditions of the system leading to where the sewer is in the network and the ambient flow conditions: dry or wet weather, time of day, day of week and season of the year; that is, the transport patterns are a combination of bed load over deposited bed and inner suspension as mentioned above, except that the discrete particles are largely organic and (unlike the definition of the dense undercurrent) appear to be in transitory contact with the boundary.

From the fieldwork in Dundee, relationships have been obtained which may be used to predict the movement of these solids near the bed in the main interceptor sewer based on antecedent and upstream conditions, hydraulic conditions and the characteristics of the transported material. These relationships may be used to determine the contribution to first foul flushes caused by the suspension of this material (Section 3.4.1). A near bed granular solids transport relationship was obtained (Arthur *et al.*, 1996) based on the modification of an existing relationship (Perrusquía and Nalluri, 1995) for transport over deposited beds. This relationship is applicable where there is known bed-load transport of granular particles, either in steep sewers or during storms (Ashley and Verbanck, 1996). For estimating the total solids moving near the bed during dry weather, an empirical equation was developed for the transport capacity C_v in the main Dundee interceptor sewer based on data collected at one site and verified using data from another, as illustrated in Figure 2.3.12.

Recently an alternative approach to the estimation of the rate of transport of the near bed solids for the Dundee data in Figure 2.3.12 has utilised the Coleman equation (2.3.6), and shown that this is effective at calculating the mass transported, although the determination of the appropriate reference concentration is problematic (Ashley *et al.*, 1999). A similar approach has been used by Verbanck (2000) to estimate the transport of the 'inner suspension' region near bed solids for some of the data collected in Hildesheim by Ristenpart (1995).

Observations of sewer sediment erosion in both Hildesheim and Dundee (Ashley *et al.*, 1993), showed the existence under a number of eroding events of a short timescale sediment peak (i.e. apparent increase in sediment depth) 'passing' the measurement sections in each sewer as illustrated in Figure 2.3.13. This peak was not found to be related to any increase in suspended solids concentration as usually associated with first foul flushes (Section 3.4.1), and was believed to be the 'pushing' of the near-bed material in front of the storm wave as it built up in the sewer network and prior to suspension of the material. Davies (1990), based on Harrison and Holmes' (1967) observations of negligible dispersion of pseudo storm water mass into an overtaken baseflow (DWF), suggests that even based on simple wave speed calculations, the effect of a rapidly varied unsteady wave progressing down a sewer (assuming no lateral input or main bed erosion) would be to concentrate the solids and pollutants moving in the ambient dry weather flow, so that the concentration in the increased flow was proportionately the same as the concentration in the base flow. Mobilisation of the near bed solids may also release the interstitial liquid trapped in the solids matrix, so contributing significantly to the pollutant load conveyed during periods of peak dry weather and rainfall induced flows, as described by McGregor *et al.* (1993) (Table 2.2.17), and contributing to foul flushes (Section 3.4.1).

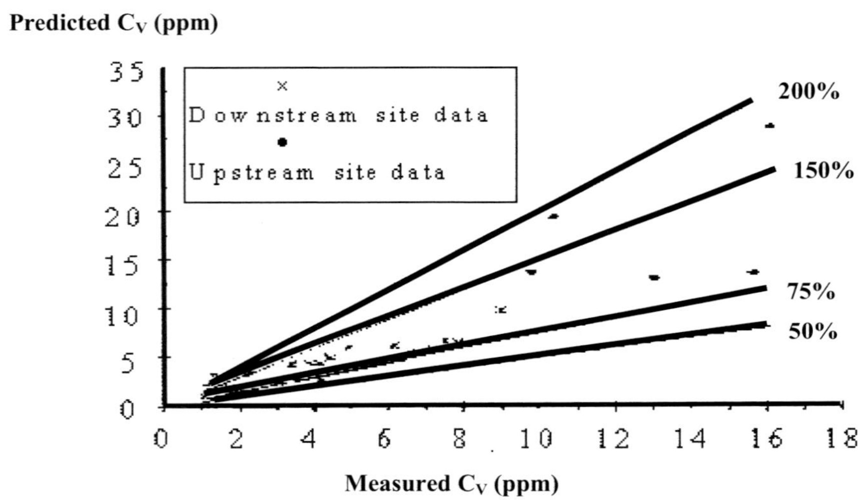

Figure 2.3.12. Near bed solids transport prediction for sites in Dundee (adapted from Arthur *et al.*, 1996).

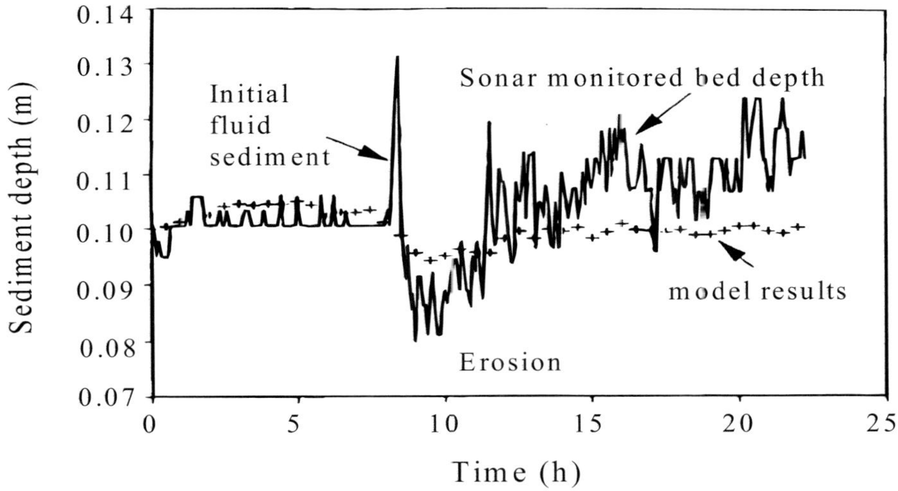

(a) DWF, 1/3/89, Dundee interceptor sewer.

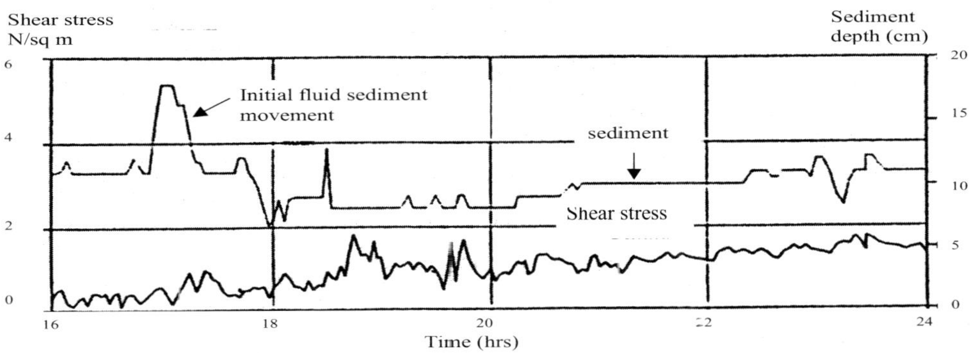

(b) Wet weather flush, 1/4/92, Hildesheim interceptor sewer.

Figure 2.3.13. Initial near bed solids flush observed in Dundee and Hildesheim.

2.3.4.2.2 Dense undercurrent

Verbanck (1995a, 2000) used measurements of solids concentrations through the depth of the water column to suggest the presence of an 'inner suspension' region (originally defined as a *dense undercurrent*). This 'inner suspension' region, over a depth of 150 mm in the Brussels main sewer, will have marked changes in the particle size characteristics and relative concentration and can contain relatively large (0.5 mm, organic content > 90 %) particles in a 'trapped' matrix of 'suspended' flow, with particles not actually in contact with the bed. The particles in this region are notably different from those found in the flow, which are finer. Under low flows the 'layer' may become stationary and appear to deposit on top of the coarser bed as a weak surficial layer, defined traditionally as a Class C sediment (Crabtree, 1989) as illustrated in Figure 2.2.11. It is believed that when there is a flow, the material may remain in a zone of fluid just above the bed in which the particles are trapped in a 'hindered settlement' matrix, never being able to actually reach the underlying bed, and supported by the strong turbulent diffusion in this zone of steep velocity gradient. Whether the material deposits on the bed or not, and because of its very weak shear resistance, the material may be considered to be constantly available for re-entrainment when bed shear stresses begin to increase.

The effects of disturbance of these 'sediments' by increasing rates of shear along the invert of the sewer are thus to mobilise this interfacial material, re-entraining the solid particles. This supposed 'release' of the dense undercurrent for dispersal into the overlying flow, is illustrated in Figure 2.3.14, for a dry weather flow in a large combined sewer in Brussels for which sewer cleaning has increased downstream shear stresses sufficiently to cause an increase in the solids in suspension. The results from Brussels suggested that bed shear stresses of 1 N/m^2 were sufficient to cause this to happen. Erosion was also observed in the German studies in Hildesheim at shear stresses slightly in excess of 0.5 N/m^2, but in this case it was not clear whether this material was in fact the 'dense undercurrent'.

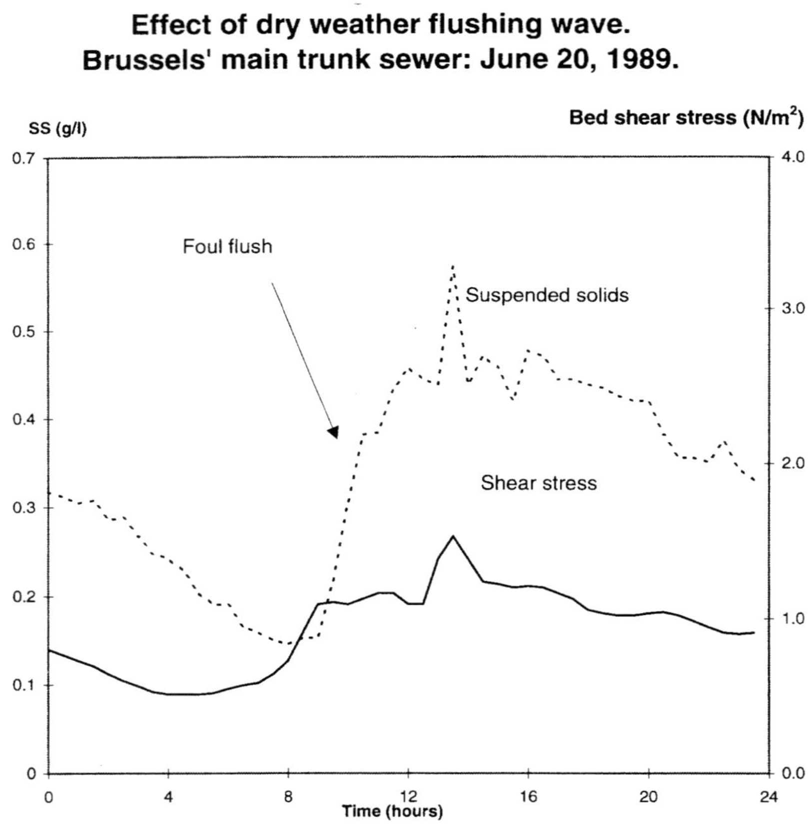

Figure 2.3.14. Increase in suspended solids concentration in Brussels main sewer due to supposed release into the flow of the dense undercurrent (adapted from Verbanck, 1995a).

2.3.4.3 *Gross solids*

Gross solids are defined by Jefferies (1992) as faecal matter, particles of paper, plastics and any other material greater than 6 mm in any two dimensions, with specific gravity close to unity (Section 2.1.4.2). They are found not just suspended throughout the depth (or floating) in the flow but also mixed in with bed deposits. Virtually all studies of gross solids have been concerned with the implications for screening at either WTPs or CSOs and hence their occurrence in suspension rather than in the bed. Gross organic solids can occupy relatively large proportionate volumes in sewage. The occurrence of gross solids in bed deposits has not as yet generally been quantified.

The conveyance of large solids in building drainage networks has been studied in the UK, Japan and the USA (Swaffield and Galowin, 1992). Of interest has been the hydraulic conditions required to ensure adequate passage of WC solids through the building drainage system and into the main sewer network. This is important for any introduction of water conservation measures via, for example, low flush WCs. Swaffield and McDougall (1996) describe the important factors for the conveyance of non-degradable solids in building drainage systems:

- operating characteristics of the WCs;
- flush volume and percentage of the volume to propel the solids;
- properties of the pipe network;
- properties of the solids being transported;
- quality of the pipe installation, e.g. joints.

Movement of solids is intermittent and progressive following each flush or any other system inflow (e.g. wash basin). Studies have investigated real faecal solids transported in shallow gradient, 100 mm diameter, pipes in hospitals, and have also utilised artificial standardised solids in laboratory tests. In building drainage conveyance systems, the individual solids are relatively large compared with the flow cross-section and hence cause an important interaction with the flow. In sewer systems, the individual solids progressively represent a smaller fraction of the flow cross-section as they pass down the system as flows increase relative to the solids. In addition, the organic solids also degrade, possibly to form the largest part of the fine-grained organic fraction found in suspensions and in bed deposits. In addition, toilet paper degrades quickly, even in building drainage, and this may be a major proportion of suspended (and wash-load) organic solids, particularly the COD load. It is clear that the larger and heavier 'discrete' solids (e.g. faeces), even when in motion in the rapidly varied 'pulse' flows in small pipes, move at a 'solid velocity' which is lower than the ambient fluid velocity or wave speed (Davies *et al.*, 1998), usually due to contact with the solid boundary. Whereas the lighter solids can move faster than the mean flow velocity in the zone of flow, remote from the bed with the high flow velocities higher than the mean. The way in which these solids actually move depends upon their individual characteristics and the relative flow depth and velocity.

A recent study of the transport of solids in sewers with intermittent flow, using prototype studies and a computer model based on the method of characteristics, has shown that there are two main mechanisms for transport (Littlewood, 2000):

1. Velocity decrement – the solid is picked up by the flow and conveyed by the flow at some proportion of the wave velocity. This occurs generally where the solids are relatively large and flush volumes small.
2. Sliding, leaking 'dam' – where the solid moves in response to wave forces on the back of the solid, remaining in contact with the pipe invert at all times. This occurs generally for larger flush volumes and small solids.

In each case the solid is transported in a series of 'hops'. The influence of toilet paper on the larger solids movement was substantial. In general 'combination' solids that included toilet paper, were found to move more easily. At low velocities, irregularities in the pipe wall (even 0.5 mm in a 100 mm pipe), were found to affect the transport of the solids, particularly at lower flow velocities (less than 0.25 m/s). It was possible to determine that sanitary sewers could be as small as 75mm in diameter provided sanitary towels are not flushed.

Verbanck (1990) monitored the 'larger' organic solids in dry and wet weather flows in the Brussels main trunk sewer and found that some 25 % of the total solids (dry weight) were larger than 125 μm (and > 90 % organic) in dry weather and in wet weather some 40 % of the 20–40 % total of suspended solids were volatile solids larger than 125 μm. It was contended that these solids moved effectively as part of the wash-load mode of transport. It is not clear, however, how the 'automatic samplers' used extracted the samples and it is possible that the largest particles were excluded from the small sampling tubes. In view of the organic nature of this material, the expression of its relative occurrence in terms of 'concentration' in flows, based on dry weight, is misleading because of the low specific gravity.

Studies carried out in Scotland of large organic solids conveyed in combined sewers (Jefferies and Ashley, 1994), used a 'gross solids sampler' with COPA sacks (mesh size 6 mm) and trash trap 'screens' to sample the entire flow volume in large (man-entry) sewers. The solids collected comprised plastic, paper and faeces, 15 %, 55 % and 30 % by mass respectively. The study showed that during dry weather the mass transport rate (g/s) of these gross solids (GSS) could be correlated quite well with the 'traditional' suspended solids load (SS). The relationship between these illustrated that considerable amounts of gross solids typically accompany the smaller suspended material during dry weather when SS ranges from 0.8–12.5 g/s. Figure 2.3.15 shows typical results.

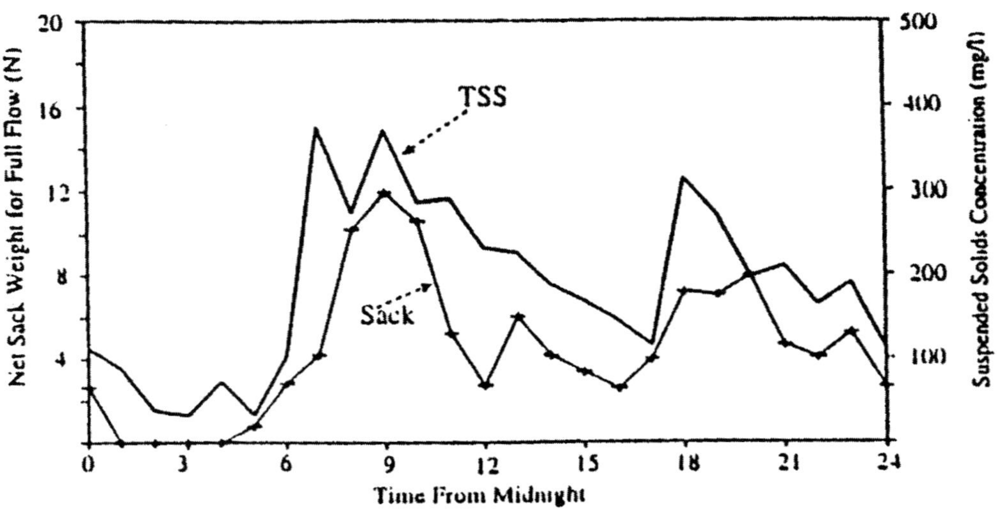

Figure 2.3.15. Variation of mass of gross solids (sack weight) as collected by gross solids sampler during DWF in a combined man-entry sewer (Jefferies, 1993).

During storm flows in combined sewers, it was found that not only does the total mass carried in suspension increase dramatically compared with dry weather, but the relative proportion of organics (by mass) conveyed appears to drop because of the entrainment into the flow of more (and denser) granular particles. It is possible that during such times, there is an increase in the amount of much larger organic gross solids, mobilised from the smaller sewers where they have been 'hopping' under the intermittent effect of the sanitary 'pulsing' dry weather flows. However, whilst predictive empirical equations were developed from this work, they have not been substantiated by measurements elsewhere. It appears from the data collected, that the 'production' of gross solids within any given catchment depends upon the nature of the sewer network (a function of population and topography) and the relative antecedent dry weather period (ADWP). Attempts to relate gross solids to other sewer sediments (e.g. Milne *et al.,* 1995) via sediment budgets have so far proved inconclusive.

Studies of gross solids and transport rates are rare, hence only recent case studies have been described here. The transferability and general validity of the results has to be further confirmed, although the significance of gross solids is apparent. Notwithstanding the relative significance, little attempt has so far been made to include these solids in the current generation of sewer flow quality models.

2.3.4.4 Granular bed load

Particles not in suspended transport (the so-called bed load particles) receive no significant upward impulses other than those due to the successive contacts between the solid and the bed (the fluid forces on the particle being essentially drag forces in the streamwise direction). Under certain circumstances in a sewer, sediment is recognisably moving as a 'traditional' bed load as granular particles being transported over a deposited bed or over a clean pipe invert in flume traction (Figure 2.3.6).

The transport of non-cohesive sediment particles as bed load in sewers can be examined by estimating the transport capacity of the flow or by examining the threshold of sediment motion. Various approaches are available, most empirically based for rigid or loose beds and for single size and mixed grain size sediments.

2.3.4.4.1 Transport rate/capacity relationships

The bed load sediment transport rate in sewers is dependent on a number of factors, primarily: hydraulic radius; shape and roughness of the conveyance; sediment concentration; grain size; threshold of motion and effective transport width. Sediment transport on rigid beds occurs with no deposit forming. Usually a clean invert is the ideal design criterion (although this is seldom realisable in practice). Transport over a sediment deposit increases the transport capacity of a pipe. The ideal of a 'no deposit' condition for engineering purposes is conservative, since it specifies high minimum gradients and flow velocities to ensure no build-up of significant amounts of sediment (Section 4.2.1).

The analysis of data from an extensive investigation of the sediment transport of non-cohesive particles as bed load without deposition on rigid beds, carried out by Mayerle *et al.* (1991) in laboratory channels with rigid, smooth and rough beds, resulted in equations which link self-cleansing velocities v_s with prevailing transport rate, grain characteristics and pipe roughness, of the form shown in Table 2.3.2.

Table 2.3.2. Bed load transport over rigid beds in laboratory channels

General for channels	$\frac{V_s}{\sqrt{gd(s-1)}} = XD_{gr}^{x_1} C_v^{x_2} \left(\frac{d}{R}\right)^{x_3} \lambda_s^{x_4}$	$\frac{(k_s - k)}{R_H} = YD_{gr}^{y_1} C_v^{y_2}$	Equations
Specific shapes:			
Circular	$14.43\, D_{gr}^{-0.14}\, C_v^{0.18} (d/R)^{-0.56} \lambda^{0.18}$	$0.0130\, D_{gr}^{0.24}\, C_v^{0.40}$	(2.3.11)
Rectangular	$11.59\, D_{gr}^{-0.14}\, C_v^{0.15} (d/R)^{-0.43} \lambda^{0.18}$	$0.0245\, D_{gr}^{0.40}\, C_v^{0.44}$	(2.3.12)

In Equations 2.3.11 and 2.3.12, C_v is the volumetric concentration (solid flux/fluid flux as volume, dimensionless), $\lambda = (8gRS / v^2)$ the dimensionless friction factor in the presence of sediment, and k_s and k are the equivalent roughness sizes with and without sediment respectively. Application of the Colebrook-White equation (Ackers *et al.,* 1996) with k_s yields λ. The equations are based on a large data set (over 400 experiments) with a fairly wide range of conditions, and are physically sound and simple to use.

Equation 2.3.11 for rigid bed circular cross section conveyances is in agreement with an alternative method due to May (1982) derived from experiments in circular cross section, laboratory flumes flowing full and part-full. Extrapolation to large diameter systems in conjunction with small particle sizes, however, results in very high flow velocities to transport the required sediment. Ackers (1984, 1991) developed a method for estimating sediment transport on loose or deposited beds. The method is based on the Ackers and White (1973) sediment transport formulae, originally developed from rectangular flume experiments and was calibrated using the measurements from May (1982). The equation relationships are:

$$\frac{v}{\sqrt{gd(s-1)}} = K\left(1 + JC_v^{1/m}\right) \tag{2.3.13}$$

where

$$K = \frac{A_{gr}\left[11.3(R/d)^{0.1}\right]^{(1-n)}}{(\lambda_s/8)^{n/2}} \tag{2.3.14}$$

and

$$J = \left[\frac{\left(\frac{A}{W_e R_H}\right)^{1-n} \left(\frac{\lambda_s}{8}\right)^{n/2}}{\left(\frac{d}{R}\right) C}\right]^{1/m} \tag{2.3.15}$$

where A is the flow area (m^2); W_e the effective transport width (m); m, A_{gr}, n and C are dimensionless coefficients in the Ackers and White transport function defined according to the dimensionless grain size as in Table 2.3.3; v is mean flow velocity; d is grain size; C_v is volumetric sediment concentration; s is specific density of sediment; λ_s is friction factor in presence of sediment.

Table 2.3.3. Coefficients in Ackers and White transport equation

Coefficients	$1 < D_{gr} < 60$	$D_{gr} > 60$
m	$1.67 + 6.83 / D_{gr}$	1.78
A_{gr}	$0.14 + 0.23 / \sqrt{D_{gr}}$	0.17
n	$1.0 - 0.56 log D_{gr}$	0.0
C	$log C = -3.46 + 2.79 log D_{gr} - 0.98(log D_{gr})^2$	0.025

The method requires unacceptably high limiting velocities in large systems when the effective transport width of 10 d_{50} (originally proposed for non deposition conditions) is considered. For design purposes, an effective transport width of 0.04 Ɔ is recommended. The uncertainties in application of these laboratory flume derived results, means that they still require verification with data from large sewer systems. Limited transport rate data collected by Rushforth (2001) in a 900 mm diameter sewer pipe indicated that both these methods can under-predict mobility (De Sutter *et al.*, 2003). This poor performance may have been due to the wide range of particle sizes present in the deposits in the latter study. However, more data are required from large sewer systems to be able to assess better the performance of this type of relationship.

Tests were carried out in Marseille utilising two bed load traps, one downstream of a steep combined sewer section, gradient 1.78 %, and the other located in a section where the gradient was only 0.1 %. In the first trap, the particles collected were virtually exclusively granular, typical size 2–3 mm (up to 10 mm), whereas in the second trap, the particles were more similar to those observed in suspension and had specific gravity variations from 2600 m^3/kg down to about 1500 kg/m^3 (Lin *et al.*, 1993). In Dundee, three trap locations were utilised. Two were on a large (1500 mm high) shallow gradient sewer (gradient 0.057 %), where ambient DWF shear stresses ranged up to 0.45 N/m^2 at one site and about 1 N/m^2 at the other, and these both collected mainly organic solids (specific gravity typically up to 1.45). The trap situated in a smaller (1030 mm high) steeper sewer (gradient 4.6 %), however, collected solids with a greater granular fraction (volatiles down to 1.4 %) and higher density. At this site the ambient shear stresses were 0.84–1.5 N/m^2. The French researchers reported that the samples collected from the first trap were influenced not by the hydraulic conditions, but by the (other) anthropogenic activities occurring during the day and hence these varied both temporally and spatially. The second French trap data did, however, correlate with flows. The evolution of the bed deposits including both erosion and deposition in the Marseille No.13 trunk sewer was shown to be predictable over a period of 1000 days with a 'bed load' relationship derived from the Meyer-Peter equation (Meyer-Peter and Müller, 1948), taking into account a mixing layer and grain size classes (Lin and LeGuennec, 1996):

$$\frac{q^*_{s,k}}{\sqrt{(s-1) g d_k^3}} = 8\left(\tau^*_k - \tau^*_{c,k}\right)^{3/2} \qquad (2.3.16)$$

where $q^*_{s,k}$ is unit-width volumetric bed load transport rate under equilibrium conditions
s is particle specific gravity
d_k is grain size (in each group)
τ_k^* and $\tau^*_{c,k}$ are respectively non-dimensional bed and critical shear stresses.

The results are shown in Figure 2.3.16.

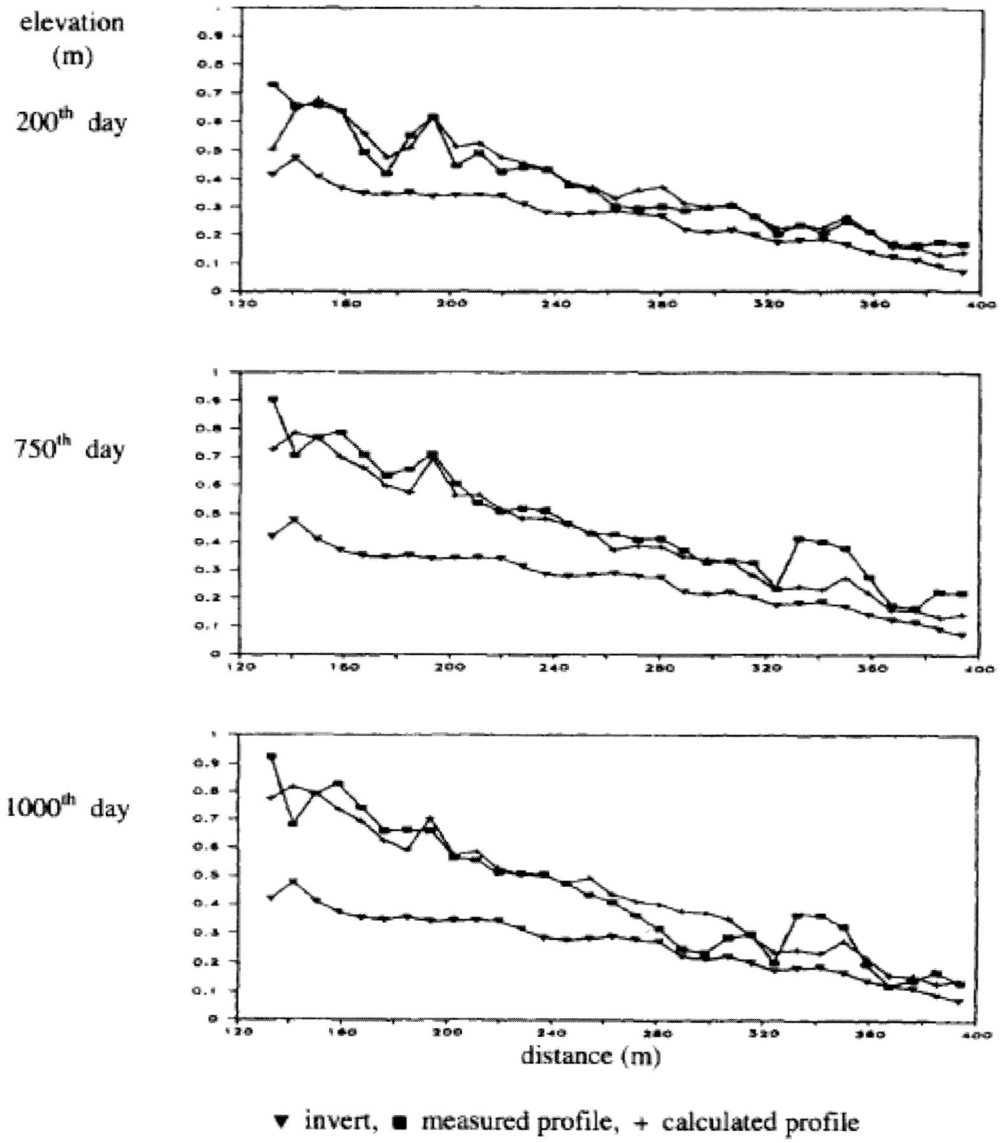

Figure 2.3.16. Comparison of observed and predicted elevation of the sediment bed in Marseille No. 13 trunk sewer (Lin and LeGuennec, 1996).

Temporal changes to transport rates at any section were determined using the relationship developed by Daubert and Lebreton (1967), which gives the rate of change of the local solids discharge $q_{s,k}$ as a function of the particle fraction settling velocity, the shear velocity and a coefficient α, where this corresponds to erosion when $q^*_{s,k} > q_{s,k}$ and conversely to deposition. For both erosion and deposition, α was recommended as 0.01.

Attempted application of a wide range of limit of deposition and other formulae developed from laboratory studies, such as May (1994), had little success for the granular fraction when applied to the Dundee results (Arthur *et al.*, 1996) for both the milder or steeper sloping sewers. The granular material, which may be viewed as a 'true' bed load will very rarely if ever, travel in true suspension in the flow. It is of major importance, because it is found ubiquitously throughout sewerage systems and when arrested is responsible for the build up of the bulk deposits within sewers (Sections 2.3.7 and 3.2.1).

2.3.4.4.2 Estimation of threshold of motion

Erosion will occur for granular particles when a critical shear stress is reached. Shields (1936) developed a relationship based on the balance between particle weight and boundary shear stress calibrated using laboratory experiments using single sized particles (Equation 2.3.17 and Figure 2.3.17). Despite the very different boundary conditions found in sewers, this relationship has been applied in a number of studies (e.g. Laplace *et al.*, 1992; Verbanck *et al.*, 1994).

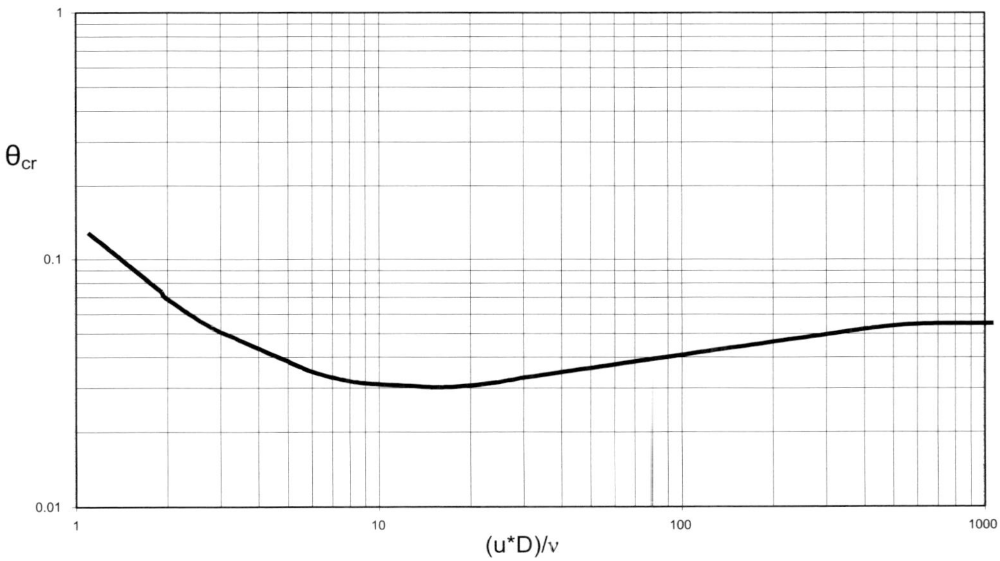

(Symbols are defined in the text.)

Figure 2.3.17. Shields diagram.

$$\theta_{cr} = \frac{\tau_0}{(\rho_s - \rho_f)gD} \qquad (2.3.17)$$

where τ_0 is boundary shear stress
ρ_s is density of sediment
ρ_f is density of fluid
g is acceleration due to gravity
D is grain size

Novak and Nalluri (1984) carried out investigations in a rectangular (305 mm width) cross section, laboratory flume with smooth and rough beds and two circular cross sections (152 mm and 305 mm diameters) flumes and based on their experimental observations estimated the critical condition for incipient motion on rigid beds as:

$$\frac{v_c}{\sqrt{gd(s-1)}} = a\left(d/R_H\right)^b \qquad (2.3.18)$$

where v_c is critical velocity (m/s)
d is particle size (m)
s is relative density of solid phase to fluid phase (-)
R_H is hydraulic radius (m)
g is acceleration due to gravity (m/s^2)
a and b are constants (-) which are a function of bed conditions and whether there are single/touching particles ($0.5 \leq a \leq 0.61$ and -0.4 b $\leq$ -0.27 respectively)

For the initiation of movement on loose or deposited beds, Shields' condition is still widely applied. Combining this criterion with Strickler's equation and representing Shields' curve as proposed by Van Rijn (1989) yields equations of the same type as Equation 2.3.18, with 'b' a constant (-1/6) and 'a' varying according to the dimensionless grain size D_{gr} between $1.38 \leq a \leq 3.91$ for $1 < D_{gr}$ to $D_{gr} > 150$.

Laplace (1991) and Bachoc (1992a) applied the Shields criterion to calculate the initiation of solids from deposits in two main trunk sewers in France. For granular solids ranging from 1 to 10 mm, and with a density of 2600 kg/m^3, the Shields relationship was adapted to determine the critical grain diameter D_c corresponding to the particles which start to move depending on hydraulic conditions.

Measurements of transport rate and boundary shear stress by Rushforth (2001) in a 900 mm diameter sewer indicated that applying Shields type relationships was not appropriate if the sewer deposit contained a wide range of grain sizes. It is apparent that for granular deposits that have a wide range of grain sizes, the use of a Shields threshold criterion will produce significant errors. This can cause large prediction errors in transport capacity relationships that require an accurate estimation of the threshold of motion (De Sutter *et al.*, 2003). In such cases an empirical 'hiding function' of a similar format to that proposed by Sutherland (1992) can be used to adjust threshold predictions for the different grain size fractions made by Shields' relationship to match those obtained by field observation.

2.3.5 Deposits in combined sewer networks

The evolution of bed deposits in sewers depends on repeated cycles of deposition and erosion. The critical parameter controlling sediment bed movement is the bed shear stress. Formerly, critical velocity criteria have been utilised, but this is now known not to be the most appropriate parameter to describe sediment transport. Shear stress also relates more readily to the unsteadiness of flows in real sewers – significant variations in bed shear stresses occur even where there are only small changes in hydraulic conditions resulting in very different bed deposit behaviour. The complexity of sediment transport phenomena in sewers and the breadth of processes which influence transport rates (and which are generally not measured) illustrates that the specification of a single threshold for defining a critical state of flow will inevitably be naive.

Section 2.3.8.1 describes how settlement will occur at a rate dependent upon the flow field, the nature of the particles and the concentration in suspension and/or near the bed.

2.3.5.1 Sediment deposition

Deposition occurs during periods of dry weather and during decelerating flows when storm runoff is receding; it has also been shown to correspond with structural and hydraulic discontinuities (Bachoc, 1992b; Chebbo *et al.*, 1995) as illustrated in Figures 2.3.18, 2.3.19 and 2.3.20.

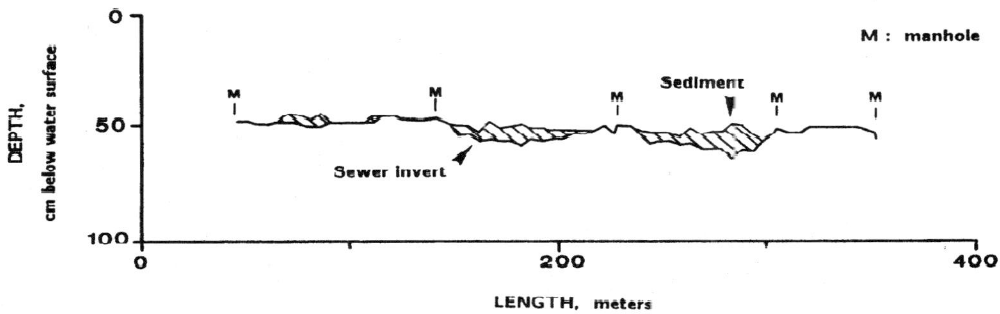

Figure 2.3.18. Typical accumulation of granular sediments in a sewer length with backfalls (Verbanck and Ashley, 1993).

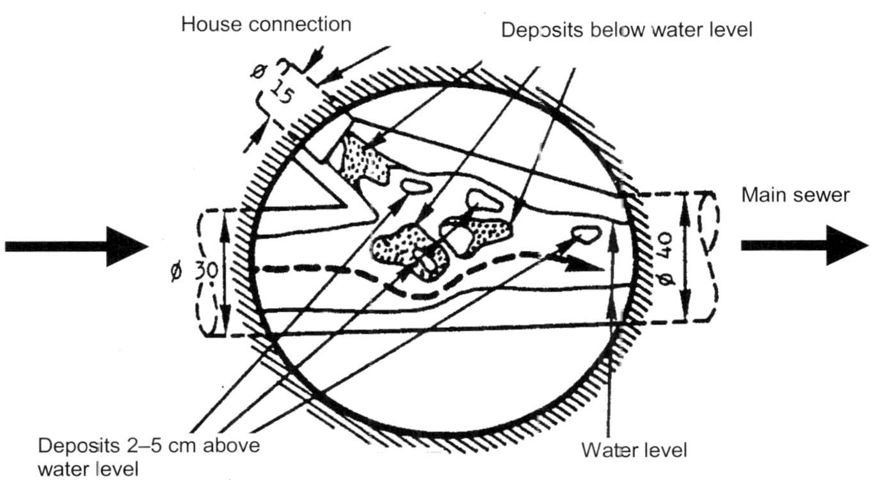

Figure 2.3.19. Sediment deposits in a trunk sewer in the vicinity of a house drain connection (adapted from Dauber and Novak, 1983).

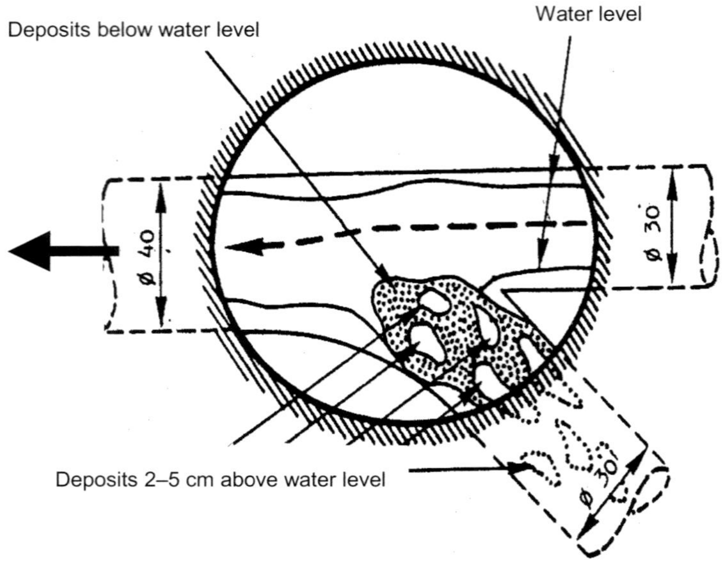

Figure 2.3.20. Deposits in the vicinity of a junction between two sewers (adapted from Dauber and Novak, 1983).

Table 2.3.4 illustrates that significant proportions of dry weather originating solids can deposit in sewer networks, together with associated oxygen demands and nutrient loads.

Table 2.3.4 Solids, carbon and nutrient generation and deposition in a catchment under dry weather conditions (Dauber and Novak, 1983)

Parameter	(a) Daily mass in dry weather inputs	(b) Mass deposited in sewer system	Ratio (b):(a)
Total solids (g)	217648	25791	1:9
Suspended solids (g)	54326	20146	1:3
COD (g O_2)	127539	12617	1:10
TOC (g C)	36351	4610	1:8
TKN (g N)	10392	783	1:13
TP (g P)	3340	184	1:18

For cohesionless particles in a weak suspension, the likelihood of deposition will depend on the particle characteristics (diameter and density) and the level of turbulence. Deposition will occur if there is a greater buoyant weight force component than the lift forces tending to retain the particle in suspension (Figure 2.3.21). However, in sewer flows, there is a continuous process of particles interchanging between suspension, the bed and the near bed region. Erosion of particles from the bed requires a larger shear stress than pertains when the same particles will deposit due to the interparticle actions, which exist even for granular material as illustrated in Figure 2.3.21.

Settlement will occur at a rate dependent upon the flow field, the nature of the particles and the concentration in suspension and/or near the bed. Erosion will occur for granular particles when a critical shear stress is reached.

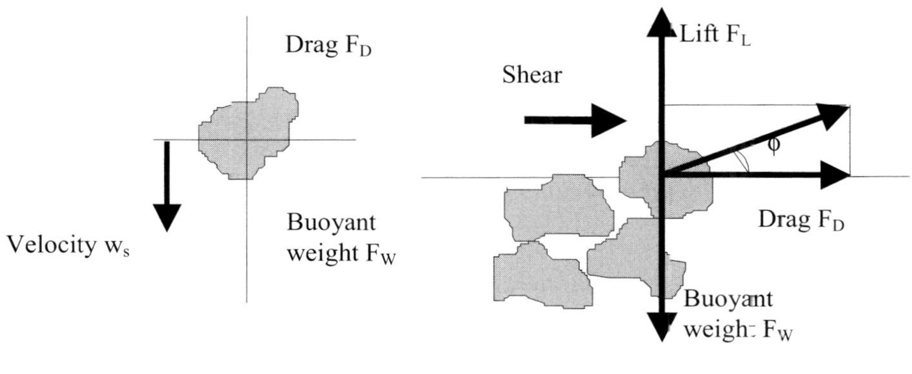

(a) deposition (b) erosion

Figure 2.3.21. Discrete particle forces acting during deposition and erosion.

A number of studies were carried out in the USA in the 1960's and 1970's to improve methods for the design of sewers to minimise deposition and also to devise strategies for sewer maintenance and cleaning. Most of these studies considered sediments in terms of idealised, small, discrete, non-cohesive, conservative particulates, and in-sewer deposition as a time-mean equilibrium phenomenon. The studies were concerned principally with the hydraulics of sewer systems rather than the polluting effects caused by the wash-out of sewer deposits, but determined build-up in smaller diameter pipes via flushing out of deposits (Table 2.3.5).

Table 2.3.5. Examples of measured rates of sediment deposition in combined sewers (Ashley *et al.*, 1992)

Sewer size (mm)	Sewer gradient	Solids deposition (g/m.d)	Population or sewer length	Comments
Flushing results				
300				Collector sewers:
	0.003–0.005	15–18	Not stated	vigorous flush[1]
	0.033	9	Not stated	vigorous flush[2]
	0.05–0.01	6.5–13	69–75 m	weak flush[3]
375	0.003–0.005	16–29	41–57 m	Collector sewer: vigorous flush[1]
300-900	0.5–1.00	None	Not stated	Trunk sewer: weak flush[3]
500	0.004	30	Not stated	Collector sewer: vigorous flush[2]
In-situ observation				
1,500	0.001	34–128	14,590	Interceptor sewer[4]
4,000	0.0004	154–176	380,000	Interceptor sewer[5]
2,750 x 1,800	0.001	500	35,000	Trunk sewer[6]

References: 1) Pisano and Queiroz (1977); 2) Larson *et al.* (1990); 3) Krejci *et al.* (1987); 4) Chebbo *et al.* (1990); 5) Techniques Sciences Méthodes (1993); 6) Laplace *et al.* (1990).

Note that flushing results are for jetting sediments from the sewer using high pressure water.

It is known that sediment deposition will occur in sewers under the conditions given in Table 2.2.8 (Section 2.2), as summarised from Bachoc (1992a), Ashley and Crabtree (1992) and Chebbo *et al.* (1995). The likelihood of deposition occurring at any given location depends upon a number of factors, of which the most important are:

- nature of solids;
- geometry, physical condition and topography of the sewer;
- nature of the upstream sewer network including the type and distribution of inputs;
- ambient flow conditions – dry weather patterns and relative magnitude of storm flows.

Figure 2.3.22 illustrates diagrammatically the deposition of sediments during dry and wet weather conditions in combined sewers.

The French studies in the Marseille No.13 trunk sewer (Figure 2.3.16) found that the most common particles deposited therein were granular and of size (2–3 mm). The bulk deposits developed principally as the result of bed-load transport of these particles. Deposit rates were found to decrease during dry weather and increase during wet weather. Presumably any sanitary solids were transported as suspended loads with little deposition during dry weather, with granular solids remaining 'stranded' in upstream deposits in the collector sewers. During storms, a combination of increased surface sediment wash-in and mobilisation of the upstream deposits caused an increase in sediment arriving (as bed-load) at the trunk sewer. For the coarsest particles the transition from 'bed' to 'bed-load' transport was predicted using Shields' criterion, whereas the Ackers approach was found valid for the suspended transport of the finer particles (Laplace, 1991; Laplace *et al.*, 1992; Lin and Le Guennec, 1996). Significantly, development of an equilibrium condition was achieved by a steepening of the sediment bed deposit gradient along the length of the sewer until a mass balance was reached for the transport of solids coming into and exiting the sewer reach.

From the studies of deposition in a large number of sewers in France (summarised in Techniques Sciences Méthodes, 1993), the following are typically observed:

- for 'small' sewers, deposits occur where slopes < 0.01 m/m, and these were up to 20 % organic;
- deposits occur in 'larger' sewers in the vicinity of discontinuities and where hydraulic gradients reduce;
- during wet weather flows, generally all particles up to 3 mm in size are moved, although not all upstream deposits are moved downstream;
- significant deposition occurs on the recession of the hydrograph and can 'mask' earlier erosion in a given sewer;
- rapidly varied flows are important as rapid fluctuations in shear (localised) and transport capacity can cause 'stalling' of the sediment transport processes, even in the middle of a storm;
- 'holes' in deposits in large sewers are found to fill rapidly with organic material during dry weather.

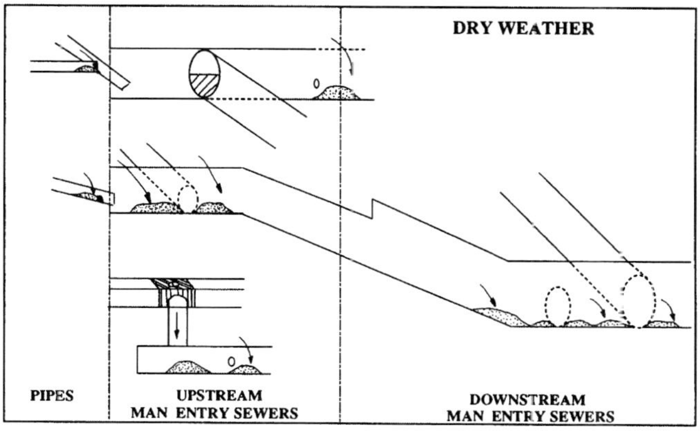

(a) Location and dynamics of deposits during dry weather, from upstream (left) to downstream (right).

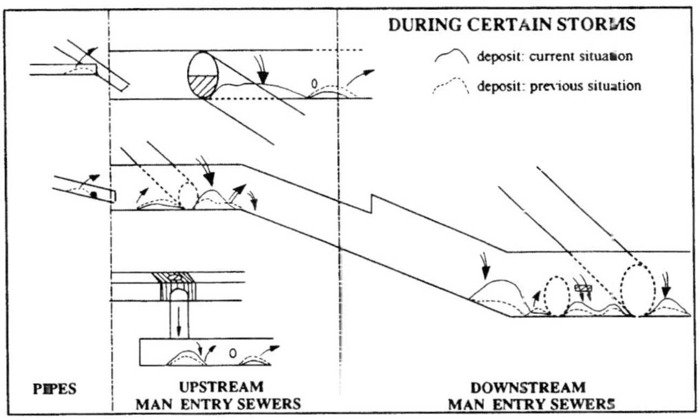

(b) Location and dynamics of deposits during storm events, from upstream (left) to downstream (right).

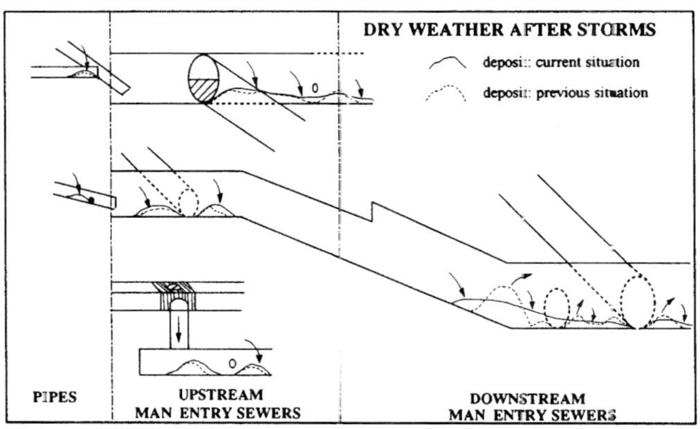

(c) Location and dynamics of deposits during dry weather after storm events, from upstream (left) to downstream (right).

Figure 2.3.22. Locations in which sediment deposits occur (adapted from Bachoc 1992a and Chebbo *et al.*,1995).

These results are reflected in studies reported elsewhere, e.g. Wotherspoon (1994), Verbanck (1995b) and Ristenpart (1995). The importance of rapidly varied flow effects on bed evolution and erosion in large sewers was observed in Dundee and also illustrated in parallel laboratory studies using synthetic sediments (Skipworth *et al.*, 1996).

From these observations and the other reported studies it is apparent that in addition to the factors given in Table 2.3.5, the following are also important for sediment deposition:

- relative sewer size: small or 'man-entry';
- distribution of discontinuities in the system (including junctions, the major source);
- temporal, and spatial variations in flows at long (gradually varied) and short (rapidly varied) time scales;
- 'bed-load' transport as the primary mode of material movement to places of deposition during dry weather (e.g. Lin and Le Guennec, 1996), and suspended load transport during wet weather.

For the smallest sewers (collectors), as studied in the field in early USEPA investigations and at laboratory scale and in buildings (Swaffield and McDougall, 1996), it is clear that the rapidly varied flow 'pulse' is the primary mechanism for solids movement and hence deposition. Whereas the transport of the materials in building drainage has been extensively studied, it is only recently that movement in the slightly larger connecting pipes has been investigated (Littlewood, 2000). The mechanisms for movement are complex and transport is not simply a matter of the amount of flow in the pipe (Littlewood and Butler, 2002). In the latter study, toilet paper was identified as a 'transport enhancing' aid for the other gross solids. Solids in the presence of paper tend to move in an efficient sliding dam mechanism, whereas those where there was no paper present stalled at a limiting solids transport distance.

Thus there must be a distinction between the conditions of intermittent flow and transport mechanics (hydraulics modelled using, for example, the method of characteristics) and the traditional suspended and 'bed-load' transport and deposition equations as generally applied to larger sewers (May *et al.*, 1996) in which the flows are assumed to be quasi-steady.

For the trunk and interceptor sewers, where flows are variable, but continuous, the traditional transport equations such as Ackers-White (Ackers and White, 1973), to check total load, or in May (1994), to check bed-load, may be appropriate, although Arthur *et al.* (1999) and De Sutter *et al.*, (2003) demonstrate that these, similar, and more recent relationships are only tenuously applicable to sewer flows. For the collector sewers, however, application of any of the traditional equations would be expected to lie outside their limits of validity. Not only are the hydraulics significantly different, but also the nature of the solids. The rate of accumulation of sediments in a range of sewers has been summarised in Ashley *et al.* (1992), as given in Table 2.3.5.

2.3.5.2 Prediction of sedimentation in larger sewers

A range of deterministic computational models have been used to predict sedimentation in sewers (Ashley *et al.*, 1999; 2000). These approach the problem usually with full-solution hydraulic models linked with transport capacity and threshold of erosion equations. The assumption is that transport occurs when capacity is available (i.e. the conditions are not sediment source limited), deposition occurs when the transporting capacity of the flow is insufficient and erosion occurs if the flow conditions exceed a certain threshold at higher flows. Thus a deposit is accumulated over time as the sediment input and flow conditions change. To be effective any model should couple the sediment bed temporal changes with the flow field and thus utilise a 'dynamic' bed for hydraulic computations.

None of the currently available models includes the effects of the near-bed solids believed to be principally responsible for foul flushes (Section 2.3.4.3). As yet, the prolonged computational times required to model annual and longer-time series rainfall events make it difficult to use detailed models for the prediction of longer-term (equilibrium) sedimentation in existing sewer systems. As a consequence, a number of simpler alternative approaches have been developed to predict sedimentation and produce control strategies.

Current methods of predicting sediment deposition include:

- surveys – usually Closed Circuit TV (CCTV);
- hydraulic analysis using characteristic criteria such as velocity thresholds;
- conceptual and risk modelling to produce a ranking for potential sediment deposition;
- use of sewer flow quality models, which include some or all of these.

Application of these methods has usually been to particular sites or catchments, without a generic approach being demonstrated. An expert system developed by Gérard and Chocat (1998) has been successfully applied to the catchment of Lyon, France. In this approach, risk of deposition was assigned to pipes and structures throughout the system based upon key structural characteristics. However, the approach was not considered flexible enough.

A simplified approach to estimating deposition was first proposed for small (< 500 mm approximately) sanitary sewers from US studies in the 1970s (Pisano *et al.* 1979):

$$Z = 40\left(\frac{\tau}{\tau_c}\right)^{-1.2} \quad \text{for } \tau > (\tau_c = 0.004) \text{ pounds/square foot} \qquad (2.3.19a)$$

$$Z = 40 \quad \text{for } \tau \le (\tau_c = 0.004) \text{ pounds/square foot} \qquad (2.3.19b)$$

where Z is percentage of dry weather solids deposited.

The relationships (2.3.19a and 2.3.19b) suggest no upper bound to deposition in any pipe length, and hence need to be modified to reflect what is observed in practice:

- Sediment deposits can become graded with time – coarser particles in the upper parts of sewer 'reaches'.
- Sediment beds attempt to ameliorate the effects of the sewer which cause them to deposit initially, i.e. where there are gradients which are too shallow, the deposits will tend to form an artificial steeper bed – thus becoming 'self-equilibrating'.
- Bed roughness (particle and form) interact with the above – having the greatest effects when the bed is at half depth in a circular pipe.

These factors have led to a modification of the above equation:

$$Z = 0.889 \times \left(\frac{\tau_0}{\tau_c}\right)^{-1.2} \times \left(\frac{W_b}{W_{\max}}\right) \qquad (2.3.20)$$

where W_b is bed width and the maximum width of the pipe is $W_{\max}$

Application of relationship 2.3.20 has been used successfully to predict deposition in the sewers in Dundee (Ashley *et al.*, 2000), as illustrated in Figure 2.3.23.

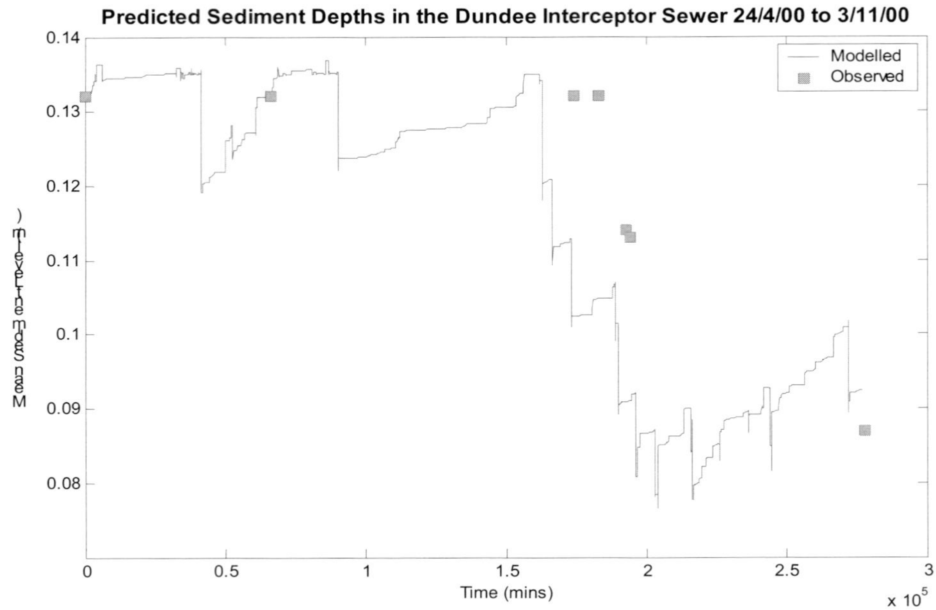

Figure 2.3.23. Predicted sediment depth in Dundee Interceptor sewer compared with observation (Ashley *et al*., 2000).

2.3.5.3 Bed deposit erosion

Erosion results in the resuspension of particles from an in-pipe sediment bed. There are a number of problems for the determination of the initiation of solids movement. In many studies, the precise moment of incipient motion is typically determined subjectively by the observer. In only a few reported studies, mainly laboratory based studies, have transport rate measurements been used for a more objective evaluation of the conditions where erosion occurs.

Nalluri and Alvarez (1992) recognised that for circular channels with flat bed deposits, critical conditions occur at lower values of mean boundary shear stresses τ_0 than those predicted by the Shields' relationship. However, analysis using 'bed partitioned' (see Section 2.3.2.2) bed shear stress τ_b have been found to be produce threshold values comparable with Shields' curve. Despite the limitations in specifying a single critical erosional shear stress, a summary of published results is given in Table 2.3.6.

A number of laboratory studies have been carried out in an attempt to evaluate the effect of cohesion on the erosion of bed deposits. Laboratory studies in Newcastle examined the erosion of synthetic cohesive sediment (Laponite clay and sand). This indicated erosion thresholds significantly higher than if no cohesion had been present in the deposit. Observations also indicated that once eroded (in small groups of particles) the sediment behaved very much as non-cohesive material (Nalluri and Alvarez, 1992). It is not known whether this erosion mechanism is also applicable for sediments in actual sewers. Some attempts have been made to understand the development of bed strength in sewers (Section 2.2.4.1.3) (Stotz and Krauth, 1986; Williams and Williams, 1989), and it is apparent that the onset of resistance to erosion is rapid (hours rather than days) once the bed has been consolidated.

Table 2.3.6. Critical erosional shear stresses $\tau_{c,E}$ - sewer type sediment

$\tau_{c,E}$ (N/m²)	Author	Notes
1.5–2	Ashley *et al.* (1993)	field observations
	Ristenpart and Uhl (1993)	field observations:
0.5–0.9		dry weather flow
1.7–3.3		wet weather flow
	Hrissanthou and Hartmann (1998)	field observations:
0.23		finer fraction of deposits (d_{30})
0.41		coarser fraction of deposits (d_{50})
2–4	Stotz and Krauth (1986)	deposits from dry weather sewage
	Nalluri and Alvarez (1992)	laboratory experiments:
2.5		weaker sediment (type C)
6–7		bed deposits (type A)
0.5–3	Torfs (1995)	laboratory experiments: mud/sand mixtures – critical erosive stress a function of fines content
0.5	Skipworth (1996)	laboratory experiments: synthetic cohesive particles
	Kamphuis (1990)	laboratory experiments:
1.3		flow with sand particles
20		clear water flow

Limited studies have been undertaken to examine the yield strength of deposits by measuring the rheological properties of sewer sediments in France and the UK utilising different rheometers (Migniot, 1968; Beyer, 1989; Williams and Williams, 1989; Wotherspoon and Ashley, 1992). It is clear that sediments exhibit an *apparent* yield stress when subjected to controlled rates of applied stress. Studies of mixed sands/muds in the laboratory and in the field (Torfs, 1993, 1995; Mitchener and Torfs, 1996) have considered the resistance to erosion (critical shear stress τ_{cr}) in terms of a relationship to bulk density ρ_b (Mehta, 1988) for a wide range of artificial mixed sediment beds:

$$\tau_{cr} = E_1(\rho_b - 1000)^{E_2} \tag{2.3.21}$$

It was concluded that for a wide range of deposited mud/sand mixtures $E_1 = 0.015$ and $E_2 = 0.73$. In general it has been found that the addition of fine-grained sediment clay/mud to a granular sand deposit increased the resistance to erosion (Torfs, 1995; De Sutter, 2001; Rushforth, 2001). There are also two principal types of erosion: surficial from the upper layer, or bulk erosion, entailing the movement of large parts of the consolidated bed. These results are important as they may be expected to apply also to the mixtures of granular and clay/mud material found in sewer deposits.

Sewer sediments have been observed to be non-Newtonian visco-elastic materials, and to have widely differing critical yield stresses ($5 > \tau_y > 2500$ N/m²) (Section 2.2.3.3) even when other physico-chemical properties appear to be similar. The differences may be attributed to the post-depositional history of the deposits and the relative importance of the organic material which, unlike in natural systems, is very heterogeneous, comprising fine grained particles, faeces, rags and fibres.

The measured yield stresses have also been considered as corresponding to the limit at which the fluidisation of the deposit occurs. This fluidisation of the bed has been considered the condition for erosion and hence the measured yield stress can be linked to the onset of erosion. Tests on samples from various locations in the Dundee sewer system, and from the Hildesheim sewer (Ristenpart, 1995), showed that the *apparent* yield strength was most closely related to the liquid content of the sample (Figure 2.3.24) The yield strength measured by an applied stress rheometer marks the onset of structural failure and may be linked to the fluidisation of the sediment.

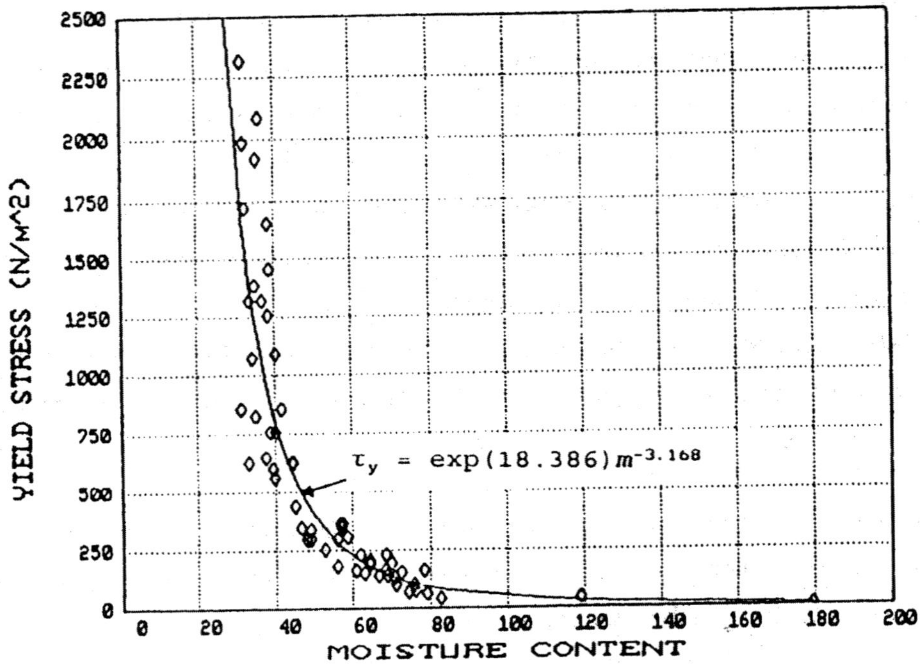

Figure 2.3.24 Relationship between critical yield strength and moisture content for sewer sediment (adapted from Wotherspoon, 1994).

The analysis of 61 sediment samples by Wotherspoon (1994) revealed a strong correlation between the yield stress, τ_y (N/m^2), and the liquid content, m (% by mass, defined as the mass of water per unit mass of solids as a percentage), of the samples (Figure 2.3.22):

$$\tau_y = 9.66 \times 10^7 \; m^{-3.1682} \qquad (2.3.22)$$

Comparison of this relationship with published results (Migniot, 1968) for a wide range of samples of silt and clay muds were good, provided the volumetric solids Cv > 0.3 (Wotherspoon, 1994). The Migniot results indicated yield strength was a function of C_V, and particle specific gravity s. Earlier studies also utilised the relationship developed for silt and clay muds and showed correlation for tests on the finest particles extracted from sewer sediment deposits (Beyer, 1989). The relationship between rheologically measured yield stress and actual critical bed shear stresses has not been conclusively investigated for sewer sediments. De Sutter (2001) used relationships developed for natural muds (Otsubo and Muraoka, 1988; Migniot, 1989) of the form in Equation 2.3.23 to investigate the link between a rheologically measured yield strength τ_y and a threshold of erosion measured in a laboratory flume τ for the same clay/sand sediment mixtures.

$$\tau = a(\tau_y)^b \tag{2.3.23}$$

where τ is the critical bed shear stress, with τ_y the yield stress measured in rheological testing
for $3\text{N/m}^2 < \tau_y < 75\text{N/m}^2$ a = 0.081; b = 1
for $0.08\text{N/m}^2 < \tau_y < 3\text{N/m}^2$ a = 0.27; b = 0.6.

Application of these relationships to tests using clay-sand mixtures showed that they produced believable results for the measured values τ_y for the low shear stresses found in the tests, whereas, application of Equation 2.3.22, for the moisture contents, followed by Equation 2.3.23, gave unrealistically low values for critical bed shear stress. It is likely that Equation 2.3.22 should not be extrapolated to very high moisture contents, or applied to finer grained sediments.

Defining when a sewer sediment deposit may behave as cohesive in erosion terms is not straightforward. For uniform sediments, particles of less than 20 μm are presumed to mark the boundary between cohesive and non-cohesive beds (Mehta *et al.*, 1994), whereas for mixed sediments it is suggested (Torfs, 1995) that the presence of small particles within the interstices of the larger granular material is the important factor in the development of a cohesive structure (Alvarez, 1990; Torfs, 1995). Some 10–20 % fines (< 63 μm) in the bed would be necessary for the onset of cohesive-like properties in mixed sands/clays; however, the required proportion of fines depends also on the relative particle sizes, with the smaller granular particles being most easily affected by the presence of even relatively low proportions of fine material, particularly if this is of clay mineral size (< 2 μm). How this information relates to sewer sediments, where the proportion of fines may not be as important as the organic content, has not yet been established, although the importance of the biofilms attached to the particles as a key component of sediment ageing and bed strength development, including consolidation processes, is apparent (Section 2.2.5.2).

Extensive field programmes investigating the erosion of sediments in sewers and taking account of their cohesive-like properties have been carried out in Dundee, Scotland and Hildesheim, Germany (Ashley *et al.*, 1993). In these studies, the deposition and erosion of the bed has been measured directly using an in-situ bed depth monitoring device. The conclusions from this work, regarding the bed deposit behaviour are as follows.

- Erosion occurs under different dry and wet weather conditions, being most closely related to magnitude and variability of shear stress (see Table 2.3.6).
- Following erosion the bed quickly (in the order of a few hours) develops a level of stability existing prior to erosion, indicating a potential rapid build-up in bed strength.
- A short time scale sediment peak has been observed for a number of events just prior to the onset of erosion; this appears to be related to a flush of concentrated solids close to the bed from upstream erosion (Section 2.3.3.3).
- Erosion may or may not occur during DWF peaks, depending upon the relative magnitude of the shear stress, the bed density and the equilibrium depth of the bed (see below).
- Storm events generally erode some of the bed, but typically not all the deposit.
- Bulk erosion of the bed has been observed to begin at the bed shear stress levels reported in Table 2.3.6.

In these studies there is an indication that the onset of erosion may occur at different initial rates of shear stress depending upon the history of sediment deposition, the antecedent dry weather period, and the rate of change of shear stress. In general, the more rapid the change in bed shear, the higher the shear stress is when erosion is first observed. In both studies there is evidence of deposition during periods of high bed shear, possibly due to the movement past the measurement point of coarser bed load material.

2.3.5.3.1 Prediction of bed erosion

The variety of sources and materials comprising sewer bed deposits make the threshold of sediment movement difficult to predict.Very few of the commercial sewer flow models deal in any way with in-sewer deposition and erosion. Typically the sediment deposits are considered to comprise a few characteristic size particles in two or three different layers. Solids are considered transferred between these layers by means of user defined empirical criteria. Release of sediment from these layers may take place when the relevant erosion threshold, usually defined as a critical bed shear stress is exceeded. Transport and subsequent deposition are estimated depending on the transport capacity of the flow. All relationships deal with the sediment as though it were granular.

In-sewer deposits can be divided into two generic groups: those deposits with little or no cohesion so that the sediments can be assumed to be entirely granular, and a second group in which cohesion is significant. In the first group, thresholds of motion are estimated using methods outlined in Section 2.3.4.4.2. Thresholds of erosion for cohesion dominated deposits, that is, deposits with significant amounts of fine-grained sediments (organic and inorganic), are dependent on many diverse mechanisms, so robust estimation is very difficult.

Correlation between resistance to erosion and the apparent yield strength of fine-grained, cohesive deposits is generally expected as they are both dependent upon similar structural factors (Otsubo and Muraoka, 1988).The relationship for estimating yield strength (Equation 2.3.22) has so far been the only direct field derived relationship to be used to develop a sewer sediment erosion model (Wotherspoon, 1994). In view of the dichotomy between peak apparent measured yield stresses ranging up to 2650 N/m^2 for the sewer sediment deposit, compared with the much lower typical hydraulic bed shear stresses of up to 25 N/m^2 found during wet weather in the Dundee interceptor sewer (Ashley *et al.*, 1992) and observed erosion at very low shear stresses (1 N/m^2), it is believed that erosion occurs as a result of progressive weakening (fluidisation) of the surficial layers of a sediment bed. This concept corresponds with the failure of estuarine sediment beds due to surficial 'swelling' or dilatation, which has also been attributed to an increase in the moisture content and a reversal of the normal consolidation process allowing transport into the flow (Torfs, 1995).

In practice, problems of application arise because of the lack of input data of the physical properties of the deposits available to operate these models. The yield strength relationship (Equation 2.3.22) has been linked to a theoretical relationship describing the variation of sediment density with depth, from estuarine studies (Mehta and Parthenaides, 1982), by Wotherspoon (1994), who also defined the erodible density to give an expression for the erodible sediment depth, h_e (m):

$$h_e = h_0\left[1-\left(\frac{\rho_F}{\xi\bar{\rho}_0}\frac{1+\left(\frac{\tau_b}{e^{18.386}}\right)^{\frac{1}{-3.168}}}{1+\frac{\rho_F}{\rho_w}\left(\frac{\tau_b}{e^{18.386}}\right)^{\frac{1}{-3.168}}}\right)^{-\frac{1}{\xi}}\right] \tag{2.3.24}$$

where h_0 is initial sediment bed depth (m)
ρ_F is sediment mass density (kg/m^3)
$\bar{\rho}_0$ is initial average bed density (kg/m^3)
ρ_w is fluid density (kg/m^3)
τ_b is bed shear stress (N/m^2)
ξ, ζ are calibration parameters (-) from Mehta and Partheniades (1982)

This relationship was compared with erosion observed at points in combined sewers in Dundee, Scotland, and Hildesheim, Germany. The results established a procedure for determination of cohesive-like sediment erosion with time and the ranges for the calibration parameters (ξ typically 0.1–0.45; ζ typically 0.65–0.80). The application of the model approach is described in Ashley *et al.* (1993). The tests carried out in Germany, where erosion was observed at stresses as low as 0.7 N/m², showed the applicability of the model approach to other locations, although it does not provide a high degree of accuracy (Ristenpart, 1996), as illustrated in Figure 2.3.25. The variations in the structure of the sediment bed over small depth changes requires further detailed examination and model improvement.

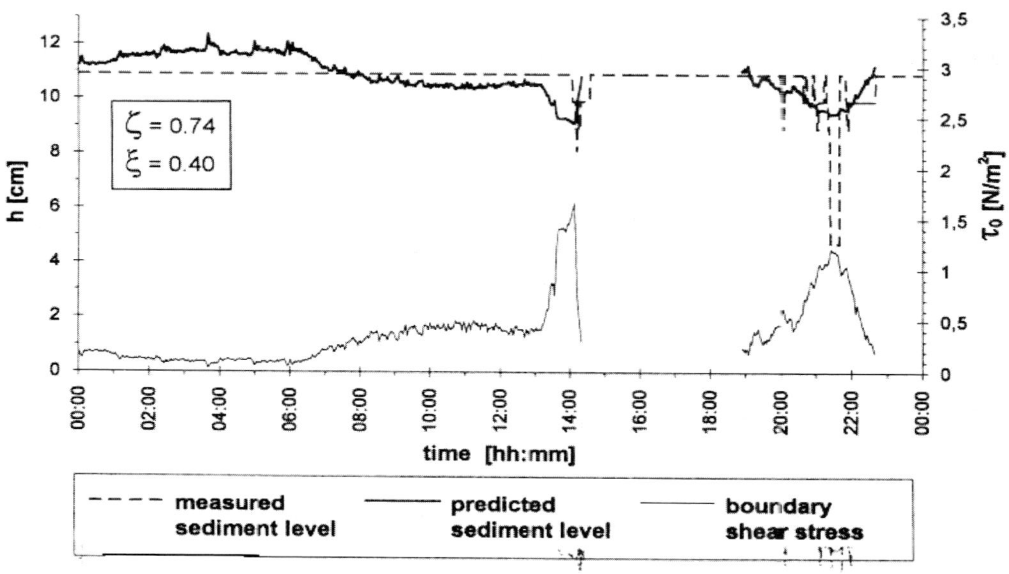

Figure 2.3.25. Sewer sediment erosion model comparison of predicted and observed results (adapted from Ristenpart, 1996).

Laboratory studies using ground olivestone as a surrogate cohesive bed material by Skipworth (1996) utilised a variable critical shear strength with depth relationship to predict erosion (Figure 2.3.26). Erosional strength of the sediment bed, τ_c, was modelled as a function of the depth of erosion, *d*, the thickness of the overlying weak layer, *d'*, the erosional strength at the surface of the undisturbed deposit, τ_{cs}, and the uniform erosional strength of the underlying bed, τ_{cu}. This was used to develop a model for erosion rate, and included the reduction in bed width and erodible area of the bed during erosion events.

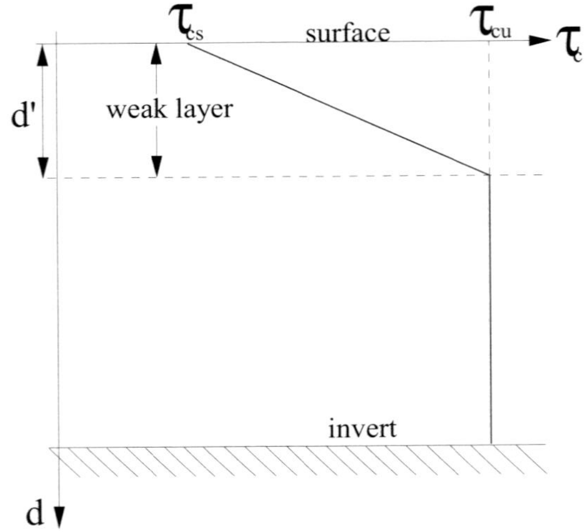

Figure 2.3.26. Erosional resistance v. depth of erosion with strength variation in weak layer assumed to be linear with respect to depth (adapted from Skipworth, 1996).

The bed strength profile illustrated in Figure 2.3.26 is similar to that suggested by Parchure and Mehta (1985) for estuarine deposits and was used to develop the concept for a unified erosion model based on sewer sediment erosion research and that undertaken in the laboratory investigating mixed cohesive/non-cohesive sediments (Torfs, 1995). The onset of erosion can be predicted using Equation 2.3.22. The rate of erosion, depending upon whether this is surficial (weak) or mass erosion (consolidated bed) is given by an excess shear stress relationship, provided the clay particle content > 3–15 % (Parchure and Mehta, 1985; Torfs, 1995):

$$\mathbf{E} = \mathbf{E}_{\mathbf{m}}\left(\frac{\tau_{\mathbf{b}} - \tau_{\mathbf{cr}}}{\tau_{\mathbf{cr}}}\right)^{\alpha} \tag{2.3.25}$$

Typical values for the erosion constant E_m for mixed muds/sands are $0.12 \cdot 10^{-3}$–$3.73 \cdot 10^{-3}$ kg/s m², with the higher values corresponding to mass erosion of the bed, with $\alpha \approx 1$ (Torfs, 1995). A similar successful application of the excess shear stress relationship to determine erosion rates has been undertaken in the laboratory experiments using beds of ground olivestone as surrogate sewer sediments, with unsteady flows (Skipworth, 1996). Unfortunately, all these modelling approaches rely on calibration data (prior to each event) to be able to simulate the change in deposit properties (water content, deposit density, critical erosional shear stress) with depth. As yet these deposit properties cannot be predicted explicitly for in-sewer deposits.

2.3.6 Biofilms

Sewer slime is a biological film with a thickness of up to a few millimetres growing mostly on sewer walls (Section 2.2.5). It contains mineral particles such as sand and the remains of dead microorganisms, as well as living cells. Slime grows mostly during periods of dry weather flow and is sloughed off during storms as well as dry weather periods due to the shear stresses at the sewer walls and the metabolism processes (desorption). The eroded slime is easily settleable (Michelbach *et al.*, 1992). In addition, sewer slime accumulates pollutants, especially heavy metals. Due to the low density of slime (biofilm), once removed from attachment to the walls and bed sediments, it is readily transported in suspension and at CSOs can be discharged into receiving waters with significant impacts.

Reiff (1992) summarised the observations regarding the abrasion of sewer slime. Combined sewage flows cause sudden erosion of sewer slime to occur due to the increase of wall shear stress. Approximately 90 % of the erodible (not total) sewer slime is lifted from the wall within the first minute just after an increase in the shear stress. The mass of sewer slime eroded depends on:

- the quantity of sewer slime present in the system before the runoff event;
- the rate of increase (gradient) of shear stress with time; and
- the hydraulic conditions under which the sewer slime has grown (i.e. 'the history' of the sewer slime). Slime that grows at a higher shear stress develops a more stable supporting matrix.

Sewer slime is never completely eroded, not even when there are considerably higher shear stresses than those for which the slime was established.

Even in times of dry weather flow sewer slime is partially and intermittently removed from the sewer walls. This is termed 'sloughing', and is due to the changing conditions inside the biofilm as a result of the metabolism processes. Continuous losses occur also due to abrasion processes caused by collision of the solids transported in the sewage flow. Significant parts of the eroded sewer slime can settle readily in sewers and contribute to the development of sewer deposits.

2.3.7 Solids transport modelling in sewers

As described in the previous sections, the movement of solids in sewerage systems involves a wide range of processes which are interlinked but which vary in relative importance according to the flow conditions and the characteristics of the solids. In order to model the behaviour of a system, it is necessary to identify the principal processes and describe them mathematically. In many cases the processes are still imperfectly understood and in others the empirical equations describing them may have been validated for only a relatively narrow range of conditions. Simplifications and assumptions therefore, have to be made in order to construct a workable model that behaves in a physically correct manner and does not suffer from problems of numerical instability. Consideration must also be given to whether the data requirements of the model can realistically be satisfied. If it contains too many factors that cannot be measured or verified, the model may be difficult to calibrate and unreliable in general use. In the following sections a general review of the problems related to solids transport modelling is given, rather than expositions of available models, as these are evolving as new knowledge becomes incorporated. For more detailed descriptions see Huber (1986); Bertrand-Krajewski *et al.* (1993); Mark *et al.* (1995) and Ashley *et al.* (1999).

2.3.7.1 Aims of models

Most of the models related to sediment transport in sewers are deterministic. Models can be of different levels of complexity according to the type of results they are intended to produce. When new gravity sewers are being designed, they are normally required to be 'self-cleansing'. The conditions needed to achieve this are often specified in terms of minimum values of flow velocity, shear stress or pipe gradient. In the past, fixed and somewhat arbitrarily selected minimum values have been used. However, an approach based on the results of recent research can now determine limiting values that are more appropriate to the particular flow and sediment conditions in each individual pipe in the system (Section 4.2.1).

More complex types of model are necessary if it is essential to simulate the behaviour of existing sewers in which sediment deposits occur because of inadequate gradients or backwater effects. During individual storm events, the changes in depth of deposited material may be quite small because net erosion or deposition in an individual pipe (considered over the whole time period) will only occur if there is an imbalance between the rates of sediment transport at the upstream and downstream ends of the pipe. Useful results can therefore still be obtained from a model that can predict movement of fine-grained or low-density sediments in suspension but in which the larger inorganic particles are assumed to form effectively fixed sediment beds (with depths determined from field surveys). Movement of pollutants associated with the suspended particles can be determined by multiplying the sediment concentrations by suitable potency factors and by assuming that the velocity of the particles is equal to that of the water or related to it by a constant of proportionality (e.g. Crabtree *et al.*, 1995).

If it is required to simulate the build-up of sewer deposits with time and their variations during dry-weather and wet-weather flow conditions, it is necessary to include a description of bed load sediment transport in the model. The movement of pollutants associated with this coarser material can be determined by using suitable potency factors and by allowing for the changes in strength that occur during the time that the sediment remains in the sewer. One notable outstanding difficulty in this respect is that of the highly concentrated near bed solids and their transport estimation, as it is unclear as yet whether this should be via 'near-bed' suspension or bed-load type formulae.

2.3.7.2 Components of models

The types of simulation model discussed above will normally consist of a core that is able to calculate time varying values of flow depth and (cross-sectional mean) velocity throughout a sewer network. Algorithms or sub-models are then added to this core to describe the movement of solids and pollutants through the system. The factors that need to be covered may include some or all of the following:

(a) Input data for sediment characteristics and pollutant loads in foul sewage.
(b) Input data for sediment characteristics and pollutant loads associated with surface water and infiltration; the former may often be predicted by a surface wash-off sub-model linked to rainfall data.
(c) Equations describing changes in the characteristics of pollutants with time, e.g. generation of hydrogen sulphide.
(d) A method of determining the hydraulic roughness of sediment deposits in pipes; this may be specified as a fixed value or may be predicted taking into account the occurrence of bed forms under certain flow regimes.
(e) A method of determining the composite roughness of pipes with deposits; allowing for the separate roughness values of the walls and the sediment bed.

(f) Equations for predicting the rate of transport of solids in suspension and a criterion for determining the onset of this mode of transport.
(g) Equations for predicting rates of bed load transport both with and without deposited beds and the conditions at which deposits begin to form and are eroded. These equations should be able to accommodate wide variations in particle size and density.
(h) A sub-model for predicting changes in the cohesive strength of deposited material with time.
(i) Methods for determining the fluid shear stress needed to erode cohesive deposits and the effect of cohesion on rates of sediment transport.
(j) A sub-model for recording quantities of deposition and erosion in each pipe and the availability of different sediment fractions to be transported.

The calibration of a model containing many different components of this type is potentially difficult because of the large number of variables that can interact with each other in very complex ways. It is therefore important to use, as far as possible, mathematical descriptions of the physical processes that have been developed from reliable data. As an example, most of the equations for sediment transport in pipes have been obtained from laboratory studies carried out with non-cohesive sediments. The ability in the laboratory to vary conditions systematically and measure them accurately has made it possible to identify the separate effects of factors such as flow velocity, water depth, pipe size, sediment depth and particle size. An initial study by Nalluri and Alvarez (1992) indicates that once the threshold of movement of a cohesive sediment bed is exceeded, the particles tend to be transported in a similar way to non-cohesive granular sediments. This suggests that the laboratory-derived transport equations may apply reasonably to real sewer sediments once they are in transport. This hypothesis can be verified by comparing the predictions of the equations with field data obtained from sewers. Although such measurements tend to be less accurate and complete than equivalent laboratory data, they should nevertheless play a key role in confirming the validity of any proposed model.

2.3.7.3 Key factors affecting model performance

2.3.7.3.1 Sediment input data

However well the physical processes are described in a simulation model, the model will not give reliable results unless the data for quantities and types of sediment entering the sewers are satisfactory. The solids load associated with the foul sewage component is more predictable but the contribution of the surface water flows can be much more variable. If a surface wash-off model is used to predict the inflows, its validity must be carefully checked. When seeking to reproduce observed patterns of deposition, it should be noted that coarse sediments transported predominantly as bed load may take a considerable time (of the order of years) to move through a sewerage system. Also, considerable quantities of sediment can enter a system as a result of specific events such as localised construction works on the surface or the collapse of a degraded section of sewer; and that these loads may not be accounted for by a generalised washoff model. At the catchment scale, the spatial variability of sediment characteristics is also significant.

2.3.7.3.2 Sediment transport equations

Transport rates can be predicted from separate equations for suspended load and bed load or alternatively from a single total-load equation. The latter approach, as typified by the Ackers-White equations (Ackers, 1991), has the advantage of being simpler to apply. However, data on the transport of widely-graded sediments in pipes are not yet available to check the validity of total-load equations. Most of these equations were developed for alluvial channels in which there is not usually such a wide range of particle densities and sizes as occurs in sewers. Organic materials and inorganic silts in sewers will usually cause deposits to exhibit cohesive-like properties, and the flow conditions needed to erode them will be significantly different from purely granular deposits. Flows will also need to be sufficient to transport the organic/inorganic sediments in suspension in quite high concentrations. By contrast, most inorganic sediments larger than about 0.5 mm will normally be transported in sewers as bed load; this has been the mode of transport in nearly all the laboratory studies so far carried out. Certain controlled experiments have investigated the erosion of granular sediment mixtures (Rushforth, 2001; Ota, 1999). These have shown the importance of the effects of variable particle sizes. An adequate description of the transport processes should include a model with three sediment components: organic, low-density particles, fine inorganic particles and coarse inorganic particles, each with characteristic particle information. Once the threshold of movement has been exceeded (allowing for cohesive effects), the first two components would move in suspension and the third as bed load. Movement of the pollutants (associated with the solids) through a sewerage system may be simulated by assigning suitable potency factors to the three sediment components. However, such constant potency factors should be replaced in the future by more complex but more realistic models accounting for biochemical processes as described in Section 2.2.

2.3.7.3.3 Cohesive effects

Further laboratory and field research is needed to quantify more precisely the effects that cohesion has on the erosion and transport of sediments from in-sewer deposits. For models used to design 'self-cleansing' sewers, it may be sufficient to specify that flows with a given frequency of occurrence should be able to erode cohesive deposits having a certain shear resistance. This approach has been adopted in new design guidelines produced by Ackers *et al.* (1996); however, in this case the shear resistance is an arbitary fixed figure for all cohesive deposits. In the case of models used to simulate long term changes, it may be necessary to take account of the change in cohesive strength that occurs with time as deposited material consolidates and transforms (e.g. Wotherspoon and Ashley, 1992; Tait *et al.*, 2003).

2.3.7.4 General assessment of contemporary models

Contemporary models for sediment transport in both suspension and at the bed have relied on a number of aspects of fundamental fluid mechanics, generally derived from laboratory studies, but there are important differences between sewer flow conditions and those found in laboratory channels and rivers or estuaries as described above. Traditional theories developed based on transport at capacity, for flows in laboratory channels using generally uniformly sized, single-density sediments in steady uniform flows, these may have only limited application to real sewer conditions. For example, the latest UK sewer design approach is novel in that it proposes for the first time the concept of 'allowable deposition' as the basis for design (May *et al.*, 1996). It is probable that, by having to make simplifying assumptions about the nature and movement of real sewer sediments, the methodology may only be generally applicable to the design of new sewers where the amounts and nature of the material entering the system can be estimated to some degree. The method may be too simplistic for the analysis of existing large sewers where the heterogeneity of the solids and the variability of the flow conditions are such that the laboratory-derived methods are beyond their valid limits.

Most modelling related to sewer sediment transport and processes in actual applied situations is carried out using one of a number of proprietary models (Ashley *et al.*, 1999). Each of these has different limitations. In general, the 'comprehensive' models cannot as yet model the processes required to properly examine the phenomena likely to be important for any quality processes, such as spill loads from overflow discharges (e.g. Langeveld *et al.*, 2003). Although these models have been reportedly used successfully for dissolved pollutants and ammonia modelling (Mark *et al.*, 1998; Holzer and Krebs, 1998), despite apparent problems noted by others (Margetts, 2000). Problems of determining in-sewer sediment transport and deposition related processes have been described recently for the most commonly used models (e.g. Bouteligier *et al.*, 2002a, 2002b, 2002c, 2002d; De Sutter *et al.*, 2003). As shown in this publication (this Section and Section 4.2), the erosion of sewer sediments is a complex process, due principally to the mixtures of types of solid, in terms of both particle size and the nature and strong variability of flow conditions. Sewer sediment beds typically have patterns of layered deposition in dry and wet weather and an active and sometimes mobile surficial organic layer, known as 'near bed solids' that contributes to any foul flush load (Section 3.4). Attempts so far to develop a generalised erosion model have been unsuccessful, other than where the pre-erosion bed can be adequately defined prior to erosion (Rushforth *et al.*, 2003). This has important implications not only for the modelling of individual storm erosion events, but also for sequences of storms used in the time series modelling that is required to determine storage volumes at overflows, as new parameters to define the state of the bed are required between each storm in the series. Currently this facility does not exist in any contemporary model approach.

Given the limitations, uncertainties and complexities described above, detailed deterministic modelling of in-sewer processes may not be achievable or even an appropriate aspiration. Various other types of model have been utilised, from conceptual, for investigating loads annually (e.g. Durschlag *et al.*, 1991), to more detailed but specific models, such as for predicting septicity (Boon *et al.*, 1998). A number of investigators have utilised simplified sewer flow quality models, such as KOSIM, which can readily be used in long-term simulation mode (Schütze *et al.*, 2002). This model is most suited to determine annual loads (chronic pollution), however, it does not include for example, backwater effects, and hence any hydraulic interactions between sewer and downstream treatment plant will not be modelled.

2.3.8 Solids movement in tanks, basins and CSO chambers

The terms 'tank' and 'basin' are often used interchangeably; in this Section, the following terminology will be employed. The term *basin* refers to large, exposed, naturally formed areas of land used for the above-ground storage of sewer flows. The term *storage tank* will be used to indicate a concrete storage structure which is enclosed underground, and which may be associated with a *CSO chamber*. *Storm tank* will be used to identify an open-topped concrete tank used at the head of sewage treatment works, whilst *settling tank* refers to a tank used for primary treatment at sewage treatment works. This Section will primarily focus on the issues relating to sediment transport and deposition in storage tanks and CSO chambers.

2.3.8.1 Solids occurrence – deposition problems

Careful consideration should be given to the question of whether or not the deposition of sediment to the bed of a storage tank actually represents a problem. Clearly, where a storage tank is used simply to provide additional storage within the sewer system, sediment deposition is best avoided. However, where a storage tank is used in conjunction with a CSO chamber, then the primary function of the combined structure is to retain sediments within the sewer system, rather than allow them to spill into the watercourse. In the latter case it may be preferable for deposition to occur, allowing for the subsequent management of the sediment. In the UK the former approach is currently favoured. The UPM manual (FWR, 1998) states: 'detention [storage] tanks should be designed to be, as far as possible, self-cleansing whilst in operation'. In Germany (Geiger, 1986) greater emphasis is placed on the operation of storage tanks as effective settling basins, and the requirement for self-cleansing operation has received little attention. A detailed description of the detrimental effects of tank deposits is given in Section 3.5.

Few detailed studies of sediment accumulation and movement in tanks have been undertaken. Geiger (1986) presented details of a Swiss storage chamber which was divided into two sections. One part of the basin filled at the beginning of an event, catching a first foul flush when one was present. The second part filled only following the filling of the first, and functioned like a settling tank. Aside from its effectiveness in pollution control, the subdivision into two parts eased maintenance requirements, as for smaller events only the first portion was filled. Geiger (1986) suggested that this type of chamber represented an optimum solution to the twin objectives of retaining the foul flush (see Section 3.4) and enabling treatment by settling.

In a survey of the sediment deposits in the Littleborough (UK) storage tank (Brownbill, 1992), it was found that deposit depths ranged from 10 to 150 mm, comprising a range of slurry to sand materials. Pollutant concentrations varied widely, and the most highly polluted samples were found in the side channels in the tank. The least polluted samples, which were also the shallowest deposits, were collected closer to the central dry weather channel.

2.3.8.2 Solids movement – settling efficiency of tanks

The sediments which enter a storage tank can be considered in two classes: gross solids and suspended solids. The gross solids include 'floaters', 'sinkers' and neutrally buoyant materials, but as the solids may alter characteristics as they are transported, faecal matter may break-up into suspended solids, whilst the absorption of water may transform floating paper products into material that sinks. The settling efficiency of a storage tank is defined by the proportion of suspended sediment entering the tank which settles to the bed of the tank. Settling efficiency is not an indication of the tanks' effect on the gross solids.

The simplest approach to estimating the efficiency of a storage tank is from classical sedimentation theory, which was first outlined by Camp (1946) to describe the operation of settling tanks. The model assumes that the tank is divided into three zones: the inflow, settling and outlet zones. In the settling zone the flow moves from the inlet towards the outlet at a velocity equivalent to the flow rate divided by the surface area of the tank, i.e. Q/A. This term is often referred to as the 'surface loading' or 'surface overflow rate'. At the same time, sediment particles are falling at a velocity w_s. A particle will only settle to the chamber bed if the settling velocity exceeds the surface loading rate. The model assumes an 'ideal rectangular continuous flow basin', in which the direction of flow is horizontal and the velocity is the same in all parts of the settling zone.

Stovin and Saul (1996) demonstrated that storage tank efficiency ξ varied as a function of the mean inlet velocity ($\bar{u}$), and suggested that the efficiency of any particular tank could be characterised by means of a $\bar{u}/\xi$ curve. Figure 2.3.27 shows an $\bar{u}/\xi$ curve derived for a laboratory scale storage tank. In this case the $\bar{u}/\xi$ curve was very similar to the efficiency curve predicted from classical sedimentation theory.

Forms of this basic model are currently used in urban drainage quality models to estimate tank efficiency. However, these models allow the user to input an *efficiency factor* for each tank modelled, recognising that the flow patterns in storage tanks are generally far more complex than the assumptions which ideal settling theory allows. Storage tanks are dynamic systems, and variations in the flow field occur in both time and space. The modelling of tank efficiency is also fundamentally dependent on the input of representative settling velocity information.

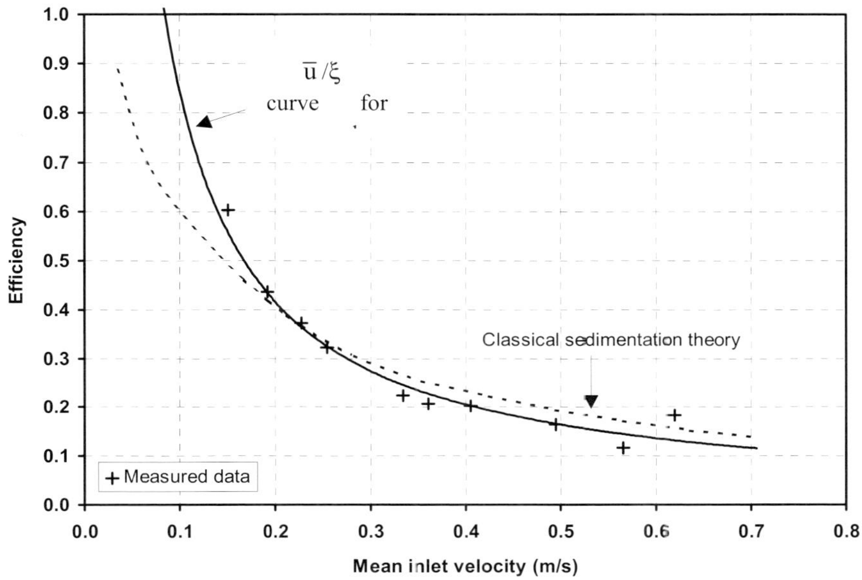

Figure 2.3.27. Laboratory efficiency-inlet velocity curve for solids in a storage tank.

Ellis (1992) and Lessard and Beck (1991) have suggested that storage tanks experience a sequence of flow regimes during each storm event. These are *filling*, *dynamic sedimentation* and *emptying*. For a tank used in conjunction with a CSO the dynamic sedimentation phase is equivalent to the *spilling* phase.

During the *dynamic sedimentation* phase, complex flow patterns develop which result in flow recirculations and the development of still zones. The flow patterns directly affect the locations where sediments are deposited to the chamber bed. Stovin and Saul (1994) have demonstrated that the location of deposited sediments relates to the positions on the chamber bed where the bed shear stress falls below a critical value, and that computational fluid dynamics (CFD) may be employed to predict the flow fields and the deposit locations. The distribution of deposited sediments may subsequently be used to provide an estimate of the tank efficiency. Stovin and Saul (1996) demonstrated that over a broad range of flow conditions, a linear relationship existed between the proportion of the chamber bed for which deposition was predicted and the overall tank efficiency.

During the *emptying* phase of operation, tanks more typically experience uniform deposition of a fine layer of sediment on the bed. Patterns of deposition observed in field situations, for example, the Littleborough tank discussed above, confirm these observations. Deposition is also associated with local obstructions, such as ladders, bends in the dry weather flow channel, columns and roof supports.

The flow patterns that develop during the *dynamic* sedimentation phase of tank operation reflect the internal geometry of a tank. The configuration of the dry weather flow channel, the longitudinal slope of the chamber and the gradient of the benching as well as the length to breadth ratio of the tank are all fundamental in determining the flow pattern that will develop and consequently the extent of the deposition that occurs.

Stovin and Saul (1996) employed computational fluid dynamics (CFD) to examine the effect of the length to breadth ratio (L/B) on chamber efficiency. FLUENT flow simulations were used to estimate the distribution of bed shear stress, from which the efficiency could be calculated. Figure 2.3.28 shows the inlet velocity/tank efficiency curves which were obtained for four different tank L/B configurations. The effects on efficiency were complex, despite the fact that only one geometric parameter was varied. The comparative performance of the different tanks varied as the mean inlet velocity changed. The efficiency predicted from classical sedimentation theory would not have shown any variation between the four configurations.

In addition to the efficiency prediction through bed shear stress analysis, Stovin and Saul (1996) also demonstrated that stochastic particle tracking with CFD could be applied to the prediction of storage tank efficiency and spatial distribution of the deposited material as illustrated by Figure 2.3.29. This plot shows the probability of sediment deposition in a simulated laboratory chamber with $\bar{u}$ = 0.4 m/s.

To describe the overall efficiency of the settling process in storage tanks, Lafond (1995) demonstrated that complex models (CFD with 2D or 3D hydrodynamic simulations) do not necessarily lead to significantly better results than simple models based on the modified Hazen or Camp theories. This is mainly due to critical hypotheses (especially the bottom condition applied to particles which reach the tank bottom) which are needed for CFD but which are difficult to verify, or to prove in the field, because of their complexity.

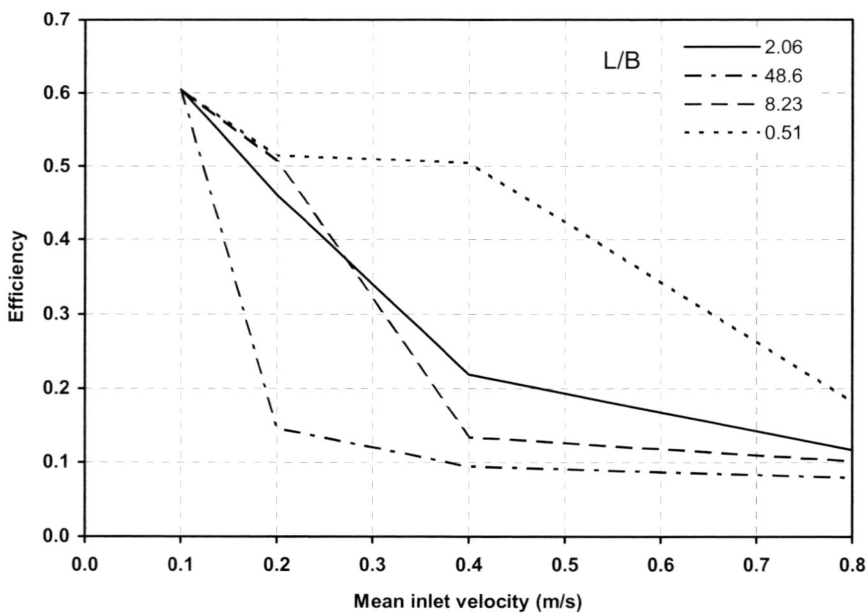

Figure 2.3.28. The effect of changes in the length (L) to breadth (B) ratio on chamber efficiency (Stovin and Saul, 1996).

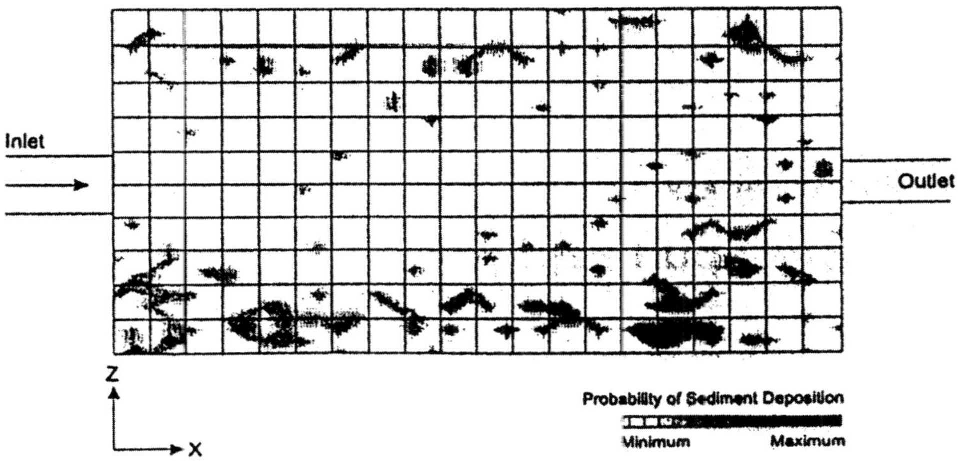

Figure 2.3.29. Probability of sediment deposition in a chamber predicted from stochastic tracking analysis (Stovin and Saul, 1998).

2.3.9 References

Ackers, P. (1984) Sediment transport in sewers and the design implications. *Int. Conf. on Planning, Construction, Maintainance and Operation of Sewerage Systems*, BHRA/WRC, Reading, England, 215–230.

Ackers, P. (1991) Sediment aspects of drainage and outfall design, *Proc. Int. Symp. on Environmental Hydraulics*, Hong Kong, December, 19–29.

Ackers, J.C., Butler, D. and May, R.W.P. (1996) *Design of Sewer to control Sediment Problems*. CIRIA project report 141. ISBN 086017 443 3. 181pp.

Ackers, P. and White, W.R. (1973) Sediment transport: new approach and analysis. *Journal of the Hydraulics Division*, ASCE, Vol. 99, No. HY11, 2041–2060.

Ahyerre, M. (1999) *Bilan et mécanismes de migration de la pollution organique en réseau d'assainissement unitaire*. PhD thesis, University Paris 6.

Ahyerre, M., Chebbo, G., Tassin, B. and Gaume, E. (1998) Storm water quality modelling: an ambitious objective ? *Wat. Sci. Tech.*, 37(1), 205–213.

Ahyerre M., Chebbo G. and Saad M. (2000) Sources and erosion of organic solids in a combined sewer. *Urban Water* N°2, 305–315.

Ali, K.H.M., O'Connor, B.A. and Nicholson, J. (1992) Theoretical and experimental study of fluid mud. *Proc. 5th Int. Symp. on River Sedimentation*, Karlsruhe.

Alvarez, E. (1990) *The influence of cohesion on sediment movement in channels of circular cross section*. PhD thesis, University of Newcastle upon Tyne, UK.

Arthur, S. (1996) Near bed solids transport in combined sewers. PhD thesis, University of Abertay Dundee, 218pp plus appendices.

Arthur, S. and Ashley, R.M. (1997) Near bed solids transport and first foul flush in combined sewers. *Wat. Sci. Tech.*, 37(1), 131–138.

Arthur, S., Ashley, R.M. and Nalluri, C. (1996) Near bed solids transport in sewers. *Wat. Sci. Tech.*, 33(9), 69–76.

Arthur, S., Ashley, R.M., Tait, S. and Nalluri, C. (1999) Sediment transport in sewers: a step towards the design of sewers to control sediment problems. *Proc. Inst. Civil Engineers, Water Maritime and Energy*, 136(1), 9–21.

Ashley, R.M., Arthur, S., Coghlan, B.P. and McGregor, I. (1994) Fluid sediment in combined sewers. *Wat.Sci.Tech.*, 29(1–2), 113–123.

Ashley, R.M. and Crabtree, R.W. (1992) Sediment origins, deposition and build-up in combined sewer systems. *Wat. Sci. Tech.*, 25(8), 1-12.

Ashley, R.M., Hvitved-Jacobsen, T. and Bertrand-Krajewski, J.-L. (1999) Quo vadis sewer process modelling? *Wat. Sci. Tech.*, 39(9), 9–22.

Ashley, R.M., Wotherspoon, D.J.J., Coghlan, B.P. and Ristenpart, E. (1993) Cohesive sediment erosion in combined sewers. *Proc. 6th Int. Conf. Urban Storm Drainage*, Niagara Falls, Vol. 1, 644–651.

Ashley, R.M. and Verbanck, M.A. (1996) Mechanics of sewer sediment erosion and transport. *J. Hydr. Res.*, 34(6), 753–769.

Ashley, R.M., Wotherspoon, D.J.J., Goodison, M.J., McGregor, I. and Coghlan, B.P. (1992) The deposition and erosion of sediments in sewers. *Wat. Sci. Tech.*, 26(5–6), 1283–1293.

Ashley, R.M., Fraser, A., Burrows, R. and Blanksby, J. (2000) The management of sediment in combined sewers. *Urban Water* 2 (2000), 263–275.

Bachoc, A. (1992a) *Le transfert des solides en réseaux d'assainissement unitaires.* Thèse de Doctorat de l'Institut National Polytechnique de Toulouse, 281 pp. + annexes.

Bachoc, A. (1992b) Location and general characteristics of sediment deposits into man-entry combined sewers. *Wat. Sci. Tech.,* 25(8), 47–55.

Bagnold, R.A. (1966) *An approach to the sediment transport problem from general physics.* Geological Survey Professional Paper 422-I, U.S. Department of the Interior, 37pp.

Batchelor, G.K. (1965) The motion of small particles in turbulent flow. *Proc. 2nd Austr. Conf. on Hydraul. and Fluid Mech.,* 19, University of Auckland.

Berlamont, J. and Torfs, H. (1996) Modelling (partly) cohesive sediment transport in sewer systems. *Wat. Sci. Tech.,* 33(9), 171-178.

Bertrand-Krajewski, J.-L., Briat, P. and Scrivener, O. (1993) Sewer sediment production and transport modelling: a literature review. *J. Hydr. Res.*, 31(4), 435–460.

Beyer, G. (1989) *Contribution à l'étude de l'érosion des dépôts en resau d'assainisement unitaire.* PhD thesis, University Louis Pasteur, Strasbourg, November.

Boon, A.G., Vincent, A.J. and Boon, K.G. (1998) Avoiding the problems of septic sewage. *Water, Science and Technology*, 37(1), 223–231.

Bouteligier, R., Vaes, G. and Berlamont, J. (2002a) Sensitivity of urban drainage wash-off models. *Journal of HydroInformatics*, 04(4), October 2002, 235–244.

Bouteligier, R., Vaes, G. and Berlamont, J. (2002b) In Sewer Sediment Transport Models. *Proceedings of 3rd International Conference on Sewer Processes and Networks*, Paris, France, 15–17 April 2002, 47–54.

Bouteligier, R., Vaes, G. and Berlamont J. (2002c) In Sewer Sediment and Pollutant Transport Models. *Proceedings of 9th International Conference on Urban Drainage*, Portland, Oregon, USA, 8-13 September.

Bouteligier, R., Vaes, G. and Berlamont, J. (2002d) Urban Drainage Quality Modelling – Case Study. *Proceedings of 3rd International Conference on Sewer Processes and Networks*, Paris, France, 15-17 April 2002, 263–270.

Bos, M.G. (ed.) (1989) *Discharge Measurement Structures*. ILRI publication 20.

Brownbill, V.R. (1992) *Tank Sediment Quality at Littleborough.* Internal report to WRc, School of Geography, University of Manchester.

Bujon, G. (1988) *Prévision des débits et des flux polluants transités par les réseaux d'égouts par temps de pluie.* Le modèle Flupol. *La Houille Blanche*, 1988, No 1, 11-23.

Camp, T.R. (1946) Sedimentation and the design of settling tanks. *Trans. ASCE*, Paper No. 2285, 111, 895–936.

Chebbo, G. (1992) *Solides des rejets pluviaux urbains: caractérisation et traitabilité.* Thèse de doctorat, ENPC, Paris, Mars 1992, 450pp.

Chebbo, G., Bachoc, A., Laplace, D. and Le Guennec, B. (1995) The transfer of solids in combined sewer networks. *Wat. Sci. Tech.*, 31(7), 95–105.

Chebbo, G., Musquère, P. and Bachoc, A. (1990) Solids transferred into sewers. *Proc. 5th Int. Conf. Urban Storm Drainage*, Osaka, Japan), 869–874.

Chebbo, G., Ashley, R.M. and Gromaire, M.C. (2002) The nature and pollutant role of solids at the water-sediment interface in combined sewer networks. *Water Sci. Tech.*, 47(4), 1–10.

Coghlan, B. (1995) *Solids transport in combined sewerage systems*. PhD thesis, University of Abertay Dundee, Scotland, 216pp. + annexes.

Coghlan, B.P., Ashley, R.M. and Jefferies, C. (1993) An appraisal of suspended sediment transport modelling methods for an interceptor sewer. *Wat. Sci. Tech.*, 27(5/6), 81-91.

Coleman, N.L. (1982) Discussion of paper 16313, *J. Hydraul. Div.*, ASCE, 108, Nr. HY1, 164–165.

Crabtree, R.W. (1989) Sediments in sewers. *J. Inst.Wat.Env.Man.*, 3(6), 569–578.

Crabtree, R.W., Ashley, R.M. and Gent, R. (1995) Mousetrap: modelling of real sewer sediment characteristics and attached pollutants. *Wat. Sci. Tech.*, 31(7), 43–50.

Daubert, A. and Lebreton, J.C. (1967) Etude experimentale sur modèle mathématique de quelques aspects des processus d'érosion des lits alluvionnaires, en régime permanent et non permanent. *Proc. 12th Congress of IAHR*, Fort Collins, Vol. 3.

Dauber, L. and Novak, B. (1983) *Quellen und Mengen der Schmutszstoffe in Regenabflussen einer stadtischen Mischkanalisation.* Eidgenossische Technische Hochschulen (ETH), Zürich. Separatum Nr. 927.

Davies, J.W. (1990) Laboratory study relating to the modelling of stormwater quality in combined sewers. *Proc. 5th ICUSD*, Niagara Falls, 1254–1259.

Davies, J.W, Butler, D., Small, J.L., Sekuloski, V. and Jefferies C. (1998) Gross solids transport and degradation. *Wat. Sci. Tech.*, 37(1), 61–68.

De Sutter, R. (2001) *Erosion and Transport of Cohesive Sediment Mixtures in Unsteady Flow.* PhD thesis, University of Gent, 283pp

De Sutter, R., Rushforth, P., Tait, S., Huygens, M., Verhoeven, R. and Saul, A.J. (2003) Validation of existing bed load transport formulae using in-sewer sediment. *Journal of Hydraulic Engineering*, ASCE, April, 325–333.

Durschlag, A., Hartel, L., Hartwig, P., Kaselow, M., Kollatsch, D., Otterpohl, R. and Schwentner, G. (1991) Total emissions from combined sewer overflow and wastewater treatment plants. *European Water Pollution Control*, 1(6), 13–23.

Einstein, H.A. (1942) Formulas for the transportation of bed load. *Trans. ASCE*, 107, 561–597, USA.

Einstein, H.A. (1950) *The Bed Load Function for Sediment Transportation in Open Channels.* Technical Bulletin 1026, USDA Soil Conservation Service, Washington D.C.

Ellis, D.R., (1992) *The Design of Storm Drainage Storage Tanks for Self Cleansing Operation*, PhD thesis, University of Manchester.

FWR (1998) *Urban Pollution Management Manual*, 2nd edition, Foundation for Water Research, Marlow.

Geiger, W.F. (1986) *Variation of Combined Runoff Quality and Resulting Pollutant Retention Strategies: Urban Storm Water Quality and Effects upon Receiving Waters*, TNO Committee on Hydrological Research, Proc. and Information No. 36, International Conference, Wageningen, The Netherlands, 71–91.

Gérard, C. and Chocat, B. (1998) An aid for the diagnosis of sewerage networks: analysis and modelling of the links between network's physical structure and risk of sediment build-up. *Wat.Sci.Tech.*, 39(9), 185–192.

Grass, A.J. (1974) Transport of fine sand on a flat bed: turbulence and suspension mechanics. *Euromech*, 48, 33–34.

Guymer, I., O'Brien, R., Mark, O. and Dennis, P. (1998) An investigation of fine sediment mixing within free-flowing and surcharged manholes. *Wat. Sci. Tech.*, 37(1), 215–222.

Hager, W.H. (1994) *Abwasserhydraulik: Theorie und Praxis*. Springer-Verlag, Berlin, Heidelberg, New York.

Harrison, A.J.M. and Holmes, D.W. (1967) *The Movement of Stormwater in Combined Sewers.* Report No. INT 73, Hydraulics Research Station, Wallingford, UK.

Harwood, R. and Saul, A.J. (1996) CFD and novel technology in combined sewer overflow. *Proc. 7th ICUSD*, Hannover, Germany, 1025–1030.

Holzer, P. and Krebs, P. (1998) Modelling the total ammonia impact of CSO and WwTW effluent on the receiving water. *Wat.Sci.Tech.*, 38(10).

Hrissanthou, V. and Hartmann, S. (1998) Measurements of critical shear stress in sewers. *Wat. Res.*, 32(7), 2035–2040.

Huber, W.C. (1986) *Deterministic Modelling of Urban Runoff Quality. Urban runoff pollution.* NATO ASI series, Vol. G10, Springer Verlag, Berlin, 167–242.

Irmay, S. (1960) Accelerations and mean trajectories in turbulent channel flow. *Transactions ASME*, December, 961-972.

Jack, A.G., Petrie, M.M. and Ashley, R.M. (1996) The diversity of sewer sediments and the consequences for sewer flow quality modelling. *Wat. Sci. Tech.*, 33(9), 207–214.

Jefferies, C. (1992) Methods for estimating the discharge of gross solids from combined sewer systems. *Wat. Sci. Tech.*, 26(5/6), 1295–1304.

Jefferies, C. (1993) *The Performance of Certain Combined Sewer Overflows with Storage.* PhD thesis, Dundee Institute of Technology, 183 pp.

Jefferies, C. and Ashley R.M. (1994) Gross solids in sewer systems: temporal and catchment based relationships. *Wat. Sci. Tech.*, 30(1), 63–71.

Kamphuis, J.W. (1990) Influence of sand or gravel on the erosion of cohesive sediment. *J. Hydr. Res.*, 28(1), 43–53.

Kazemipour, A.K. and Apelt, C.J. (1979) Shape effects on resistance to flow in open channels. *J. Hydr. Res.*, 17(2), 129–147.

Kleijwegt, R.A. (1992) *On Circular Transport in Circular Sewers with Non-Cohesive Deposits.* Ph.D. thesis, Delft University of Technology, The Netherlands.

Kline, S.J., Kim, H.T. and Reynolds, W.C. (1971) The production of turbulence near a smooth wall in a turbulent boundary layer. *J. Fluid Mech.*, 50(1), 133–160.

Krejci, V., Dauber, L., Novak, B. and Gujer, W. (1987) Contribution of different sources to pollutant loads in combined sewers. *Proc. 4th Int. Conf. Urban Storm Drainage*, Lausanne, 34–39. Edited by B.C. Yen, pub. Secretariat of XXII Congress IAHR Laboratoire d'hydraulique, Ecole Polytechnique Fédérale Lausanne, Switzerland.

Lafond, J.-M. (1995) *Comparaison de modèles de transport en suspension. Application à des oeuvrages de stockage-dépollution.* PhD thesis, Cemagref, Lyon, 240 pp.

Langeveld, J.G., Clemens, F.H.L.R. and van der Graaf, J.H.J.M. (2003) Increasing wastewater system performance: the importance of interactions between sewerage and wastewater treatment. *Wat.Sci.Tech.*, 45(3), 45–52.

Laplace, D., Sanchez, Y. and Dartus, D. (1990) Trunk sewers clogging development observation and first interpretations. *Wat. Sci. Tech.*, 22(10/11), 259–266.

Laplace, D. (1991) *Dynamique du dépôt en collecteur d'assainissement.* PhD thesis, Institut National Polytechnique de Toulouse, 202 pp. + annexes.

Laplace, D., Bachoc, A., Sanchez, Y. and Dartus, D. (1992) Trunk sewer clogging development: description and solutions. *Wat. Sci. Tech.*, 25(8), 91–100.

Larson, M., Berndtsson, R., Hogland, W., Spanberg, A. and Bennerstedt, K. (1990) Field measurements and mathematical modelling of pollution build-up and pipe deposit wash-out in combined sewers. *Proc. 5th Int. Conf. Urban Storm Drainage*, Osaka, Japan, 325–332.

Lessard, P. and Beck, M.B. (1991) Dynamic simulation of storm tanks. *Water Research*, 25(4), 375-391.

Lin, H. (1993) *Le transport solide en collecteurs unitaires d'assainissement et sa modélisation.* PhD thesis, ENPC, Paris, 281 pp. + annexes.

Lin, H.S., LeGuennec, B., Dartus, D. and Bachoc A. (1993) Bed load transport in the No.13 sewer trunk of Marseille. *Proc. 6th Int. Conf. Urban Storm Drainage*, Niagara Falls.

Lin, H.S. and LeGuennec, B. (1996) Sediment transport modelling in combined sewer. *Wat. Sci. Tech.*, 33(9), 61–67.

Littlewood, K. (2000) *Movement of Gross Solids in Small Diameter Sewers*, unpublished PhD thesis, Imperial College, University of London.

Littlewood, K. and Butler, D. (2002) Movement mechanisms of gross solids in intermittent flow. *Proc. 3rd International Conference on Sewer Systems and Processes*, Paris, April, 13–20.

Lorenzen, A., Willms, M. and Dinkelacker A. (1992) The Gottingen Boat. *Wat. Sci. Tech.*, 28(8), 37-46.

Macke, E. (1982) *Über den Feststofftransport bei niedrigen Konzentrationen in teilgefüllten Rohrleitungen.* Mitteilungen des Leichtweiß-Instituts für Wasserbau der TU Braunschweig, Heft 76, 1-151.

Margetts, J. (2000) Sewer sediment modelling: dishing the dirt. *Proc. Wastewater Planner Users Group*, Blackpool, UK, Nov.

Mark, O. (1995) *Discussion of Numerical Modelling Approaches for Sediment Transport in Sewers*. PhD thesis, University of Aalborg, Denmark.

Mark, O., Appelgren, C. and Larsen, T. (1995) Principles and approaches for numerical modelling of sediment transport in sewers. *Wat. Sci. Tech.*, 31(7), 107–115.

Mark, O., Hernebring, C. and Magnusson, P. (1998) Optimisation and control of the inflow to a wastewater treatment plant using integrated modelling tools. *Water, Science and Technology*, 37(1), 347–354.

Markofsky, M. (1980) *Strömungsmechanische Aspekte der Wasserqualität.* Schriftenreihe GWF Wasser, Abwasser, Heft 18, ISBN 3-486-24251-2, Oldenbourg Verlag, München.

May, R.W.P. (1982) *Sediment Transport in Sewers.* HR Wallingford, Report IT 222.

May, R.W.P. (1994) Transport of sediments in pipes: application to design of self-cleansing sewers. *European Water Pollution Control*, 4(5).

May, R.W.P., Ackers, J.C., Butler, D. and John, S. (1996) Development of design methodology for self-cleansing sewers. *Wat. Sci. Tech.*, 33(9), 195–206.

Mayerle, R., Nalluri, C. and Novak, P. (1991) Sediment transport in rigid bed conveyances. *Journal of Hydraulic Research*, IAHR, 29(4), 475–495.

McGregor, I., Ashley, R.M. and Oduyemi, K.O.K. (1993) Pollutant release in sewer systems and their release into receiving waters. *Wat. Sci. Tech.*, 28(8–9), 161–169.

McTigue, D.F. (1981) Mixture theory for suspended sediment transport. *Journal of Hydraulics Division, ASCE*, Vol. 107, No. HY6, 659–673.

Mehta, A.J. (1988) Laboratory studies on cohesive sediment deposition and erosion. *Physical Processes in Estuaries*, edited by J. Dronkers and W. Van Lewsen, Springer-Verlag.

Mehta, A.J., Lee, S.-C. and Li, Y. (1994) *Fluid Mud and Water Waves: A brief review of interactive processes and simple modeling approaches.* Report UFL/COEL/MP-94/01, Coastal and Oceanographic Eng. Dept., University of Florida, Gainesville.

Mehta, A.J. and Parthenaides E. (1982) Resuspension of deposited cohesive beds. *Proc. 18th Conf. Coastal Engineering*, 2, 1569–1588.

Meyer-Peter, E. and Müller, R. (1948) *Formulas for Bed-Load Transport.* IAHR-Meeting Stockholm, Zahlentabellen zu den Geschiebetriebsversuchen der Versuchsanstalt für Wasserbau und Erdbau an der ETH Zürich, unpublished supplement.

Michelbach, S., Striebel, T. and Wöhrle, C. (1992) Absetzbarkeit von schwermetallbelasteten Fest-stoffen im Mischwasser, *GWF Das Gas- und Wasserfach (Wasser/Abwasser)*, 133(8), 404–410.

Migniot, C. (1968) Etude des propriétés physiques de différents sediments tres fins et de leur comportement sous des actions hydrodynamiques. *La Houille Blanche*, 1, 591–620.

Migniot, C. (1989) Tassement et rhéologie des vases. *La Houille Blanche*, 1–2, 11–28, 95–111.

Milne, D.A., Jefferies, C. and Ashley, R.M. (1995) Pollutional aspects of gross solids and their interaction with sewer sediments. *Wat. Sci. Tech.*, 33(9), 31–37.

Mitchener, H. and Torfs, H. (1996) Erosion of mud/sand mixtures. *J. Coast. Eng.*, 29 (1–2) 1–25.

Myers, W.R.C. (1982) Flow resistance in wide rectangular channels. *J of Hydr. Div., ASCE*, 108(4), April, 471–482.

Nalluri, C., Ab Ghani, A. and El-Zaemey, A.K.S. (1994) Sediment transport over deposited beds in sewers. *Wat. Sci. Tech.*, 29(1–2), 125–133.

Nalluri, C. and Alvarez, E.M. (1992) The influence of cohesion on sediment behaviour. *Wat. Sci. Tech.*, 25(8), 151–164.

Nalluri, C. and Novak, P. (1973) Turbulence characteristics in a smooth open channel of circular cross section. *J. of Hydr. Res.*, 11(4), 343–368.

Nalluri, C. and Novak, P. (1975) Turbulence characteristics in smooth fixed bed channels. *Proc. XVI IAHR Congress*, Sao Paulo, Brazil.

Novak, P. and Nalluri, C. (1984) Incipient motion of sediment particles over fixed beds, *J. of Hydraulic Research*, IAHR, 22(3), 181–197.

O'Brien, M.P. (1933) Review of the theory of turbulent flow and its relation to sediment transportation. *Transactions AGU*, 27–29 April, 487–491.

Oms, C., Gromarie-Mertz, M.-C. and Chebbo, G. (2002) Observation of sediment using an endoscope. *Proc. 3rd International Conference on Sewer Systems and Processes*, Paris, April, 5–12.

Ota, J.J. (1999) *Effect of Particle Size and Gradation on Sediment Transport in Storm Sewers*. PhD thesis, University of Newcastle upon Tyne, UK.

Otsubo, K. and Muraoka, K. (1988) Critical shear stress of cohesive bottom sediments. *J. Hydr. Engg. ASCE*,114(10), 1241–1256.

Parchure, T.M. and Mehta, A.J. (1985) Erosion of soft cohesive sediment deposits. *J. Hydr. Engg. ASCE*, 111(10), 1308–1326.

Perrusquía, G. (1988) *Part-full flow in pipes with a sediment bed.* Chalmers University of Technology, Department of Hydraulics, Report A:17, Göteborg, Sweden.

Perrusquía, G. (1990) Flow resistance in storm sewers with a sediment bed. *5th International Conference on Urban Storm Drainage*, Osaka, Japan, 873–878.

Perrusquía, G. (1991) *Bedload Transport in Storm Sewers*. PhD thesis, Report A:22, Chalmers University of Technology, Department of Hydraulics, Göteborg, Sweden.

Perrusquía, G., Lyngfelt, S. and Sjöberg, A. (1987) Flow capacity of sewers with a sediment bed. *4th International Conference on Urban Storm Drainage*, Lausanne, Switzerland, 331–336. Ed Yen B C. pub. Secretariat of the XXII Congress IAHR Laboratoire d'hydraulique Ecole Polytechnique Federale Lausanne, Switzerland.

Perrusquía, G. and Nalluri, C. (1995) Modelling of bedload transport in pipe channels. *Proc. Int. Conf. on the Transport and Sedimentation of Solid Particles*, Prague.

Perrusquía, G., Pedersen, O. and Larsen, T. (1995) Influence of sewer sediments on flow friction and shear stress distribution. *Wat. Sci. Tech.*, 31(7), 117–126.

Pisano, W.C. (1996) Summary: United States 'sewer solids' settling characterization methods, results, use and perspective. *Wat. Sci. Tech.*, 33(9), 109–115.

Pisano, W.C. and Queiroz, C. (1977) *Procedures for Estimating Dry Weather Deposition in Sewerage Systems*. US EPA Report EPA-600/2-77-1, 20 July.

Pisano, W.C., Connick, D., Queiroz, C. and Aronson, J. (1979) *Dry Weather Deposition and Flushing for Combined Sewer Overflow Pollution Control*. EPA 600/2-79-133.

Raudkivi, A.J. (1990) *Loose Boundary Hydraulics.* 3rd edition, Pergamon.

Read, G.F. (ed.) (1997) *Sewers: Rehabilitation and new construction (Repair and renovation).* Arnold, London.

Reiff, H. (1992) *Wachstum und Abtrag der Sielhaut in Mischwasserkanälen.* Heft 9 der Schriftenreihe Wasser – Abwasser – Abfall, Verein zur Förderung der Fachgebiet-siedlungswasserwirtschaft an der Universität Kassel.

Ristenpart, E. (1995) *Feststoffe in der Mischwasserkanalisation: Vorkommen, Bewegung und Verschmutzungpotential*, PhD thesis, University of Hannover, Germany.

Ristenpart, E. (1996) Modelling of sediment erosion in a combined sewer. *Proc. 7th Int. Conf. Urban Storm Drainage*, Hannover, Germany, 851–856.

Ristenpart, E., Ashley, R.M. and Uhl, M. (1995) Organic near-bed fluid and particulate transport in combined sewers. *Wat. Sci. Tech.*, 31(7), 61–68.

Ristenpart, E. and Uhl, M. (1993) Dynamic behaviour of sewer sediments. *Proc. 6th Int. Conf. Urban Storm Drainage*, Niagara Falls, Canada, 748–753.

Rouse, H. (1937) Modern conceptions of the mechanics of sediment suspension. *Transactions, ASCE*, 102, 463–543.

Rubey, W.W. (1933) Settling velocities of gravel, sand and silt particles. *American Journal of Science*, 5th series, 25(148), 325–338.

Rushforth, P.J. (2001) *The Erosion and Transport of Sewer Sediment Mixtures*. PhD thesis, University of Sheffield, UK.

Rushforth, P J., Tait, S J. and Saul, A.J. (2003) Modeling the erosion of mixtures of organic and granular in-sewer sediments. *Journal of Hydraulic Engineering*, ASCE / April, 308–315.

Saul, A.J., Skipworth, P.J., Tait, S.J. and Rushforth, P.J. (2003) Movement of total suspended solids in combined sewers. *Journal of Hydraulic Engineering*, 129(4), April, 298–307.

Schmidt, W. (1925) *Der Massenaustauch in freier Luft und verwandte Erscheinungen.* Probleme der kosmischen Physik, Band 7, Hamburg, Germany.

Schütze M., Butler, D. and Beck, M.B. (2002) *Modelling, Simulation and Control of Urban Wastewater Systems*. Springer Verlag, London.

Shields, A. (1936) *Anwendung der Ähnlichkeitsmechanik und der Turbulenzforschung auf die Geschiebebewegung.* Mitteilungen der Preußischen Versuchsanstalt für Wasserbau und Schiffbau, H. 26.

Skipworth, P.J. (1996) *The Erosion and Transport of Cohesive-Like Sediment Beds in Sewers*. PhD thesis, University of Sheffield, UK.

Skipworth, P.J., Tait, S. and Saul, A.J. (1996) Laboratory investigations into cohesive sediment transport in pipes. *Wat. Sci. Tech.*, 33(9), 187–193.

Stotz, G. and Krauth, K.H. (1986) Deposition in combined sewers and their flushing behaviour. *Proc. Int. Symp. on Comparison of Urban Drainage Models with Real Catchment Data*, UDM 1986, Maksimovic, Dubrovnik, 443–453.

Stovin, V.R. (1996) *The Prediction of Sediment Deposition in Storage Chambers Based on Laboratory Observations and Numerical Simulation.* PhD thesis, University of Sheffield, UK.

Stovin, V.R. and Saul, A.J. (1994) Sedimentation in storage tank structures. *Wat. Sci. Tech.*, 29(1–2), 363–372.

Stovin, V.R. and Saul, A.J. (1996) Efficiency prediction for storage chambers using computational fluid dynamics. *Wat. Sci. Tech.*, 33(9), 163–170.

Stovin, V.R. and Saul, A.J. (1998) A computational fluid dynamics (CFD) particle tracking approach to efficiency prediction. *Wat. Sci. Tech.*, 37(1), 285–293.

Sumer, B.M. and Deigaard, R. (1979) *Experimental Investigation of Motions of Suspended Heavy Particles and the Bursting Process.* Series paper No. 23, Inst. of Hydrodyn. and Hydraul. Eng., Tech. Univ. Denmark.

Sutherland, A.J. (1992) Hiding functions to predict self-armouring. *Proc. Int. Seminar on Grain Sorting*, Mitteilungen 117, ETH Zurich, 273–298.

Swaffield, J.A. and Galowin, L.S. (1992) *The Engineered Design of Building Drainage Systems*. Ashgate.

Swaffield, J. and McDougall, J.A. (1996) Modelling solid transport in building drainage systems. *Wat. Sci. Tech.,* 33(9), 9–16.

Techniques Sciences Méthodes (1993) Special Issue: *Les solides en réseaux d'assainissement*. October, ISSN 0299-7258.

Tait, S., Rushforth, P.J. and Saul, A.J. (1998) A laboratory study of the erosion and transport of cohesive-like sediment mixtures in sewers. *Wat. Sci. Tech.*, 37(1), 163–170.

Tait, S.J., Marion, A. and Camuffo, G. (2003) Effect of environmental conditions on the erosional resistance of cohesive sediment deposits in sewers. *Water Sci.Tech.*, 47(4), 27–34.

Torfs, H. (1993) Erosional behaviour of partly cohesive, partly non-cohesive sediments in uniform flows. *Proc. 6th. Int Conf. Urban Storm Drainage*, Niagara Falls, Canada, ISBN 1-50056-253-3, 754-759.

Torfs, H. (1995) *Erosion of mud/sand mixtures.* PhD thesis, Katholieke University of Leuven.

Vanoni, V. and Brooks, N. (1957) *Laboratory Studies of the Roughness and Suspended Load of Alluvial Channels.* California Institute of Technology, Report E-68, USA.

Van Rijn, L.C. (1989) *Sediment Transport by Currents and Waves*. Delft Hydraulics Report H 461.

Velikanov, M.A. (1954) Gravitational theory for sediment transport. *Academy of Science Bulletin*, USSR, Geophysics, 4, 349–359 (GTS Translation No. 62-15004).

Verbanck, M.A. (1990) Sewer sediment and its relation with the quality characteristics of combined sewer flows. *Wat.Sci.Tech.*, 22(11), 247–257.

Verbanck, M.A. (1993) Identification of the depth-dependent transportation of particulate solids in dry-weather sewage flows. *Proceedings of 6th International Conference on Urban Storm Drainage*, edited by J. Marsalek and H.C. Torno, Niagara Falls, Canada, 742–747.

Verbanck, M.A. (1995a) Capturing and releasing settleable solids: the significance of dense undercurrents in combined sewer flows. *Wat. Sci. Tech.*, 31(7), 85–93.

Verbanck, M.A. (1995b) *Transferts de la charge particulaire dans l'égout principal de la ville de Bruxelles*. PhD thesis, Université Libre de Bruxelles, May 1995, 193pp.

Verbanck, M.A. (2000) Computing near-bed solids transport in sewers and similar sediment-carrying open-channel flows. *Urban Water*, 2(4), 277–284.

Verbanck, M.A. and Ashley, R.M. (1993) International workshop on origin, occurrence and behaviour of sediments in sewer systems: outline of technical conclusions. *Wat.Sci Tech.*, 27(12), 173–176.

Verbanck, M., Ashley, R.M. and Bachoc, A. (1994) International workshop on origin, occurrence and behaviour of sediments in sewer systems: summary of conclusions. *Wat. Res.,* 28(1), 187–194.

Water Research Centre (WRc) (1994) *Sewerage Rehabilitation Manual*. 3rd edition

Williams, P.R. and Williams, D.J.A. (1989) Rheometry of concentrated cohesive sediments. *J. Coastal Res.*, Special Issue No. 5.

Wöhrle, C. and Brombach, H. (1991) Probenahme im Abwasserkanal. *Wasserwirtschaft*, 81(2), 60–65.

Wotherspoon, D.J.J. (1994) *The movement of cohesive sediment in a large combined sewer*. PhD thesis, Dundee Institute of Technology, UK.

Wotherspoon, D.J.J. and Ashley, R.M. (1992) Rheological measurements of the yield strength of combined sewer sediment deposits. *Wat. Sci. Tech.*, 25(8), 165–170.

3

The effects of sewer solids on the performance of wastewater systems

3.1 OUTLINE OF THE EFFECTS OF SEWER SOLIDS

The presence of solids in sewers can cause a variety of problems, as illustrated in Table 3.1. Many reported problems, however, have limited supporting evidence and may be regarded as anecdotal. Deposits cause blockages which may eventually lead to undesirable surcharging and even flooding during both dry weather and storm conditions, whilst microbial processes contribute to the production of hydrogen sulphide, causing odours, dangers to sewer workers and corrosion, as well as acute and chronic pollutant effects in receiving waters. Deposits in manholes, tanks and blinding, and partial blockage of screens at CSOs and treatment plants, cause more localised problems. Large organic and inorganic gross solids also escape from sewers, even when screens are installed and cause both aesthetic and (potential) health risks when deposited in receiving environments. The problems may be summarised as follows.

1. Accumulation of sediments in sewers inhibits the designed conveyance of flow; this has two effects:
   - surcharging of the sewer causing premature operation of CSOs;
   - increases the frequency of flooding.
2. Sewer solids can overwhelm 'control' systems such as screens or treatment plants when washed out.
3. Sediments act as stores/accumulators of pollutants due to two processes:
   - gradual accumulation with time;
   - provision of the opportunity for biochemical activity within the deposit.

© 2004 IWA Publishing. *Solids in Sewers*. Edited by R.M. Ashley, J.-L. Bertrand-Krajewski, T. Hvitved-Jacobsen and M. Verbanck.. ISBN: 1 900222 91 4.

In the latter case, the problems occur due to acute in-sewer pollutant build-up, releasing gases for example, and when re-eroded, by impact on either the environment via direct discharge, or as a shock load at a treatment plant.

Sewers should be designed so that the flow velocities are sufficient either to convey sediments through the system or to re-erode any deposits. Because design and construction practices in the past were based on limited knowledge, existing sewers suffer extensively from deposition. As a consequence there is also a need to define appropriate maintenance practices to minimise the blockage and pollutant build up problems which occur. Current knowledge, however, is also limited, and the design of 'sediment free' sewers is generally not possible, as sedimentation occurs due to the inevitable intermittent and variable nature of the flows and the inputs to sewer systems and also due to the spatial variations in sewerage layouts (Section 2.3). The treatment of sewer solids, the impact of solids on wastewater treatment plants (WWTP), tanks and CSOs and their disposal are thus of primary importance in sewage handling and sewerage design practices.

Table 3.1. Effects of sewer solids on wastewater and other systems

Effect caused by solids	Description of solids cause
Reduction in hydraulic capacity, increase in surcharging, flooding	Deposition of solids in inverts, permanent or semi-permanent
Blockage	Deposition in inverts, build-up on walls (progressive or sudden)
Premature operation of CSOs	As above (downstream of CSO)
Enhanced pollutant washout from CSOs	All types of solids, and associated pollutants during wet weather
Gases, odours, explosions	Generated from biological degradation in bed deposits (hydrogen sulphide, methane and other odorous substances)
Sewer corrosion	As above, in moist atmosphere
Pump impeller abrasion	Inorganic solids in flow (typically washed through system in wet weather)
Screen blockages and damage	Large solids (inorganic and organic)
Shock loads to treatment plants	Foul flushes and bed erosion, releasing both solids and associated pollutants
Rodents (rats)	Source of food (organic solids)
Health risks to sewer workers	Increased hazards, infections: gases (asphyxiation, toxicity), rodents (disease transmission)
	Access and maintenance problems increased by solids presence
Fat and grease deposits – can reduce capacity or get washed out in 'lumps'	Build up on sewer walls, particularly around ambient surface levels, can also develop into balls

3.2 EFFECTS OF SOLIDS ON CONVEYANCE AND BLOCKAGE OF SEWERS

Sediment deposition is considered in detail in Section 2.3.4.5.1, and the effects of sediments on hydraulic resistance in Section 2.3.2.3; this Section therefore deals with the consequences of deposition.

3.2.1 Hydraulic resistance changes caused by sewer solids

The evaluation of hydraulic resistance to flow is reviewed by Yen (2002). The presence of solids in the invert or attached to the walls of a sewer (slimes, biofilms) changes the boundary resistance to the flow; this has been known for some time (e.g. Horton, 1901). The effect of the solids is usually to increase the relative roughness and reduce the flow capacity. However, wall sliming may reduce the resistance to flow in smooth-walled pipes. For example, Perkins and Gardiner (1982) demonstrated that the roughness of various pipes, once a slime layer had been established in a few days, was prone to decrease, albeit temporarily, as the slime layer sloughed off from time to time. Several studies (e.g. Ackers, 1961; May, 1982; Perkins and Gardiner, 1982; Henderson, 1984a) have investigated the effects of solids on roughness in laboratory scale prototype sewers and from field data. Those by Henderson (1984a) and May (1982) have been used for the UK Sewerage Rehabilitation Manual (Water Research centre, 2001) as guidance for the selection of sewer boundary roughness for hydraulic modelling purposes. The guidance is based on the equivalent k_s, as used in the Colebrook-White equation.

The important factors for solids effects on hydraulic resistance are (Henderson, 1984b):

- flow regimes – flow depths and velocity ranges;
- sewer pipe material and initial wall roughness when new;
- age and condition for used pipes and brick sewers;
- presence of sediment, and bed forms.

There is evidence that the pipe-full hydraulic roughness in pipes with fully developed biofilms is a function of the maximum daily velocity in the pipe. With k_s ranging from 1-6 mm at a maximum velocity of about 0.6 m/s down to 0.5–1.5mm at 1.7 m/s. In fact k_s will vary with flow and flow depth, being usually largest at lower depths (proportional depths of 0.2–0.4). The sewer would be expected to have differing boundary roughness characteristics at points on the circumference.

Where there are permanent or semi-permanent bed deposits, the occurrence of bedforms is important due to 'form roughness' as well as the sediment 'grain roughness' (May, 1993), as illustrated in Figure 3.1. Where the sediment bed is flat, Henderson (1984b) suggests:

$$k_s \approx 1.5 d_{65} \text{ (mm)} \tag{3.1}$$

where d_{65} (mm) is bed particle size for which 65% of the sample is finer.

Where there is part-full flow, the presence of bedforms is more likely, particularly where the Froude number is 0.25-0.5 (see also Figure 2.3.6). In the presence of bedforms May (1993) suggests the use of a simple relationship for the equivalent bed roughness k_b for pipe-full conditions:

$$k_b = 2.41\, D^{0.61} d_{50}^{\;0.39} \text{ (mm)} \tag{3.2}$$

where D is pipe diameter
d_{50} is median particle size

Equation (3.2) is pessimistic as it presumes the presence of bed forms.

Using the value of k_b, from Equation (3.2), the bed friction factor may be determined from the Colebrook-White equation (rough turbulent flow). A composite roughness may then be determined by weighted summation of the pipe boundary (bed and walls) based on the distribution of wetted perimeter.

Figure 3.1. Regular bedforms in a Dutch stormwater sewer.

3.2.2 Small sewers

Full pipe blockage is a feature particularly characteristic of small sewers at the extremities of sewer networks, and in these parts of main sewer networks (typically defined as pipes < 150 mm in size) where the flows are similar to those found in building drainage systems. Flow is intermittent, with pipes possibly empty for considerable periods of time. Even when flowing, the pipes are likely to be carrying water at only a shallow depth. Close to the point of connection of the individual dwelling to the sewer, the discharge is often in the form of a wave, especially from the WC. It is at these points that the depth of flow is greatest. As the flush wave travels along the pipe, its velocity and height decrease; this attenuation depends on the gradient, pipe diameter and roughness as well as the occurrence of simultaneous discharges from other appliances connected to the system. Faeces, toilet paper and other sanitary refuse transported in the water do not necessarily progress steadily through the whole length of the drain. If the velocity and depth of water drop below certain limits, the solids in transport will stop until the next discharge, or series of discharges, when further progress along the pipe should occur (Swaffield and Gallowin, 1992).

If, for example, the pipe or pipe joint is rough or the number of discharges are infrequent allowing drying out and adhesion of the material to the pipe wall, or the flow depth is small, then the solids may not be transported by following discharges and additional solids may further accumulate around the initial deposit. In this way a blockage may start to develop, trapping water upstream of the solid congestion. If the water discharge upstream of the blockage is greater than the rate of leakage past the deposit then the head of water will build up until eventually the material is moved, or the flow is ejected from a manhole. As it moves through the system, the solid will either be degraded into more easily transported sizes or will move into a sewer with sufficient continuous flow to be carried in suspension or along the bed. This behaviour model presumes therefore that 'mini-blockages' or stoppages occur regularly in small sewers with intermittent flows, but where constructional and flow conditions are adequate (that is, as designed) these stoppages do not develop into major problems but are dispersed (Lillywhite, 1976).

3.2.2.1 Blockage formation

Lillywhite and Webster (1979) conducted a field study of blockages and their apparent causes in 100 mm and 150 mm drains. They found that defective pipe joints and manholes were the major triggers of blockage formation (Table 3.2). It was suggested that these defects were generally due to poor construction. Defects of this type tend to restrain larger solids such that the build-up of head behind the solid does not cause movement, even after the coalescence of several solids, thereby leading ultimately to full blockage.

An early study in the UK (Institution of Sanitary Engineers, 1954) tried to establish the types of solids associated with such blockages. At that time some 37 % of blockages were found to be associated with sanitary towels.

An interesting feature of Lillywhite and Webster's (1976) study is their finding that blockage did not seem to be influenced by sewer gradient (Table 3.3). For example, one drain surveyed was laid at a gradient of 1:1200, with no history of blockages. Similarly, drains in very hilly regions were also not particularly associated with high incidences of blockage. Lillywhite (1976) found that most of the drains prone to blockage were very lightly loaded hydraulically. Very few recurring blockages were found in systems constructed with modern materials and jointing methods.

Table 3.2. Summary of features of small diameter drains subject to repeated blockages (adapted from Wise and Swaffield, 1995)

Feature observed	Sample (%)
Faulty joints	30
Broken pipes	10
Backfalls	11
Deposits, root present	24
Manhole faults	23
Others	2
Total	100

Table 3.3. Proportion of surveyed 100 mm clayware drains subject to repeated blockages (adapted from Wise and Swaffield, 1995)

Gradient	Drains blocking (%)
0.050-0.030	3
0.030-0.014	11
0.014-0.002	4

3.2.2.2 Blockage frequency

The estimation of blockage rates in small-bore sewers is difficult due to the lack of comprehensive records. In an Institution of Sanitary Engineers (1954) report, average blockage levels in Local Authority owned housing areas was 8.8 blockages/100 houses/year. Assuming an occupancy of 3.5, this amounts to 0.025 blockages/capita/year, with the highest rate up to 0.06 blockages/capita/year. A more recent study in Scotland (Stokes and

Morris, 1996) suggests widely varying figures, depending on location, but with the highest rates noted as 0.15–0.2 blockages/capita/year. Unpublished data from various UK locations gives rates, expressed in terms of length of sewer, as typically 0.2–0.4 blockages/km/year.

3.2.3 Large sewers

In large sewers the mechanisms of blockage formation and the implicated solids are quite different to small sewers. The most troublesome solids in large sewers tend to be the larger inorganic sediments, typically of surface origin (Section 2.1). The nature of the problems induced by sediment blockage are described in Section 3.2.1.3.2.

Due to the time-varying nature of the flow in urban drainage systems it is not possible to prevent the deposition of sediment. In well designed and maintained systems, flow velocities are attained at frequent intervals which are sufficient to re-erode deposited sediment and transport it either to the outfall of the system, or further downstream. However, if the rate of input of the sediment increases or the hydraulic conditions deteriorate, the flow may not be able to transport the sediment through the system and 'permanent' deposition will occur.

3.2.3.1 Blockage formation

If an initially clean sewer flowing part-full is subjected to a flow with a sediment load, but conditions are not sufficient to prevent deposition, the development of a sediment bed will increase the resistance and cause the depth of flow to increase and the velocity to decrease. There are two possible conditions which may develop as a result (Butler *et al.*, 1996).

1 The reduction in velocity may cause a reduction in the sediment-transporting capacity of the flow and lead to further deposition. Ultimately, the depth of sediment may increase until the pipe is flowing full and under pressure; if sufficient head is available, an equilibrium would finally be reached when the rate of sediment transport in the partially blocked pipe matches the rate of sediment input.
2 The presence of the deposited sediment bed may allow the flow to acquire a greater capacity for transporting sediment which more than compensates for the reduction in velocity caused by the roughness of the deposited bed. Ultimately, the increased deposited bed width and the increased sediment transport capacity which is associated with it may reach an equilibrium with the incoming sediment load and prevent further deposition.

The first condition has been observed in early laboratory studies (Laursen, 1956) and is consistent with certain field observations of progressive blockage to the point of surcharge. During more recent measurements under a limited range of conditions, however, May *et al.* (1989) have observed the second condition. It is evident that the occurrence of either condition will be influenced by the degree to which the sediment load exceeds the transporting capacity of the flow, and various hydraulic considerations, such as the ratio of the discharge to the pipe-full capacity.

3.2.3.2 Extent of problems

The extent of problems due to blockage by sediment deposition varies widely. A survey in the UK (CIRIA, 1986) found that, amongst the authorities contacted, almost 80 % acknowledged having sewers containing sediment. One-third of the areas surveyed had more than 25 % of the sewers in their systems containing sediment. Over 45 % of the systems were regularly surcharged during non-storm periods and 95 % surcharged during storm conditions. At the time, the sedimentation problems were affecting 25,000 km of the UK sewers, at an estimated annual cost of £60 million. Generally, however, very little quantitative data exists for the costs associated with sewer sediment deposits.

A study in Sweden (Perrusquía *et al.*, 1987) showed that the reduction in hydraulic capacity in sewers containing sediment can be considerable, reducing by more than 50 % even for low levels of sediment (20 % of pipe diameter) as illustrated in Figure 2.3.3. Figure 3.2 gives a similar illustration of the significant effect of deposits on flow capacity using data provided from CIRIA (1986).

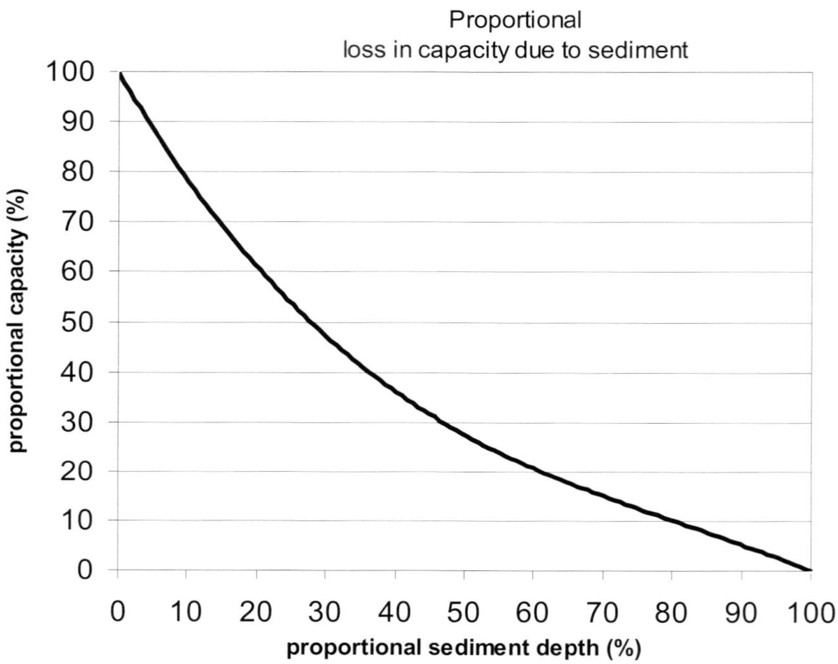

Figure 3.2. Illustrative example of effect of sediment deposits on pipe-full capacity.

In a review of the management of flooding from sewers, May *et al.* (1998) identified that proactive sewer cleaning was a low-cost option. Nonetheless, many sewer operators do not conduct programmed cleaning operations, responding only reactively to problems caused by blockages. Problems occur due to the diversity of the solids deposits, from large organics and inorganics to fine clay and organic particles. Subsequent methods of removal (Section 4), require to be diverse to deal with this variety and disposal may be problematic in terms of the waste designation as 'special' or 'hazardous'.

Problems are different in developing countries compared with the 'traditional' first industrialised countries. The problems of solid waste management, and poor quality of services, often result in large solids being dumped into storm drainage systems, or being washed there during rainfall as outlined in Section 2.1.8. (e.g. Kolsky, 1998). These effects dominate the performance of the drainage systems, as they result in rapid and widespread blockages, often invalidating the storm drain network. Studies in Indore, India, have shown that large storms during the monsoon do not clear the solids from the open storm drains, which remain typically at least one-third full of solids, and even after cleaning, the rate of re-deposition during dry weather averages some 5 % of available depth of drain per 100 days.

Whilst covered drains (sewers) could reduce these problems, they bring other difficulties in countries like India:

- potential for inlet blockage;
- more difficult to clean;
- greater difficulty in identifying where problems are when blockages occur.

Kolsky (1998), amongst others, recommends more usage of roads and paths as drainage conduits, to overcome some of these problems.

3.3 SULPHIDE FORMATION AND EFFECTS

3.3.1 Hydrogen sulphide formation

Hydrogen sulphide formation is a microbial process taking place under anaerobic conditions. Sulphide in sewers is primarily produced in the biofilms and the sediments and is thereby closely related to the presence of sewer solids. Only limited activity occurs in the bulk water due, for example, to detached biofilm particles (USEPA, 1974). When dissolved oxygen and nitrate in the wastewater of sewer systems is depleted, sulphate reducing bacteria may use sulphate as the electron acceptor in their transformation processes of wastewater organic matter (organic carbon), resulting in the formation of sulphide (hydrogen sulphide). Schematically, the sulphide producing process is:

$$SO_4^{--} + \text{organic carbon} \rightarrow \text{new bacteria} + CO_2 + H_2S \qquad (3.3)$$

Sulphide production in pressure mains requires anaerobic conditions, which are particularly dependent on the dissolved oxygen concentration of the incoming wastewater. This production may start after 10 to 30 minutes. Even at low temperatures, problems associated with sulphide formation may be clearly observed when the residence time of the wastewater exceeds 0.5–2.0 hours, (Hvitved-Jacobsen *et al.*, 1995; Hvitved-Jacobsen, 2002).

In gravity sewers, sulphide production takes place mainly at relatively high temperatures in slow-flowing pipes with insufficient reaeration. The presence of sulphide in the bulk water of gravity sewers in Northern Europe is rare due to the moderate temperatures even during summer periods. Although sulphide is produced in the biofilms and the sewer sediments, aerobic conditions in the bulk water phase typically result in a corresponding oxidation of the sulphide produced, thereby preventing the occurrence of sulphide in the water phase. Elsewhere, where temperatures are higher, such as in the countries in Central and Southern Europe, in the Southern USA and Australia, sulphide problems under gravity sewer conditions are common.

Sulphide concentrations of 0.5, 3 and 10 mgS/l may be considered as low, moderate and high respectively. In countries with long pressure mains or high temperatures in wastewater of gravity sewers, significantly higher concentrations than 10 mgS/l are reported (e.g. Thistlethwayte, 1972; Pomeroy and Parkhurst, 1977).

Anaerobic conditions, i.e. no presence – or limiting concentration – of dissolved oxygen or nitrate, is the basic requirement for sulphate reduction. Under such conditions the most important factors determining sulphate reduction rates are as described below.

3.3.1.1 Presence of sulphate

Sulphate is typically found in all types of wastewater in concentrations > 5–10 mgS/l, i.e. in concentrations which are not limiting for sulphide formation in biofilms. In sediments, however, where sulphate may penetrate deep into the deposits, the potential for sulphate reduction may increase with increasing sulphate concentration in the bulk water phase.

3.3.1.2 Quantity and quality of biodegradable organic matter

Biodegradable organic matter is available in wastewater as substrate for sulphate reduction. However, in wastewater from food industries, for example, with easy biodegradable organics, the sulphate reduction rate may be higher than in wastewater from households.

3.3.1.3 Temperature

The temperature dependency of sulphate reduction for sulphate reducing bacteria is high, corresponding to a temperature coefficient of about 1.13 per °C. Because diffusion of substrate into biofilms or sediments is typically the limiting process, the temperature coefficient is reduced to about 1.03 per °C.

3.3.1.4 pH

Sulphate reducing bacteria may exist between pH 5.5 and 9. Only when the pH exceeds about 10, will the sulphate reduction rate be significantly reduced.

3.3.1.5 Prediction of sulphide formation

Several empirical equations have been developed to predict the sulphide production rate r_a from the biofilm surface in pressure mains (in gS/m^2h, Table 3.4). Equations 3.4, 3.5 and 3.6 assume that the sulphate concentration is high and non-limiting, which is usually the case when the concentration is above 4–5 mg SO_4-S/l (Nielsen and Hvitved-Jacobsen, 1988). Typical concentrations in wastewater are usually above this value. Equations 3.7 and 3.8 estimate maximum possible rates and they are a conservative result of measurements on many pressure mains. Equation 3.7 is useful for predicting variations in the sulphide build-up in a pressure main when the actual type of wastewater is known. This has been shown to be suitable for estimating diurnal and seasonal variations in the sulphide production rate in many Danish combined or separate pressure mains (Hvitved-Jacobsen *et al.*, 1988). Sulphur compounds such as sulphite and thiosulphate are also reduced to sulphide if present in wastewater. Organic sulphur compounds are usually insignificant sources for sulphide production.

In gravity sewers there is usually no sulphide present in the bulk water when the oxygen concentration is above 0.2–0.5 mg/l (USEPA, 1974). When the oxygen concentration is lower, the empirical equation 3.6 in Table 3.4 can be used successfully, although the rate constant is usually lower than 1×10^{-3} (USEPA, 1974). Moreover, the other equations which are normally used for pressure mains accord with actual measurements. An estimation of the sulphide concentration in the bulk water in gravity sewers, however, has to take loss to the atmosphere and chemical and biological oxidation in the bulk water into account (Pomeroy and Parkhurst, 1977; Tchobanoglous, 1981; Wilmot *et al.*, 1988).

Table 3.4. Empirical equations for forecasting sulphide production rates in pressure mains (3.4–3.7) and oxygen consumption in gravity sewers and pressure mains (3.8)*

Equation	Source	No.
$r_a = 0.5 \times 10^{-3} u C_{BOD}^{0.8} S_{SO4}^{0.4} 1.139^{(T-20)}$	Thistlethwayte, 1972	(3.4)
$r_a = 0.228 \times 10^{-3} C_{COD} 1.07^{(T-20)}$	Boon and Lister, 1975	(3.5)
$r_a = 1 \times 10^{-3} C_{BOD} 1.07^{(T-20)}$	Pomeroy and Parkhurst, 1977	(3.6)
$r_a = k^{**} \times 10^{-3} (S_{COD}-50)^{0.5} 1.07^{(T-20)}$	Nielsen and Hvitved-Jacobsen, 1988	(3.7)
$r_a = 5.3 S_O (su)^{0.5}$	USEPA, 1974	(3.8)

* Units for r_a are $g/(m^2\,h)$. Can be expressed as $g/(m^3\,h)$ by division with the hydraulic radius.

**k = 1.5 for typical Danish domestic wastewater, 3 for wastewater with some contribution from food industry and 6 for easily degradable wastewater from food industry (see Figure 3.3, A (k = 1.5), B (k = 3) and C (k = 6))

where:

C_{BOD}	is biochemical oxygen demand (mg O_2/l)
C_{COD}	is chemical oxygen demand (mg O_2/l)
S_O	is dissolved oxygen concentration (mg/l)
S_{COD}	is soluble COD (mg O_2/l)
S_{SO4}	is sulphate concentration (mg SO_4/l)
k	is rate constant (-)
T	is temperature ($^\circ$C)
s	is slope (m/m)
u	is flow velocity (m/s)

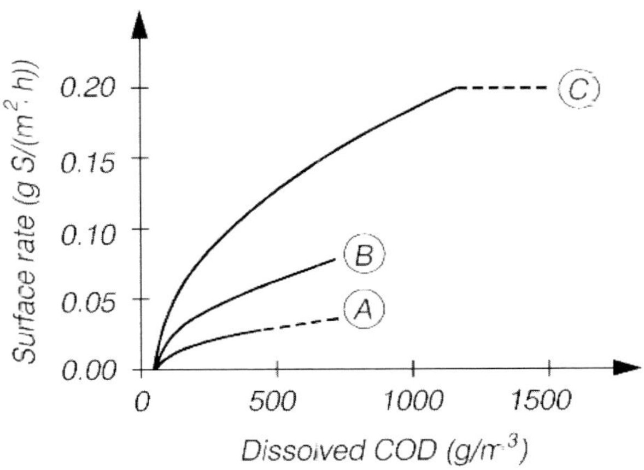

A (k = 1.5), B (k = 3) and C (k = 6)

Figure 3.3. Prediction of sulphide formation in pressure mains (see Table 3.4).

3.3.1.6 Release of hydrogen sulphide gas from wastewater to the atmosphere

The release of hydrogen sulphide gas from wastewater into the overlying atmosphere is of fundamental importance for the effects of sulphide such as corrosion and odour problems. As long as the sulphide remains in the water phase these effects do not occur.

Only the molecular form, H_2S, and not the dissociated forms of sulphide, HS^- and S^{--}, can be released from the wastewater phase to the atmosphere. Hydrogen sulphide (H_2S) is a weak acid which dissociates as follows:

$$H_2S \Leftrightarrow H^+ + HS^- \Leftrightarrow 2H^+ + S^{--} \tag{3.9}$$

An increase of the wastewater pH will, at a constant total sulphide concentration in the wastewater, reduce the hydrogen sulphide concentration in the overlying atmosphere.

For processes typical of sewer conditions only the equilibrium between H_2S and HS^- is of importance. S^{--} will only play a role if pH > about 12. In a closed system, an equilibrium between the molecular form (H_2S) in the water phase and the overlying atmosphere is established. The relationship demonstrating this phenomenon is described by Henry's law, which will depend on pH and temperature (see Figure 3.4).

The partial pressure curves shown in Figure 3.5 are essential for the evaluation of the potential risk for microbial corrosion and odour problems in an actual situation. Turbulence in the wastewater and the establishment of equilibrium is important. Release of H_2S into the atmosphere at e.g. pumping stations and hydraulic jumps may be substantial and therefore cause correspondingly high corrosion rates and health risks. The curves show the H_2S concentration at equilibrium in the atmosphere per unit concentration in the water phase.

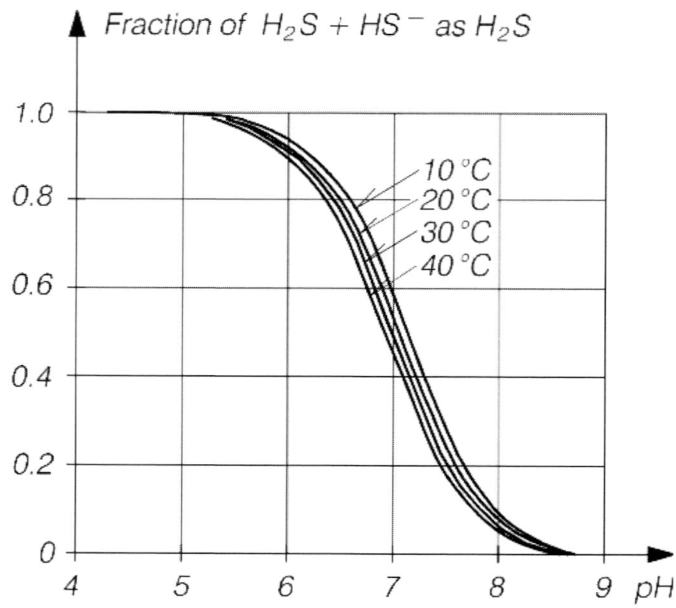

Figure 3.4. Dissociated equilibria for H_2S in aqueous solution.

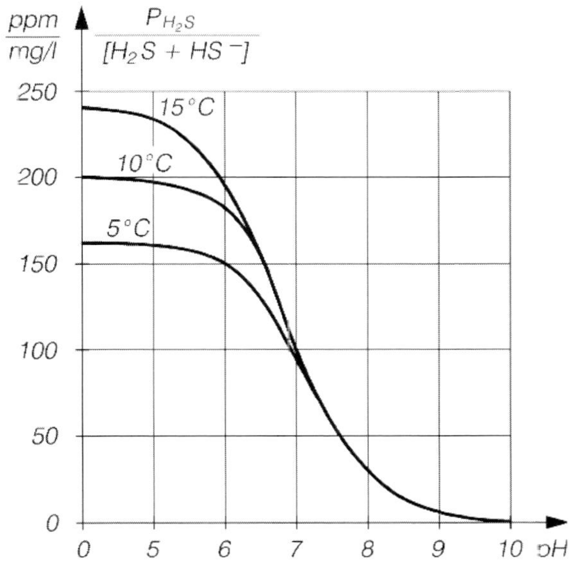

Figure 3.5. Partial pressure of H_2S measured in ppm volumetrically in the atmosphere in equilibrium with a water phase of sulphide.

3.3.2 Corrosion

The corrosion process is depicted in Figure 3.6. Those areas where corrosion may be especially noticeable are often close to where hydrogen sulphide is released, e.g. close to an area with permanent anaerobic conditions and the water surface or point of turbulence, or at points where moisture may be easily adsorbed. It is important to appreciate that corrosion does not take place under the water surface.

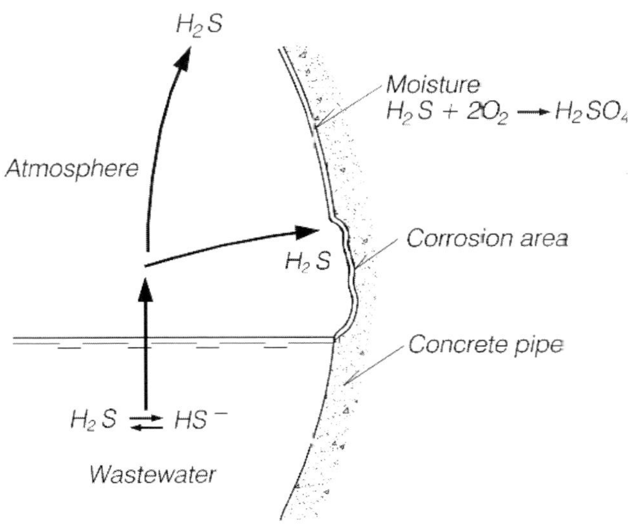

Figure 3.6. Principle for release of hydrogen sulphide and corrosion.

When H_2S has been released to the atmosphere, the second step of the corrosion process is absorption of hydrogen sulphide at moist surfaces. After absorption the following microbial process, carried out by aerobic bacteria of the *Thiobacillus* family may take place:

$$H_2S + 2O_2 \Rightarrow H_2SO_4 \quad (3.10)$$

Species of the *Thiobacillus* family, e.g. *Thiobacillus concretivorus*, have been observed active at solutions with up to 7 % sulphuric acid. Although hydrogen sulphide may corrode, for example, metals by producing metal sulphides, the main corrosion problem of significance in sewers is related to attacks on concrete surfaces, caused by the sulphuric acid produced.

If the rate of sulphuric acid production is low, almost all acid produced on concrete surfaces will react with the cement in the concrete and only loose bound, inert components will be left. If the rate of sulphuric acid formation is high, part of the acid may be washed away before further reaction with the concrete takes place. The formation of sulphate ions is associated with acid corrosion which may cause sulphate attack on concrete surfaces.

The principle of the concrete corrosion processes are given by the expression:

$$H_2SO_4 + CaCO_3 \text{ (cement)} \Rightarrow H_2O + CO_2 + CaSO_4 \text{ (gypsum)} \quad (3.11)$$

Based on this formula and the density of concrete, a simple formula for the corrosion rate c in units of mm/year is:

$$c = 11.4\ f/A \quad (3.12)$$

where f is rate of hydrogen sulphide absorption on a concrete surface (g/m^2h).
A is alkalinity of concrete material in units of $CaCO_3$.

In cases of severe concrete corrosion in sewer systems, a corrosion rate of 1–5 mm/year may be observed.

3.3.3 Particular problems in pumping mains and gravity sewers

Except for problems associated with discharges of sulphide into treatment plants or combined sewer overflows into receiving waters, the particular problems of sulphide in wastewater are associated with its potential release into the overlaying atmosphere (cf. Section 3.3.1.6). Therefore, as long as hydrogen sulphide remains in, for example, a pressure pipe, no problems exist. It is at the outlet of a pressurised pipe system and above the surface of the wastewater in a gravity sewer that problems can occur.

In addition to the corrosion problems, the odour and health related effects of hydrogen sulphide must be understood. Hydrogen sulphide can be detected by smell at very low concentrations down to about 0.1 ppm and increasing risk to health can be observed at increasing levels of concentration. Potential risk of hydrogen sulphide in sewer systems is shown in Table 3.5.

Table 3.5. Odour and health related effects of hydrogen sulphide in the atmosphere

Odour or human effect	Concentration in atmosphere (ppm)
Odour limit	0.1–0.2
Unpleasant smell	3-5
Recommended maximum criterion for working	10
Effect on eyes	50–100
Inactivation of smell	150–250
Serious water accumulation in lungs	300–500
Deadly impact on nerve system	500–1000
Immediate cessation of respiration	1000–2000

The density of hydrogen sulphide is slightly higher than the atmosphere (relative density 34/29). Therefore, hydrogen sulphide has a tendency to accumulate in low areas such as pumping stations and manholes. As hydrogen sulphide typically will not be detected by smell if the concentration exceeds about 200 ppm, instruments or alarm systems for monitoring must be used whenever working in sewer systems.

3.4 CONTRIBUTIONS OF SOLIDS TO FLOW QUALITY AND CONCEPT OF FOUL FLUSH

3.4.1 Introduction

During storm events, significant loads and concentrations of solids (usually expressed as total suspended solids (TSS)) are frequently observed in both separate and combined sewer systems. These are often termed 'foul flushes' (e.g. Krauth, 1970; Thornton and Saul, 1987). From Sections 2.1 and 2.2, the four main sources of the solids and associated pollutants in flushes are:

1. runoff from catchment surfaces;
2. dry weather sewage;
3. deposits in sewers;
4. slime on sewer walls.

During any storm event, the TSS load measured at any point in a sewer system contains solids from the above four sources, but in variable proportions depending on various factors, such as:

- dry weather flow rate and concentrations;
- period of the day when the storm event occurs;
- rainfall intensity and duration;
- antecedent dry weather period;
- amount and type of deposits in the upstream sewers and on catchment surfaces;
- amount and growth rate of slime (biofilm);
- age and condition of the sewer fabric (pipe corrosion or infiltration entry);
- sewer maintenance and cleaning practices;
- sewer geometry, size and slope, land use, street cleaning practice, etc.

Temporal and spatial variability of these factors leads to complex and unpredictable conditions in the sewer flow and, even for the same site, from storm event to storm event conditions may vary, depending upon initial and antecedent conditions.

As sewer flow pollutant control strategy is dependent upon knowledge about the nature and occurrence of flushes, many studies have been carried out to investigate the phenomena as described in the next section.

3.4.2 Relative contributions

Assessing and measuring the respective contributions of the four main sources of solids is a complex problem. As direct measurements based on tracers or radioisotopes are very expensive and difficult, evaluations have mainly been by means of mass balances from indirect measurements, the transposition of models and results, and from direct measurements in both sub-catchments and locations where each contribution could be more easily identified or ignored. Three examples are presented hereafter, in chronological order.

Dauber and Novak (1983) undertook one of the first documented studies, monitoring the total solids load transported in wet weather in combined sewers in Zürich, Switzerland. By deducting the dry weather flow component measured during dry periods, it was possible to show that the erosion of deposits during wet weather produced significant contributions to the total load in transport. An example of this work is shown in Figure 3.7, which illustrates that of a total of 56.2 kg of total solids (dissolved plus suspended) being transported, some 36.5 % may be attributed to the bulk deposits and 23.2 % to the eroded slimes.

Krejci *et al.* (1987) also confirmed that the contribution from previously deposited in-sewer solids to the quality of flow during a storm event can be significant. They found, in more detailed investigations of a small 12.7 ha residential catchment in Zürich, Switzerland, that a significant solids and related pollutant contribution originated from in-sewer erosion of bed deposits and slimes. The average contributions for four storm events are given in Figure 3.8.

Bachoc (1992) determined that 30 to 44 % of the SS loads measured during two heavy rain events at the outlet of a combined sewer in Marseilles originated from sewer sediments. More recently, French studies in a large combined sewer in the Paris city centre (Gromaire-Mertz *et al.*, 1998) concluded that approximately 50 % of the solids transported in suspension during storms originated from previously deposited sediments and near-bed solids, as illustrated in Table 3.5, for five monitored storm events.

Table 3.5. Contribution of in-sewer solids to suspended load during storms (slime was not measured and, if present, is included in the contribution attributed to sewer sediments)

Rainfall (mm/h)	Contribution to load of solids in suspension (%)				
	Sanitary sewage	Roof runoff	Yard runoff	Street runoff	Sewer sediment
4.5	28	6	5	15	47
4.7	9	10	4	17	59
35.3	7	10	7	10	66
5.6	6	23	10	10	50
1.8	37	3	3	11	45

It is apparent therefore that previously built-up deposits and wall slime in combined sewer systems can contribute significantly to the total solids loads transported during storm events, with contributions ranging from 35 to 55 % by mass. However, these results should not be considered universally applicable to any sewer system and to all storm events. Due to the complexity of the origins, catchments, causes and, hence, difficulties of quantification of these effects, site-specific measurements are necessary to determine when, where it may occur and the significance of the phenomena. There are, however, significant problems in measurement of all the solids in transport. Single depth sampling from one height above the sewer bed using vacuum samplers will not account for the variability in concentration and the nature of the particles with height; nor will the solids moving near the bed be sampled.

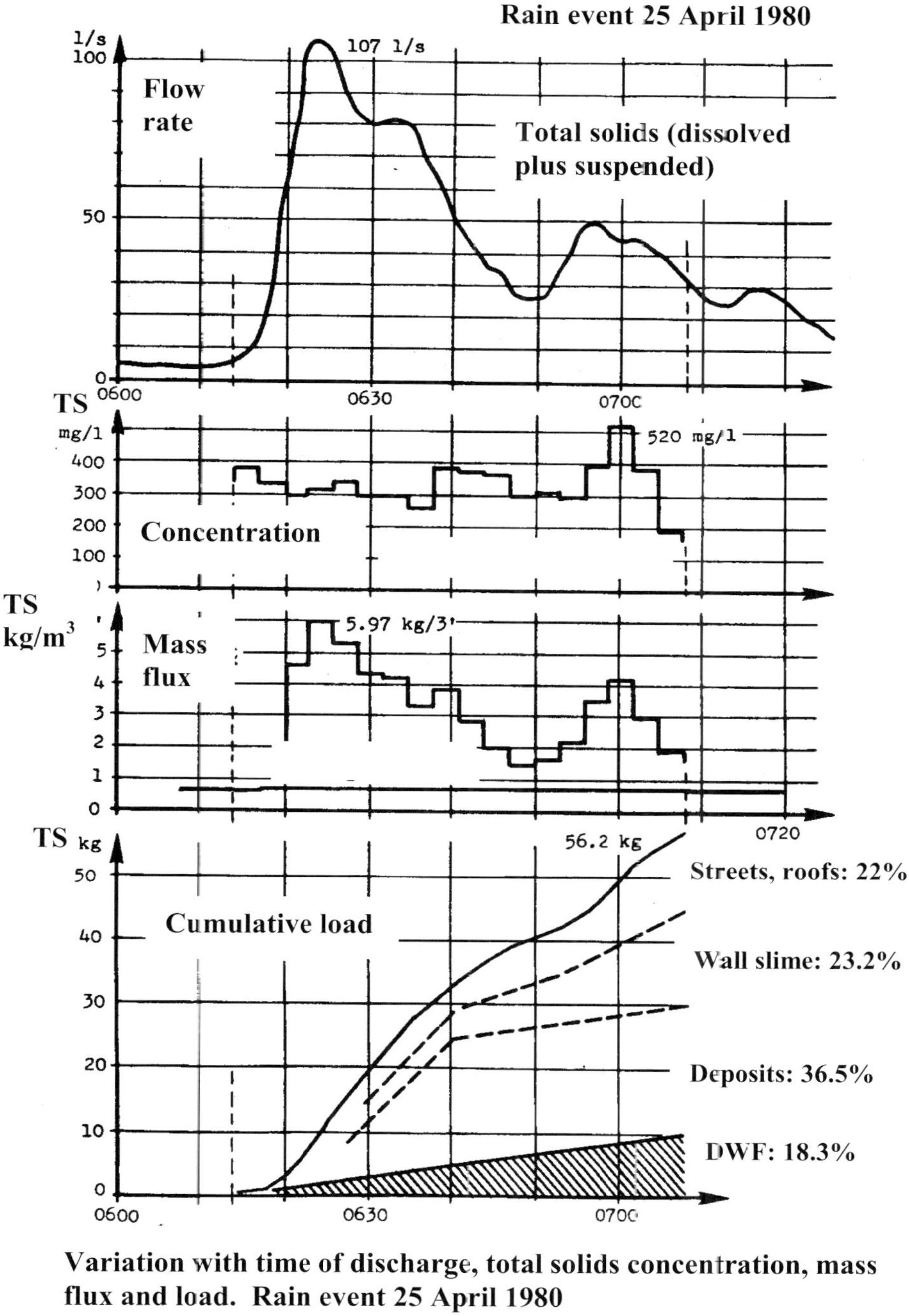

Figure 3.7. Relative contributions of total solids loads from various sources in a combined sewer in Zurich (adapted from Dauber and Novak, 1983).

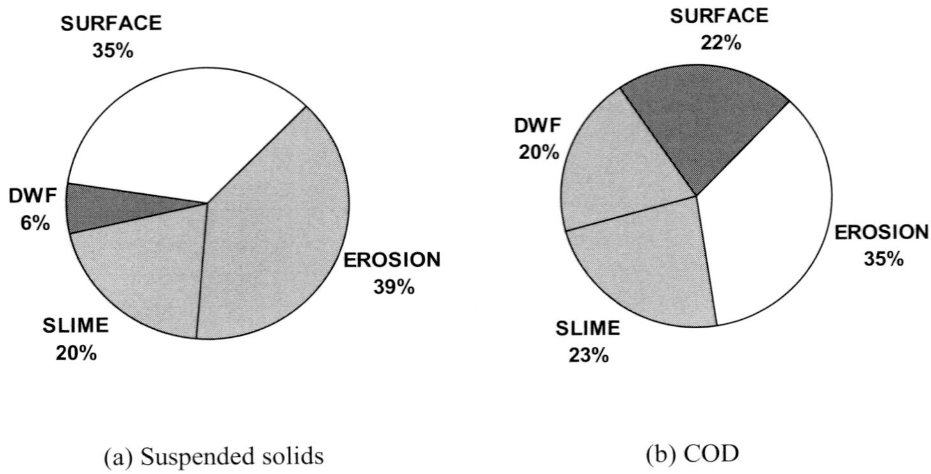

Figure 3.8. Solids and pollutant sources in combined sewage flows (data taken from Krejci *et al.*, 1987).

In perhaps the most extensive experimental programme yet undertaken, Gromaire *et al.* (2001a) measured experimentally the pollutant loads from different types of runoff, from sewage and the total load at the catchment area outlet, for the same rain events and in Le Marais catchment in Paris. Mass balance calculations over thirty rainfall events showed that 30 to 80 % of the mass of SS, VSS, COD and BOD_5 masses originated from in-sewer sources. When calculations are extended over a scale of one year (Gromaire *et al.*, 2001b), the main source of wet weather flow pollution was the sewage (40 to 60 % of the SS, VSS, COD and BOD_5 loads), with in-sewer erosion a major source of pollutants (30 to 45 % of the pollutant loads). Runoff inputs comprised only a minor part of the particle and oxidisable matter transported to the outlet of the network during storms. The significant contribution from sewage was explained by the predominance, over a whole year, of storms of a low intensity.

Any observations showing that the solids being transported during wet weather in a combined sewer are highly organic are an indirect indication of the likelihood of a significant contribution of previous bed deposits and wall slime. This is because the latter solids have a higher organic content than the solids transported in separate storm sewer systems, even during very heavy storm events generating rapid surface runoff. In such cases, the proportionate contribution of the dry weather sewage (the main origin of organic solids in sewers) is not sufficient to explain the high organic content.

3.4.3. Concepts of flush

3.4.3.1 Origins of the first flush concept

The concept of a 'first flush' to describe the increase in solids and other pollutant concentrations in an accelerating flow in a sewer has been extensively used since the end of the 1960s. This was originally based on measurements in a small and steep catchment near Stuttgart, Germany, where sewer deposits were flushed by the increased flow generated by storm events (Krauth, 1970). After this flush, the sewer was presumed to be clean and subsequent events had peak solids concentrations that were found to be lower. The term 'first flush' (denoted 'foul flush' or 'first foul flush') has been used subsequently to describe this phenomenon.

Many studies have been carried out, mainly in Germany, France, UK and Denmark, in the late 1970s and in the 1980s, in order to verify the concept in a range of sewers and to develop explanatory and predictive ideas (e.g. Göttle, 1978; Stotz and Krauth, 1984, 1986; Johansen, 1985; Pearson *et al.*, 1986; FNDAE, 1986; Davies, 1987; Thornton and Saul, 1987; Philippe and Ranchet, 1987; Hvitved-Jacobsen *et al.*, 1987). Reportedly high TSS concentrations occurring early in the flood wave during storm events in combined sewers are common (example in Figure 3.9), if deposits previously accumulated during dry weather periods are present and eroded by the flow.

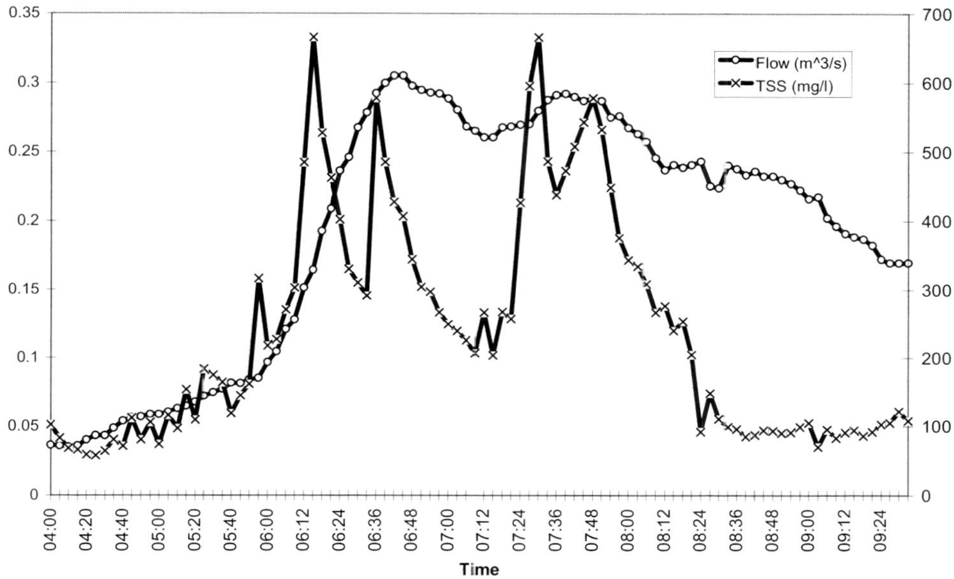

Figure 3.9. Typical foul flush of solids (main Dundee Interceptor sewer).

Hence it is apparent that the antecedent dry weather period is a significant factor implicated in the origins of these flushes. Catchment area and slope have also been claimed as secondary contributing factors. For example, larger catchments are claimed to lead to less noticeable downstream flushes.

Solids do not only come from previously deposited in-sewer deposits, as illustrated in Section 3.4.2. Consequently, flushes may be attributable either to surface source inputs and/or to the scouring of in-sewer deposits, and may also occur in sewers where the dry weather flow peak is sufficiently large to cause erosion of deposits. Dilution (lower concentration compared with normal dry weather concentration) may occur when the concentration of pollutants in the sewer is reduced by the ingress of runoff water, and is common when there is a lack of pollutant supply (i.e. sewer deposits), or when the rainfall event is of low intensity. In some cases, a second flush is also observed (see Figure 3.9), and is generally attributed to surface sediment wash-in and the relative time of concentration for the system. This second flush observed in combined sewers is comparable to the first flush observed in stormwater sewers, where solids are mainly coming from catchment surfaces (see an example of first flush in a separate sewer in Figure 3.10).

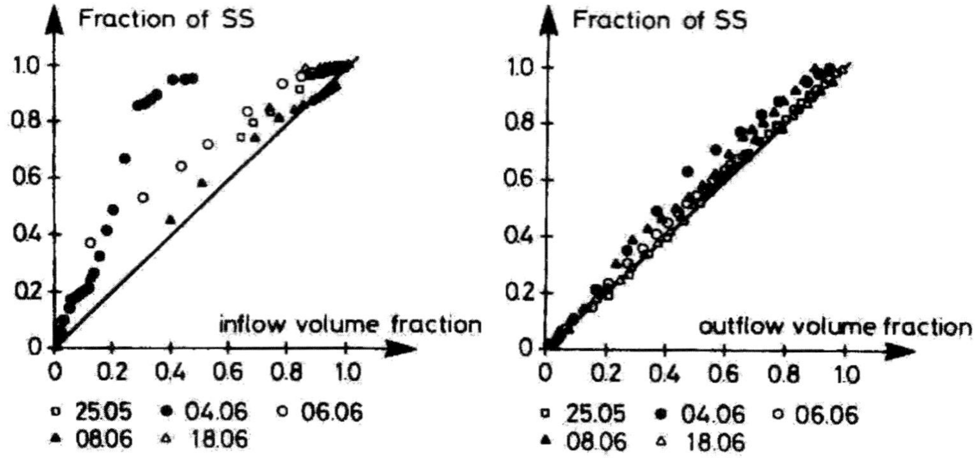

Symbols refer to storms on different dates in 1986

Figure 3.10. Foul flush from storm sewer observed entering and leaving a pond in Danish studies (adapted from Hvitved-Jacobsen *et al.*, 1987).

Concurrently, wall slimes and biofilms attached to the solids particles may also be released into the flow (Section 2.3.4.6), adding to any flush effects due to the bulk deposit erosion. Descriptions of the effects of sewer slime erosion in the USA are given by Tomlinson *et al.* (1980); Hoffmann and Meighan (1984). There have been only few studies investigating the effects of sewer slime (e.g. Dauber and Novak 1983, Michelbach 1995). A fictitious sewerage system was established (Reiff, 1992) in order to be able to estimate the degree of pollution produced by sewer slime eroded during storm flows. Calculations of the proportional pollutant load from sewer slime due to (equivalent) rainfall of 100 l/(s.ha) showed that typically in excess of 90 % of the COD load arose from this source, with 0.051 kg COD/metre sewer length being provided by the slime. Dauber and Novak (1983), with a much smaller equivalent rainfall intensity of only 18 l/(s.ha), nonetheless estimated that 30 % of the total COD load in the combined flow was attributable to sewer slime, and 38 % to bed deposits during a runoff event.

Heavy metals are mainly associated with sewer slimes and not with the larger (bed) solids in sewers, or in the dissolved phase in bulk flows, as shown by Michelbach and Wöhrle (1992). Sewer slimes show proportionately higher heavy metals concentration than settled solids in sewers. In this study it was found that the heavy metal concentrations were typically very high during the first flush of combined sewage during storms. It was contended that the first flush consisted mainly of eroded sewer slime and associated pollutants (Figure 2.2.20).

The implications of these erosion and flush observations are the occurrences of 'pulses' of highly concentrated suspensions of solids and associated pollutants (notably COD and slime associated heavy metals), which may be conveyed into watercourses by spills from CSOs. The considerably elevated concentrations of heavy metals found in receiving water sediments in areas adjacent to CSOs can be largely attributed to sewer slime erosion and subsequent deposition. Flushes may also be responsible for shock loads at treatment plants, and can disturb the equilibrium operation of these plants where there are no balancing facilities (Ashley *et al.*, 2002). Because of these effects, a number of regulators specifically require wastewater system operators to provide storage systems to retain first flushes within sewer systems. This requirement is often responsible for operational misunderstandings and paradoxical or apparently contradictory conclusions, due to differing approaches and definitions, as described in Section 3.4.3.2 below.

3.4.3.2 Definition of the first flush

Provision of an adequate in-sewer or other storage volume to store the highly concentrated first foul flush is not alone sufficient to ensure adequate receiving watercourse protection during CSO spills. Although the relative concentration of solids and associated pollutants may decrease after an initial flush during a prolonged storm, the protracted elevated concentration and the sustained high flow volume may result in a higher load (volume x concentration) still being discharged from the CSO to the receiving water.

In published documents dealing with foul flushes and their effects, the use of two alternative definitions for 'first flush' can cause confusion. The expression is used to denote either:

1 any initial TSS concentration peak (i.e. as observed) – Figure 3.9; or
2 a relative fraction of the total TSS load (i.e. it is operational criteria used to define a control strategy for stormwater pollutant loads) – Figure 3.10.

Few references to 'foul flushes' provide an adequate definition of what is meant by the term. Unfortunately the lack of definition can lead to contradictory conclusions: a significant initial concentration peak (definition 1 above) does not necessarily lead to a significant initial relative fraction of the total load (definition 2 above). Precise definition attempts have been proposed, but not universally adopted due to their often conflicting nature (see Bertrand-Krajewski *et al.* 1998 for a review and discussion). The various definitions fall into two groups:

Group 1. The first flush is defined as high pollutant concentrations in the first part of the increased flow due to storm inputs (note that the terms 'high concentrations' and 'first part' of increased flow are often not explicitly defined). Two typical examples illustrate the approach. Pearson *et al.* (1986) proposed the following definition: 'The initial period of a storm flow during which the concentration of pollutants is significantly higher than those observed during the latter stages of the storm event.' Two types of flush were defined: Type A for which suspended solids and COD concentrations were lower than for the ambient DWF conditions, and for which the time of the peak concentration preceded the flow peak. Type B had a pollutant concentration in excess of those found in DWF conditions. The latter were found to occur for the larger storms and after an antecedent dry weather period (ADWP) of at least 3 days, and the timing generally corresponded with the flow peak. In each case, a second flush could also occur coincident with the flow peak. Stotz and Krauth (1984) proposed another approach where the first flush is defined by the fact that 'the maximum pollution load (in kg/min) appears before the maximum water flow'.

Group 2. The first flush corresponds to a specific proportion of the total load transported within a certain proportion of the total flow volume. This type of definition has the advantage of setting a more precise specification to the variables and allowing comparisons between storms and catchments, but it still relies on a subjective specification as to the rate of change of the cumulative pollutant load (as a proportion of the total during the storm) compared with the rate of change of the cumulative flow. Three typical examples are given here. According to Geiger (1984, 1987), 'there is a first flush if the cumulative load/flow curve, also named *m(v)* curve, (see Figure 3.11) has an initial slope at the origin greater than 45°.' If the TSS concentration remains constant during the event, then the curve slope is at 45°. For an initial slope less than 45°, there is a dilution effect. A more restrictive definition based on *m(v)* curves has been proposed by Saget *et al.* (1996), who defined the first flush as occurring only where 'at least 80 % of the pollution load is transferred in the first 30 % of the (flow) volume' (also termed the 30/80 LFF). Using this definition it was concluded from assessments of 197 events measured in twelve French separate storm and combined sewer systems, that there was only one which produced a first flush. Note that the first flush shown in Figure 3.10 would not be classified as such according to the Saget *et al.* (1996) definition. Another, but similar definition, by Gupta and Saul (1996) proposed that 'the flush occurs during that part of the storm up to the maximum divergence between the dimensionless cumulative percentage of pollutants and the cumulative percentage of flows when plotted against the cumulative percentage of time', as shown in Figure 3.12.

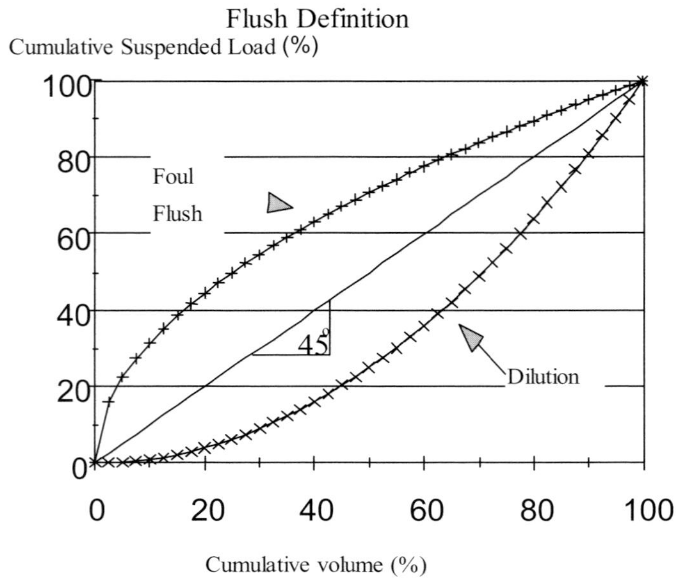

Figure 3.11. Cumulative mass/flow definition of first foul flush as specified by Geiger (1987).

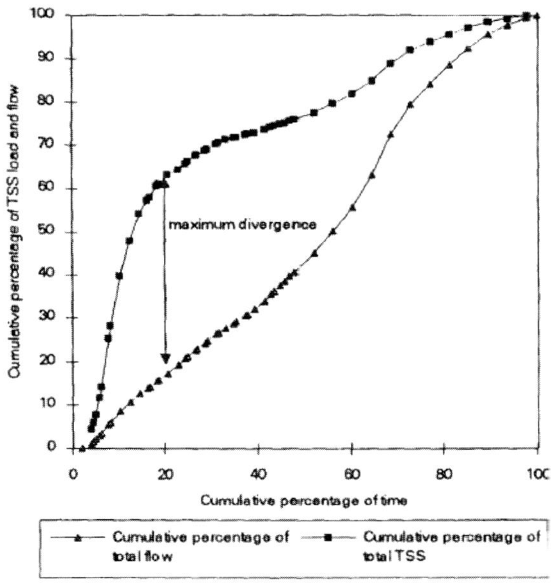

Figure 3.12. Cumulative mass and flow/time definition of foul flush (Gupta and Saul, 1996).

A practical problem with the second group of definitions is the definition as to when the 'event' is deemed to start and end, as this influences the computation of the total loads, flows and the slopes of the lines. In addition, the Saget *et al.* (1996) definition suggests that foul flushes are very rare, at least in French sewers. From a practical point of view, this conclusion is unrealistic as wastewater system operators do observe elevated concentrations and loads of solids in their systems and need to ensure that these are managed effectively. They also require to be able to make reference to these events using recognised and acknowledged terminology. The above examples demonstrate that there is a need to adopt clear and common definitions that are understood and accepted in both the scientific and operational communities to prevent misunderstanding and help to provide useful management tools. Consequently it is proposed here to provide two new definitions for the way in which the phenomena may be described:

> The **Concentration First Flush** (CFF) is that which occurs where the solids concentration is observed to increase above the ambient DWF concentration during the early part of a storm event.
>
> The relative **Load First Flush** (LFF) is used to describe the occurrence of a significant proportion of the cumulative flush load with a corresponding lesser proportion of the cumulative flow volume. The precise definition may be expressed in terms of the relative flow to load percentages assumed to define the occurrence of a flush. Most commonly this is 20/80, meaning 20 % of the cumulative flow volume carries 80 % of the cumulative load of solids.

Another important difficulty preventing comparative assessment of first flushes between studies is the spatial scale considered. Some studies have been carried out at very local scale (some tens or hundreds of metres of sewer), whilst others have been carried out at a larger scale (from several to some thousands of hectares). If a local concentration first flush (CFF) of the type shown in Figure 3.9 has been observed and interpreted in terms of local hydraulic and solid transport characteristics, the consequential extrapolation to larger scales is typically not feasible as this would depend on a wide range of parameters, including concentration and transport times, mitigation effects and the complex combination of numerous (superposed) local concentration first flushes (Ashley and Bertrand-Krajewski, 1993).

The complex spatial distribution, superposition and propagation of a flush through a catchment was originally hypothesised by Göttle (1978) who verified the conceptual model using extensive field experiments. The main result is summarised in Figure 3.13, where the relative sediment mass observed at the outlet of the catchment is shown related to the relative flow volume. Assuming that the catchment is made of *m* subcatchments, where significant load first flush (LFF) occurs (first line of the figure), the resulting effect at the outlet shows a decreasing LFF effect when the number *m* increases (last line of the figure). A similar conclusion was also drawn later by Stotz and Krauth (1984).

3.4.3.3 Modelling of first flushes

Techniques and models for predicting CFF and LFF in runoff are given in Section 2.1.3. It should be noted that these currently are largely semi-empirical and need to be calibrated for each site. Deletic (1998), defining the LFF as the total load transported in first 20 % of storm volume, concluded that it could not arise from catchment surface inputs. Comparable results have also been obtained by Stanley (1996) using a 20/80 LFF definition previously suggested by Stahre and Urbonas (1990).

Catchment-based approaches to the prediction of LFF in sewers have been attempted using empirical models. For example Gupta and Saul (1996), using the definition given in Figure 3.12, related the cumulative pollutant load ($LOAD_{ff}$, kg), carried by the flow in a catchment in North West England to the storm duration (STDURN, min), the rainfall intensity ($RINT_{max}$, mm/h) and the antecedent dry weather period (ADWP, h):

$$\text{summer: } \boldsymbol{LOAD_{ff}} = 1.35 \times \boldsymbol{STDURN}^{0.61} \times \boldsymbol{RINT}_{\max}^{0.71} \times \boldsymbol{ADWP}^{0.23} \text{ (kg)} \tag{3.13}$$

The peak concentration of total suspended solids TSS_p (mg/l) was shown to be related to the peakedness of each individual storm event (PEAKEDNESS: the ratio of maximum and average rainfall intensity) and the ADWP (h):

$$\boldsymbol{TSS_p} = 123 \times \boldsymbol{PEAKEDNESS}^{0.64} \times \boldsymbol{ADWP}^{0.17} \text{ (mg/l)} \tag{3.14}$$

Although the CFF is largely dependent upon the source material and the characteristics of the storm event, other factors are also believed to be important:

- total volume of DWF;
- antecedent dry weather period length (ADWP);
- duration and magnitude of previous storm events;
- sewer physical characteristics (size, shape, gradient, etc.);
- time at which during the diurnal DWF pattern the storm occurs.

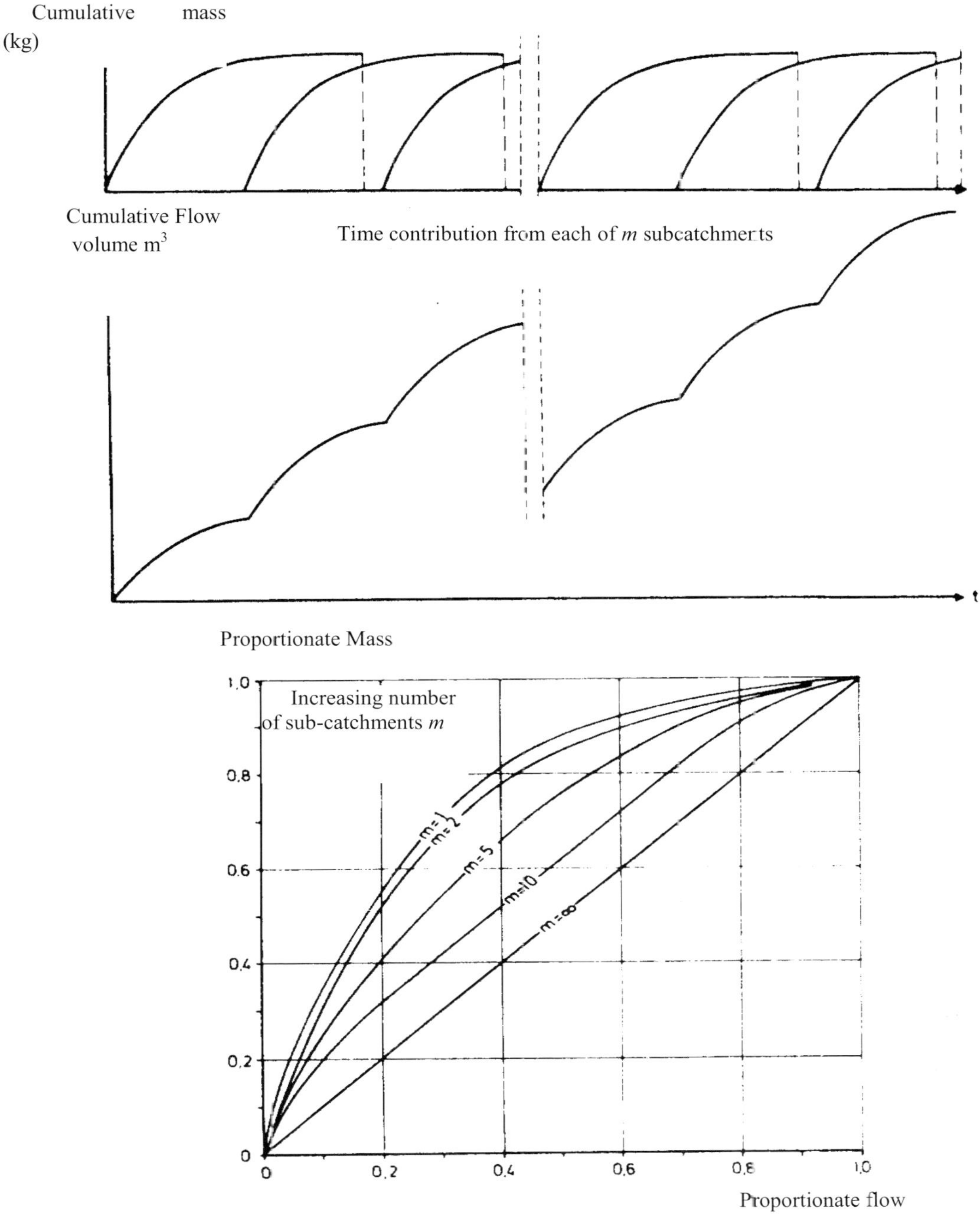

Figure 3.13. Modelling of a large catchment by m subcatchments and consequence for first flush phenomena with increasing values of m (adapted from Göttle, 1978).

Each of the empirical relationships proposed for the prediction of storm pollutant loads rely principally on rainfall hyetographs and the ADWP. More recently these relationships have been reviewed and modified (Saul *et al.*, 2003), although the fundamental approach has been maintained. The importance of the rainfall hyetograph characteristics has been highlighted by many other researchers investigating the pollutant potential of urban runoff events (e.g. Jewell and Adrian, 1982; Geiger, 1984; Philippe and Ranchet, 1987; Lindholm and Aaby, 1989). However, when dimensionless definitions from group 2 (i.e. LFF) are considered, the hyetograph characteristics are not always significant and it is very difficult to identify common explanatory factors (Bertrand-Krajewski *et al.*, 1998) because the relative dimensionless quantities used to define the first flush cannot be easily related to any physical factors.

3.4.3.4 Role of sewer sediments and near bed solids

In an attempt to explain the physical processes involved in flushes, more detailed studies have investigated the effects of disturbance of the bed sediments by increasing rates of shear along the invert of sewers. It is believed that erosion mobilises the surficial material, re-entraining particles and releasing the interstitial liquid trapped in the solids matrix, so contributing significantly to the pollutant load conveyed during periods of peak dry weather and rainfall induced flows (see Section 2.3.4.5).

Detailed models for the near bed solids contribution to the first flush have been developed using field data (Verbanck, 1995a, 1995b; Arthur and Ashley, 1998) and surrogate laboratory experiments with cohesive sediment erosion (Skipworth *et al.*, 1996 and Tait *et al.*, 1998; Skipworth *et al.*, 2000; Saul *et al.*, 2003), being related to a flush of suspended solids (Section 2.3.9).

When a storm wave progresses through a combined sewer reach, the flow regimes may be considered segregated into three discrete sections each with a different ambient velocity, (Davies, 1990b; Davies *et al.*, 1996; Arthur, 1996) as illustrated in Figure 3.14. The front part of the storm wave consists of overtaken baseflow, i.e. DWF that is forced to travel faster than normal due to the surge wave. It is possible to presume only limited mixing in this region. In experiments using a saline baseflow it was observed that there was negligible dispersion of pseudo-storm water mass into overtaken baseflow (Harrison and Holmes, 1967) (Section 2.3.4.5). In the overtaken region, the velocity is increased, resulting in the solids moving near the bed being accelerated and subsequently entrained into the flow column. Hence the total mass in transport in this region comprises the sum of the mass transported in suspension in dry weather plus the near bed solids mass. Evidence for this comes from observations of sewer sediment bed levels in Hildesheim, Germany and Dundee (Ashley *et al.*, 1993), which show the existence under a number of eroding events of a short timescale peak in bed depth passing the measurement sections in each sewer, as illustrated in Figure 2.3.13. This transient peak was not found to be related to any increase in suspended solids concentration as usually associated with first flushes, and was believed to be the 'pushing' of the near bed solids in front of the storm wave as this built up in the sewer network and prior to entrainment of the solids. Hence this phenomenon is distinct from that illustrated in Figure 3.9, which indicates an increase in solids concentration in the suspended phase. It is believed that the former is a 'near bed' solid pulse, which ultimately may become entrained into the main flow. Alternatively, if the solids in the near bed phase are large and dense, then they may move more in a bed load transport mode.

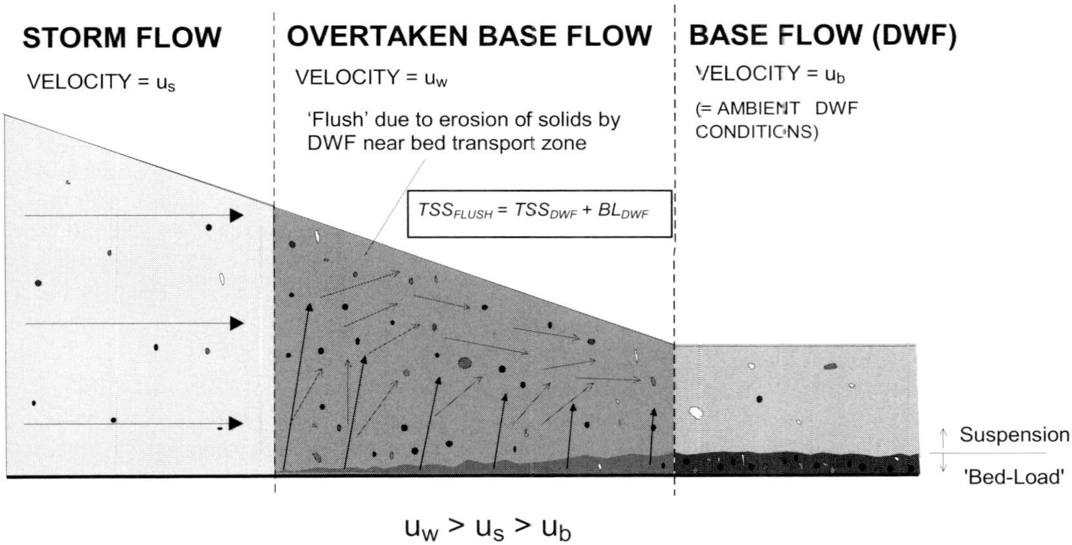

TSS_{FLUSH} = solids concentration of the 'flush'

TSS_{DWF} = ambient DWF solids concentration

BL_{DWF} = near bed solids in transport in ambient DWF conditions

Figure 3.14. Idealisation of storm hydrograph progressing through a combined sewer (adapted from Ackers *et al.*, 1968 and Davies, 1990a, 1990b).

The near bed solids convey a disproportionate pollution load, via very high concentrations (Section 2.2.6.2). These may settle to (and enter) the bed, be conveyed through sewer networks to treatment plants (normal fate) or be discharged into the environment via CSOs, depending upon changing hydraulic conditions and the relative density of the particles. However, their entrainment high enough up into the flow column to reach the weir of a CSO would often require extreme hydraulic conditions. The ranges of quality of the solids and associated pollutants collected in traps in Dundee for more than 200 samples of NBS is illustrated in Table 3.6, summarised from Table 2.2.11 for the main interceptor sewer.

Table 3.6. Near bed solids characteristics and associated pollutants – Dundee interceptor sewer

Solids concentration (g/l)	Volatile solids (g/l)	COD (g/l)	BOD (g/l)	AmmN (g/l)	Specific gravity (dry solids)
0.4–165	0.33–135	85-328	1–685	0.068–2.3	< 1–1.45

The potential erodibility of this material and associated pollutant release has been examined by carrying out a series of controlled laboratory tests (McGregor *et al.*, 1993). These considered the potential for pollutants remaining in suspension once eroded, in solution or attached to fine or coarse sediment fractions, and the potential for subsequent deposition (Section 2.2). The test revealed that of the total 100 % available, some 52 % of the COD, 53 % of BOD, 56 % of total solids, 53 % volatile solids and 71 % of the AmmN were 'released' by the erosion of the near bed solids. The major proportions of each of these released pollutants were either associated with the finest particles or were in the dissolved phase and were hence unlikely subsequently to re-deposit within the sewer. Thus the erosion of the NBS would appear to be a primary source of flushes in sewers and can contribute to the polluting loads discharged via CSOs into the environment.

Application of these concepts to the data collected in Dundee (Arthur and Ashley, 1998) shows that if all the material travelling near the bed in a reach of the combined interceptor sewer were to move into suspension when the baseflow was overtaken, the DWF suspended sediment concentration would increase only by between 0.5 and 9.0 %, which represents a relatively modest increase in solids transport. However, the biochemical data indicate that the increase in COD and BOD_5 concentrations will be much more significant at 77 % and 120 % respectively. Figure 3.15 shows the COD concentration flush for the same event as illustrated in Figure 3.9.

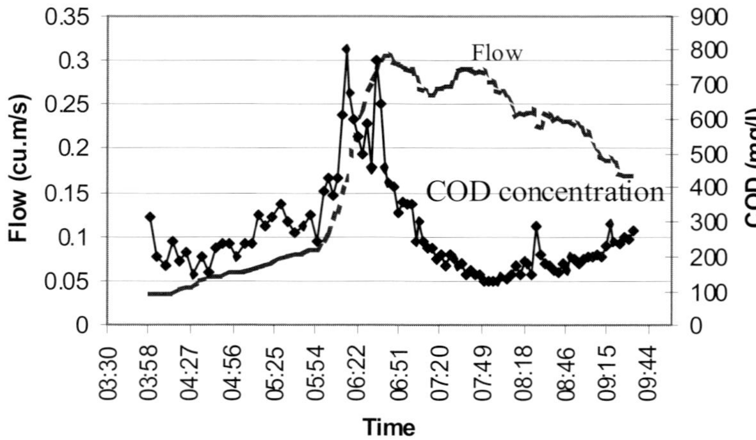

Figure 3.15. Flush of COD in Dundee Interceptor sewer due to rainfall (same event as Figure 3.9).

The studies in Le Marais catchment in Paris (Section 2.3.4.5) have linked the erosion of the weak interfacial solids with the observed solids in transport in suspension at the catchment outlet. This material, which is found extensively throughout the catchment (Ahyerre, 1999), is eroded by low shear stresses (> 0.4 N/m^2). Ahyerre *et al.* (2000) carried out flushing experiments on a 150 m sewer length in the Le Marais catchment. Before flushing, the mass and the pollutant load of the solids in the different types of deposits along the sewer were precisely analysed. These were then compared with the mass and pollutant loads of particles eroded during artificial hydrant flushing. This comparison indicated that the organic layer at the water-sediment interface was the main source of eroded solids. The organic layer was the only type of sediment having pollutant loads of the same order of magnitude as the eroded particles. Type A sediments were too mineral and biofilms had metallic loads that were too high. Moreover, the mass of eroded particles was consistent with the mass accumulated in the surficial 'organic layer', estimated by taking into account the antecedent dry weather period. Similar conclusions were reached from mass balances carried out for a sequence of storms and intermediate dry weather periods. Recently the excess shear stress erosion model developed by Skipworth, 1996 has been applied to the prediction of the flushes (Tait *et al.*, 2003) due to NBS erosion (see below).

A similar approach as applied by Ahyerre was also applied over the whole Le Marias catchment area. The particles eroded during rainfall events (Gromaire 1998) were compared with the particles present in the sewer sediments (Ahyerre 1999) using two criteria: their pollutant loads and mass. This showed that the mass of type A deposits was considerable (184 tonne) in the entire catchment, but these sediments had too high a mineral content

(4 % volatile) to contribute to the suspended particles, as they would not travel high enough in the flow column with typical storm hydraulic conditions. The biofilm mass (56 kg) was much lower than the mass of the observed eroded particles (409 kg). Hence, biofilms could not have been the source of the eroded particles during rainfall events. The only identified type of deposit that could have contributed to the suspended load was thus the interfacial organic layer. Indeed, the solids in this layer had mean organic (volatile solids: 64 %; COD: 1.6 g/g and BOD_5: 0.28 g/g) and heavy metal loads that were of the same order of magnitude as the observed particles being transported in suspension. Moreover, the mass of the organic layer estimated for the different antecedent dry periods was similar to measured eroded masses during rainfall events.

Thus the role of the NBS in CFF and LFF is clearly of major importance, but flushes may also comprise significant amounts of eroding deposits from the bulk of the deposited bed. Data from Brussels suggest that local shear stresses at the level of the interface of only 1 N/m^2 are sufficient to cause bed entrainment contributions to flushes, as illustrated in Figure 2.3.15 (Verbanck, 1995a) and hence this can occur in dry as well as wet weather. Wotherspoon (1994) monitored the evolution of type A sediments in the interceptor sewer of Dundee, using a sediment surface sonar device. Significant erosion of sediments was observed during rainfall events, the erosion being initiated for $\tau > 1.8$ to 2 N/m^2. However, it was not known whether the sediments which were eroded into suspension were from the NBS or bulk bed. Thus it was not possible to define which material contributed to the CFF. Transport as bed load was not monitored, although computations using the Ackers-White formula were used to show that a mass balance could be achieved between the eroding bed and increasing SS in transport.

Laboratory tests in a discrete length pipe channel (Skipworth *et al.*, 1996, Tait *et al.*, 1998) have investigated the link between the prevailing hydraulic conditions and the erosion of fine grained (cohesive-like) organic material from in-pipe deposits into suspension. In the tests the discharge was increased at a uniform rate and then held constant. As the discharge increased the suspended sediment transport rate and concentration were found to increase rapidly to a maximum (Section 2.3.4.5), emulating a flush (Figure 2.3.26). The peak values of suspended sediment concentration and transport were observed as the peak discharge was attained. Following this, the suspended sediment transport rate diminished significantly despite the peak discharge being maintained. Examination of the rate of change in suspended sediment transport in relation to the time varying flows indicated that there was a link between the rate of change of flow discharge and peak solids transport rate. The values of these peaks in suspended sediment appeared to be strongly dependent on the rate of change of the flow discharge and the characteristics of the weak surface layer. The model developed to describe the erosion and subsequent transport rate of solids utilises a two-layer bed (Section 2.3.4.5.3), and a transport rate equation based on excess shear stress (similar in form to Equation 2.3.24).

Further controlled experiments utilising sediment mixtures are attempting to make these results more generally applicable to sewer sediments (Tait *et al.*, 1998; Tait *et al.*, 2002; Saul *et al.*, 2003), and may lead to a better model for flushes.

An example of prediction of flushes at laboratory scale is shown in Figure 3.16. The flow has been increased uniformly with time for 6 and 10 minutes and then maintained constant. Transport is all in suspension. It can be seen that the rate of change of flow (acceleration) has an effect both on the total amount of solids transported during an 'event' and also on the magnitude of the peak transport rate. The bed became more resistant to erosion with depth eroded until a point was reached at which the resistance of the bed to erosion was effectively constant. Modelling presumed a two-layer bed structure, with a weaker surface layer, having an increasing resistance to erosion overlying a stronger layer. More recently this has been applied to erosion events observed in the main Dundee interceptor sewer and in the

Le Marias sewer in Paris. In the former case, sediments at the surface were weak Class C, underlain by coarser Class A material, whereas in Paris the erosion was primarily of the near bed solids (Saul *et al.*, 2003).

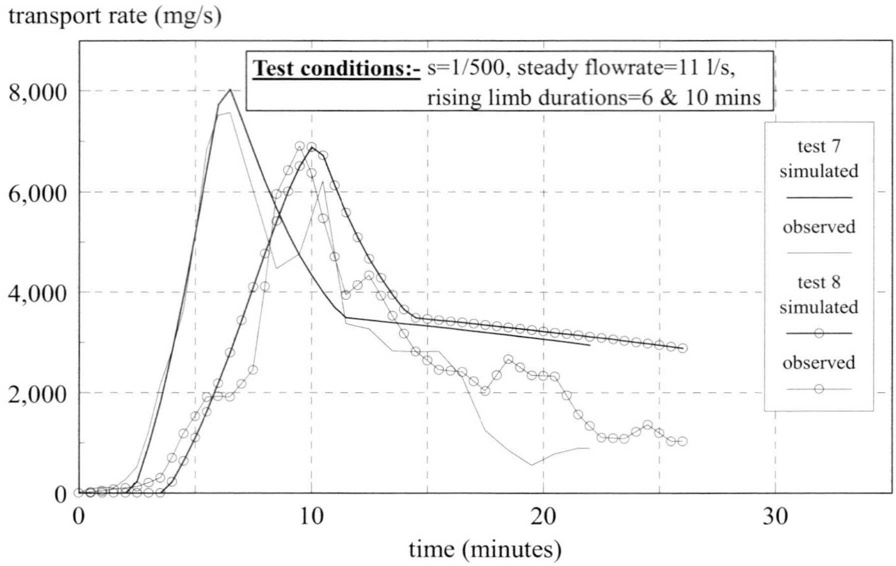

Figure 3.16. Observed and simulated transport rates for ground olivestone eroding from a bed in a circular pipe channel for a ramped hydrograph (Skipworth, 1996).

For this application:

$$E = M\left(\frac{\tau_b - \tau_c}{\tau_c}\right) \tag{3.15}$$

$$\tau_c = \frac{d^{1/b}(\tau_{cu} - \tau_{cs})}{d'^{1/b}} + \tau_{cs} \tag{3.16}$$

where:

τ_{cs} is erosional resistance at the bed surface
τ_{cu} is uniform layer resistance to erosion
τ_c is critical shear stress at point of erosion
τ_b is applied bed shear stress
d′ is depth of the weak layer
d is depth of erosion
b is coefficient to describe the resistance of the weak layer to erosion in relation to depth
E is erosion rate at bed shear stress τ_b
M is erosion rate when $\tau_b = 2\tau_c$

The weak/stronger layer is illustrated in Figure 2.3.26.

Figure 3.17 shows the results for two Dundee storms. The optimised coefficients for the storms using the relationships (Equations 3.15 and 3.16) are given in Table 3.7.

Table 3.7 Optimised bed description parameters for Storms A1 and A2 (Figure 3.17)

Parameters	Storm A1	Storm A2
τ_{cs} (N/m^2)	1.96	0.46
τ_{cu} (N/m^2)	10	10
d′	14.1	2.92
b	0.049	0.330
M (mg/s)	1900	850

As yet, however, linkage of the detailed studies of local sediment erosion to flushes in real sewer catchments as a whole has not been achieved, although simplistic erosion criteria (such as critical bed shear) for sewer flow quality models have been utilised to demonstrate the ability to reproduce observed flushes (e.g. Gent *et al.*, 1996). The linking of sediment erosion to consequential effects such as flushes is in theory possible, and is reported for certain 'physically based' commercial models. Of concern is the apparent ease of prediction given that these models utilise typically only one or two sediment fractions, and the limited state of knowledge about critical erosive stresses and supposed bed erosion (not observed).

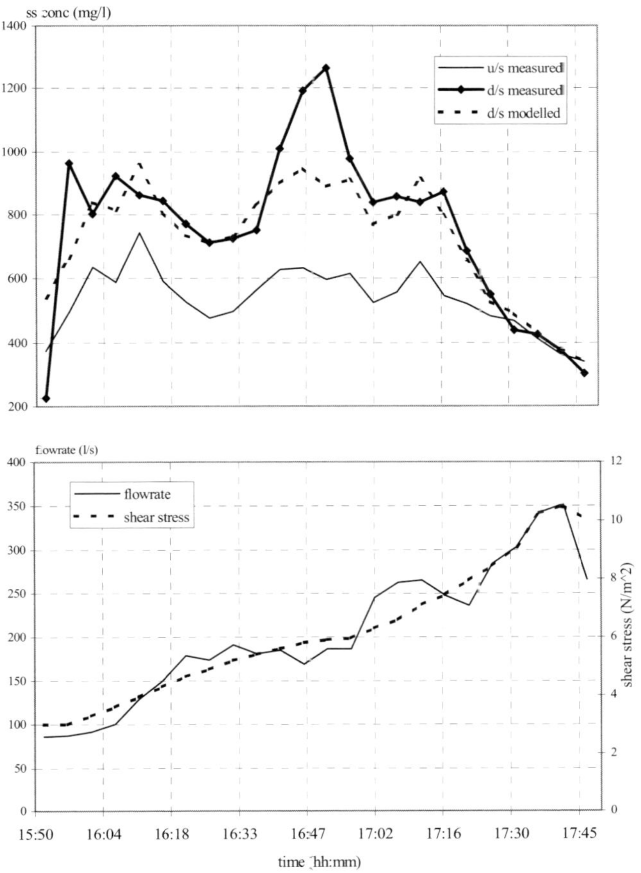

Figure 3.17a. Storm A1

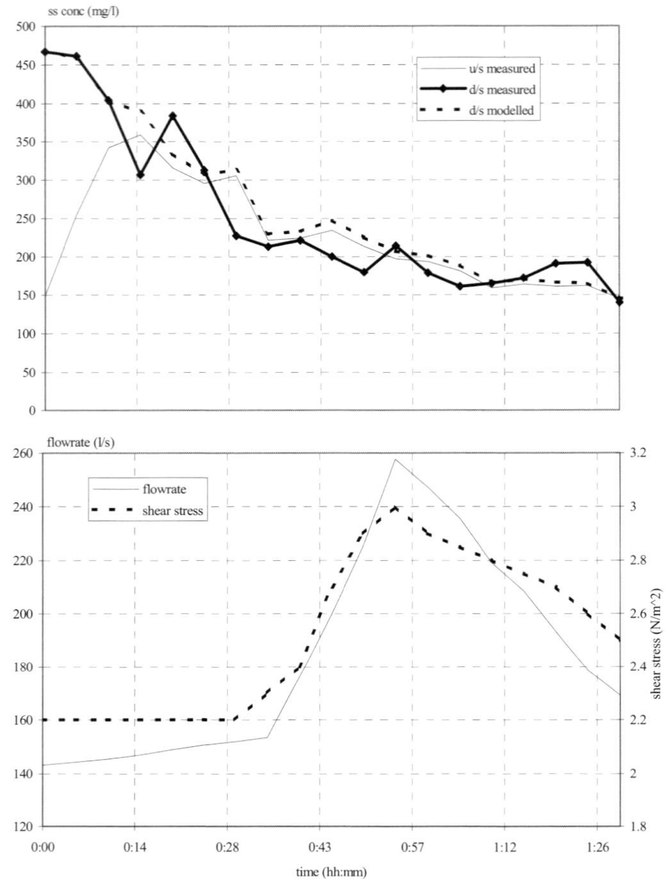

b. Storm A2

Figure 3.17. Comparison of predicted downstream TSS concentration based on upstream measurements using the excess shear stress model (see also Ashley *et al*., 2003).

3.5 EFFECTS OF SEWER SOLIDS ON PERFORMANCE OF ANCILLARIES AND TREATMENT PLANTS

3.5.1 Tanks and CSOs

This section deals with the effects of sewer solids on the hydraulic performance of tanks and CSOs; the separation of solids is not considered. It is therefore appropriate to consider suspended solids and gross solids separately. Where sediment deposits are allowed to accumulate they begin to reduce the hydraulic efficiency of the tank, and may result in blockages, surcharging, flooding and premature overflow operation. In addition, tank deposits are highly organic and their pollutant potential may increase with age. They therefore represent a store of pollutants with consequent increased threat to water quality, and risk of odour problems.

3.5.1.1 *Suspended solids*

Suspended solids in the concentrations usually found in combined sewers do not, in general, result in any detrimental impacts on the hydraulic performance of ancillary structures. When suspended solids settle out in manholes (on benching), tanks or chambers they usually occupy only a small proportion of the structure's capacity, and there are no reported instances of blockages or impaired hydraulic performance due to deposited sediments in chambers. In the UK it appears that most operators aim to construct storage chambers and CSOs which are self-cleansing. This is achieved through the incorporation of sloped benching and other design features which are discussed in Section 4.3. In practice, some sedimentation will occur, although it is generally not significant.

It should be noted that deposited sediments in tanks may create problems of odours, explosion risks, and hazards to sewer workers, in addition to only limited effects on hydraulic performance. Routine maintenance programmes generally ensure that any deposited sediments are cleared from chambers at regular intervals.

3.5.1.2 *Gross solids*

Screens and specific proprietary devices at CSOs and treatment plants have varied effectiveness at retaining gross solids. Retention efficiency is primarily a function of orifice size and it is clear that the nature of the solids is an important influence on efficiency (Balmforth, 1997; Saul *et al.*, 1998); this is likely to vary between catchments reflecting both the nature of the sewerage system and the socio-economic communities contributing to it (Section 2.1.4). A perfectly retentive screen does not exist and, for example, polypropylene-based cotton buds ('cue tips' in the USA) are notorious for orientating themselves in such a way as to escape through even very small orifices. As these are not biodegradable they persist within wastewater systems and, once in the environment, and can even cause damage to drum screens.

Sewage Related Debris (SRD) found in watercourses and on beaches is both aesthetically unpleasant and, under certain circumstances, may also be a health hazard, if from a population in which infectious disease is present. SRD items are evident on coast lines around the world. When comparing UK beach pollution from SRD for 1995 and 1996, the UK Marine Conservation Society found the prevalence of solid waste items on UK beaches to be increasing (up by 68 %), whilst most SRD was found to be reducing, the difference being due to the presence of increasing numbers of cotton buds.

The need to install screens at combined sewer overflows (CSOs) to prevent aesthetic pollution of receiving watercourses (and comply with the Urban Wastewater Treatment Directive in Europe) is now accepted and has prompted investment in developing screen technology and understanding the effectiveness of the various screens and CSO structures available (e.g. Riding *et al.*, 1994); a range of novel screens and screen equivalents are now available (Section 4.3).

In summary, two types of operational problems need to be considered with respect to the larger gross solids.

1 When unusually large gross solids (including for example, rocks, tree branches, traffic cones and other extraneous matter) enter the sewer, these frequently become trapped across outlets, meshes or screens. These tend to produce rapid changes in the hydraulic performance of the system, which may include reverse flows, surcharging of the system and the premature operation of CSOs. This problem may be addressed through the control of inputs into the sewer system, in the design of

gully pots and manholes (Section 4.1), and should be considered in the development of maintenance programmes (Section 4.3).

2 In ordinary operational conditions the gross solids associated with domestic sewage may gradually accumulate across meshes and screens, resulting in a gradual deterioration in the hydraulic performance of the system. Routine maintenance (see Section 4.3) can be scheduled to prevent such deterioration in performance from ever becoming significant.

3.5.1.3 Operation of CSOs

This Section considers the effects of sediments in terms of in-sewer deposition and consequent CSO discharges. There are three effects to consider:

- the 'designed' operation of CSOs;
- the premature operation of CSOs due to downstream sediment deposition;
- the effects of improving CSO discharges by renovation and the consequential effects on the downstream sewerage system.

In Denmark in 1999 there were some 7000 combined sewer overflow (CSO) structures. In the UK in 1993, the Environment Agency estimated there were some 25 000 CSOs, with 20-30 % deemed unsatisfactory. The UK government has specified that in the period 2000-2005, some 4000 currently unsatisfactory CSOs in England and Wales be upgraded. Details of the role of sewer solids in the impact of discharges via CSOs into receiving waters are still not fully understood, although a number of studies have linked solids washout with acute and chronic impacts (e.g. Rees and White, 1993; Ellis and Hvitved-Jacobsen, 1996). There are implications for both the immediate (acute) effects and the longer term chronic effects in terms of the behaviour and assimilation of the solids and the associated pollutants (Section 2.2). In a year-long study of CSO discharges to the Manchester ship canal in the UK, Rees and White (1993) concluded that the main impacts arose from resuspension of sewage derived sediments.

Work from Denmark has shown that the rate of hydrolysis of organic matter into readily biodegradable substrate is highly dependent on the nature of the waste and its attachment to sediments (Vollertsen *et al.*, 1998) (Section 2.2). Under all normal flow conditions particles of very small size or low density may remain in suspension and therefore be transported as a *washload*. Although these particles have a negligible effect on the hydraulic capacity of the sewerage system, they may have a significant impact on the pollutant loading, both in the flow, and at discharge points such as CSOs, where they are spilled without separation. Sediments with higher settling velocities may form deposits under low flow conditions but these may become re-entrained into the flow column with higher velocities due to storms or diurnal variations in flow. In contrast, larger or denser particles may only be transported (as bed load) under peak flow conditions which may be relatively infrequent and therefore may form permanent or semi-permanent, deposits near their point of entry (Ackers *et al.*, 1996). Significant oxygen demands exist within sewerage systems during spill events, leading to an immediate DO sag in receiving waters, and there is an additional longer-term 'downstream' oxygen demand due to sediment, although the disposition of this between 'new' and 'old' sediment is as yet indefinable. The timescale effects are illustrated in Figure 3.18.

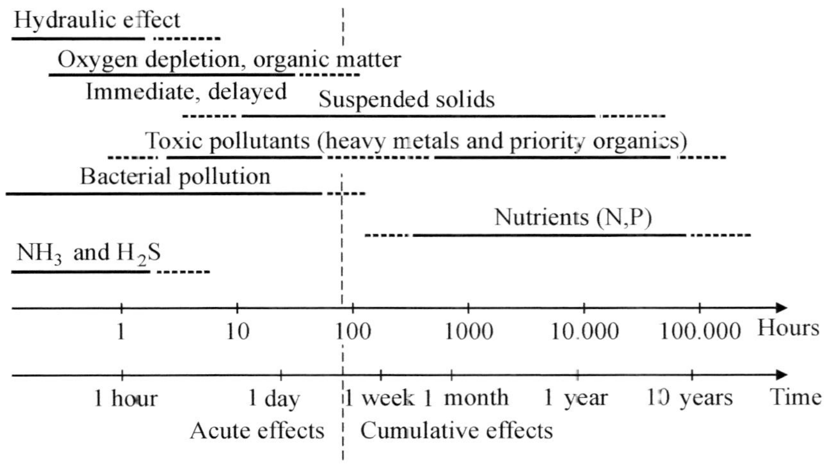

Figure 3.18. Timescale for receiving water effects from intermittent pollutant discharges (Ellis and Hvitved-Jacobsen, 1996).

The current assessment of receiving water impacts, based on CSO discharges of soluble and particulate fractions of COD, is an approach with limitations. The soluble fraction includes substances with different biodegradation characteristics, i.e. inert, readily biodegradable and rapidly hydrolisible COD, and the particulate fraction consists of different types of solids in terms of adsorption characteristics, settling rates and biodegradability, (Vollertsen *et al.*, 1998). Basic microbial characteristics, i.e. biomass growth and corresponding substrate utilisation, is, in the traditional approaches currently used, not taken into account from a process point of view. Furthermore, the description is at present based on loadings at the sewer/receiving water interface and is not linked to the complex solids erosion and washout phenomena in the sewer (Sections 2.2 and 2.3). Figure 3.19 illustrates a result from a simulation model for the performance of a CSO structure during a storm for a downstream loss in pipe cross section of 40 % due to sediment. There is a significant increase in the overflow spill volume and frequency (37 compared with 18 spills per annum, and 6345 m^3 compared with 1621 m^3).

Any system changes which include enhancements to CSO structures reducing either spill volumes or the amounts of solids discharged will inevitably increase downstream loads in the sewer network. Traditionally this potential problem is ignored in design, due to the lack of information available to quantify it. Even where the downstream sewer had an efficient transport capacity, the effects of retaining more solids in the sewer systems are to provide more opportunity for biodegradation and to overload downstream treatment plants.

3.5.2 Pumping stations

Pumping stations are very sensitive to solids. The accumulation of deposits in either the contributing sewers or pump sumps, can lead to blockages or mechanical damage of pump impellers. The mixtures of organic and inorganic solids in sump deposits can also lead to septicity (Section 3.3). Even without such acute effects, the recurrent transport of solids through pump stations can lead to premature abrasion and ageing of pumps. In addition, the large gross solids can accumulate into balls, blocking inlets. Grease and fats can exacerbate these problems (Section 3.6.1). Nonetheless it is recommended that the solids should normally be kept in the flow and the pump station designed accordingly.

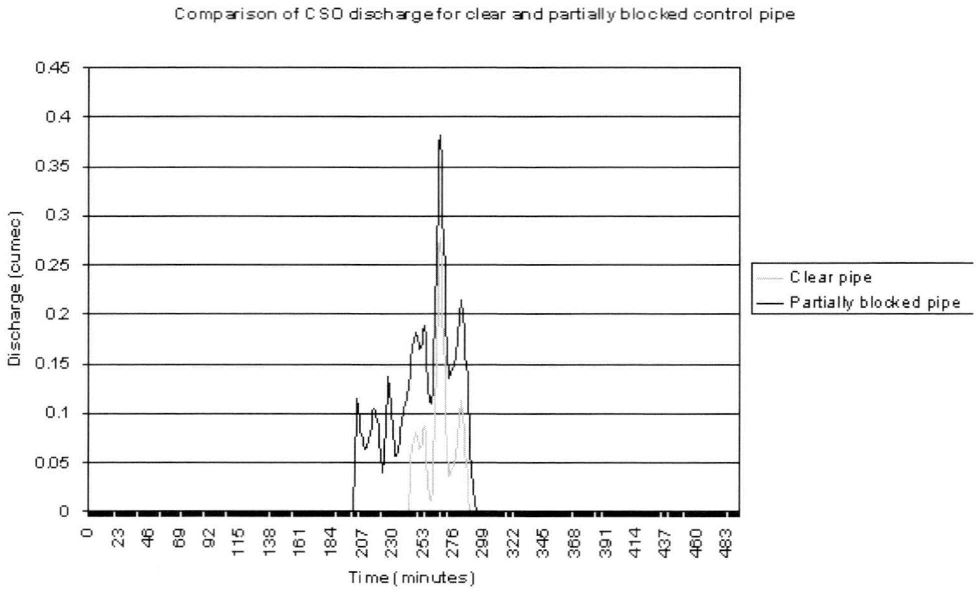

Figure 3.19. Relative spill volumes from a CSO with and without downstream sedimentation in the sewer network.

Alternatively, to protect pumping installations (CEN, 1997), the following devices can be used:

- screening (from decimetric to millimetric mesh size);
- pre-pump station settling basins (with automatic and regular sand extraction);
- grit chambers (Section 4);
- covered and slotted bed load sediment traps (Section 4).

Since the beginning of the 1990s, improved hydraulic design to avoid settling in pump sumps has been developed using pilot-scale studies and by CFD modelling (e.g. Hydraulic Models, 1995).

3.5.3 Treatment plants

During dry weather periods, screening and degritting devices control the solids entering wastewater treatment plants. These devices are usually designed for maximum dry weather conditions, and work well until these conditions are exceeded. Problems can occur, however, during storm weather conditions. These may arise due to flushes of solids overloading the plant, and also where there is significant in-system storage, resulting in dilution of drain down flows following rainfall, with consequent biomass die-off due to lack of nutrients. The latter, however, is typically not a sediment related problem (Ashley *et al.*, 2002).

Understanding of the changes in flow quality which occur in sewer systems prior to arrival at end of pipe treatment plants has advanced considerably in the last decade (Section 2.2.2). Under aerobic conditions in sewers with retention times of at least 1 hour there may be significant removal of the readily biodegradable COD, although in pressure mains, where conditions are typically anaerobic (Section 3.3) this does not occur. In the former case, the COD is transformed into particulates, potentially increasing the suspended solids load. It is possible, however, that where sediment transport capacity is low, significant solids may deposit within the sewerage system prior to arrival at the treatment plant (Section 2.3). From pumping mains, the higher hydrogen sulphide content is a triggering factor for the development of filamentous microorganisms in the treatment plant, potentially leading to sludge bulking, and inhibiting the nitrification process (Hvitved-Jacobsen *et al.*, 1998). Other problems occur from odours arising when anaerobic conditions prevail (Section 3.6.2).

3.5.3.1 Screening

During storm weather, due to the wash-off of solids accumulated on catchment surfaces and the erosion of solids deposited in the sewers, the mass of solids collected by screens can be 5 to 10 times greater than during dry weather periods (Section 2.1.4) (Sidwick, 1984; Kassner, 1987). Consequently it is necessary to adapt the cleaning frequency of screens to avoid blockage.

3.5.3.2 Degritting

The solids loads arriving at wastewater treatment plants during storm periods can be significantly greater than during dry weather periods. Londong (1990) indicates grit volumes ranging from 20 to 200 litres/1000 m^3 with a mean value of about 60 litres/1000 m^3. The highest values are reached during storm events, especially in summers and after long dry periods. It is interesting to compare these values to usual French design ratios for degritting devices: 5–12 litres per 1000 m^3 (Degrémont, 1989).

There is a risk the degritting device will become overloaded and block. Similarly to screens, it is necessary to extract the grit more frequently during or just after storms. For particular cases where loads during storm periods are very high (i.e. steep catchments with runoff on natural soils connected to the sewers), it is recommended that a specific grit chamber or sediment trap be installed upstream of any preliminary treatment unit. This device may be equipped with a sand screw conveyor.

The emptying and cleaning of online or offline storage tanks, with or without sedimentation, should also be taken into account; the discharges from these tanks can lead to large masses of solids entering wastewater treatment plant (Bertrand-Krajewski *et al.*, 1995) (see Figure 3.20).

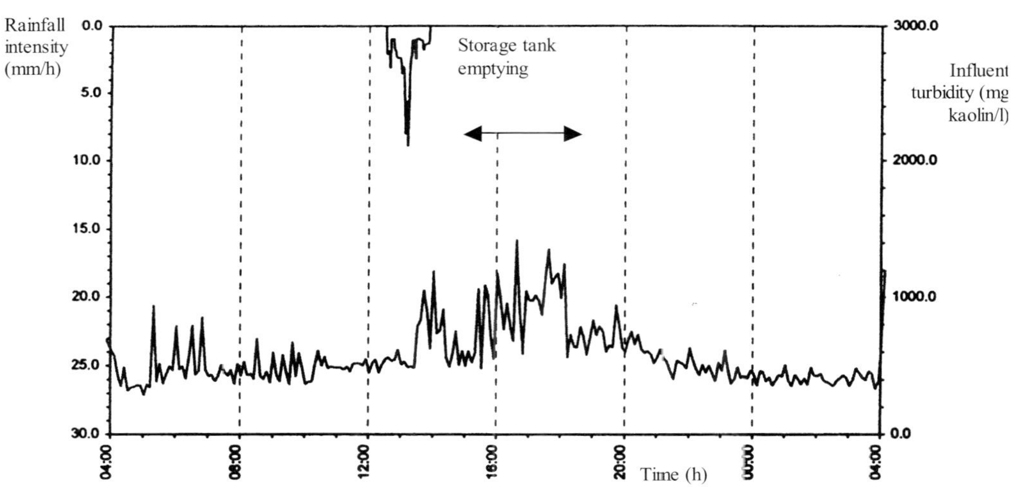

Figure 3.20. Turbidity variation at the inlet of a treatment plant (Boran-sur-Oise, France).

The solids that are not intercepted in any pretreatment stage can lead to mechanical impacts. The following have been observed:

- premature pump impeller abrasion;
- deposition and clogging of pipes;
- deposits in the treatment tanks, which progressively reduces the capacity and leads to mixing problems and interactions with biological processes
- deposits in digesters.

3.5.3.3 Primary clarification

During storms the production of primary sludge increases considerably, and sludge removal must be adapted accordingly to avoid consequential disturbances. The increasing mineral content of these solids, due to the storm conveyance of greater masses of grit, results in improved thickening characteristics of the sludge; whilst the quantity of sludge as a proportion of the total volume increases significantly, the quantity of inert solids may increase proportionately by a greater amount. Kassner (1987) gives an example for which an observed sludge volume increased by a factor of two and was associated with a five-fold increase in the volume of inert solids, due to a better thickening efficiency.

This increase in mass and in mineral content should be taken into account by the plant operator, especially when biological digestion is being utilised (Fayoux and Herremans, 1995). If the sludge extraction rate is not adapted to this increase, the resulting unavoidable storage of sludge in the primary clarifier will lead to detrimental anaerobic conditions. Oversizing the primary clarifier to take heavy storm periods into account should be avoided, otherwise the residence time will be excessive during dry weather periods. This leads to settlement of too high fractions of organic particles, with resultant problems for the primary sludge treatment process due to anaerobic conditions. A better solution for storm weather periods is to install a separate primary clarifier or a specific plant for storm water treatment (see Section 4.3).

3.5.3.4 Pollutant removal

During storm events, the increase in masses of solids entering the wastewater treatment plant can lead to problems affecting the biological processes. As the greatest part of the COD or BOD_5 is attached to the solids, the normal equilibrium of removal processes between biological growth and hydrolysis and adsorption is changed, with more adsorption. As adsorption is rapid in comparison with residence time in the aeration tank, the overall efficiency of pollutant removal may be sustained during wet weather, at least for plants working with low BOD_5/MLVSS/day ratios.

If the influent mass of solids is large, and the mineral fraction high, the volatile solids/total solids ratio in the activated sludge can decrease, leading to less efficient biomass activity per unit mass of suspended solids of activated sludge during the hours or days following the rainfall event. Unfortunately, this negative effect is usually reinforced by the normal daily 'excess' sludge extraction which is often increased at such times to maintain the total solids concentration of activated sludge at a specified level. A better option would be to maintain a constant concentration of active biomass. Fortunately problems from this 'imbalance' in mineral compared with organic solids coming into treatment plants appears to be limited for typical plants.

Nitrification and denitrification processes are sensitive to toxics, such as heavy metals ions. However, these metals do not significantly disturb the biological processes if their concentrations are lower than 1 mg/l. As the heavy metals in surface runoff are mainly associated with the fine solids (Section 2.1) or are not easily bioavailable, they exert little impact on treatment plant biological processes. There is no evidence that metal inputs from sewers during storm conditions ever reach levels which impair wastewater or sludge treatment processes (Lijklema *et al.*, 1993).

Denitrification requires easily available or soluble carbon. During storm events, changes in influent characteristics compared with dry weather flows occur:

- larger fractions of COD and BOD_5 are attached to solids;
- the COD/BOD_5 ratio increases, indicating that the carbonaceous pollutants are less easily biodegradable.

As a consequence denitrification efficiency may be reduced.

3.5.3.5 Sludge treatment and dewatering

During storm events sludge production in treatment plants typically increases significantly in response to increasing influent solids. As biological treatment processes operate most efficiently with stable inputs, this can cause disruption. The resultant effects include too much sludge being retained in treatment tanks and/or clarifiers.

Aerobic stabilisation is typically very sensitive to influent load variations, and these should be avoided where possible. Anaerobic digestion is less sensitive to variations in inputs, although daily load increases should not exceed 25 or 30 %.

Specific and separate sludge treatment units should be utilised where sludge loads increase significantly during storm weather periods. If the increase is not significant, specific temporary storage should be installed in association with less sensitive and more reliable sludge treatment processes such as aerobic thermophilic stabilisation (which can operate intermittently) or two-phase anaerobic digestion (Fayoux and Herremans, 1995). However, these problems are not fully understood, and further research is necessary to improve knowledge and develop reliable solutions.

Solids load increases during storm events should also be taken into account for the design of dewatering devices (band or press filtration, centrifugation, etc).

3.5.3.6 Sludge disposal

Many pollutants attached to solids such as heavy metals, are stored in sludge following storm periods. If concentrations are too high to meet the legal requirements for agricultural use, the sludge must be disposed of in alternative ways, such as in landfill for toxic residues or incinerated where facilities are available.

3.6 OTHER EFFECTS

3.6.1 Fats and greases

Fats, oils and greases have always been a problem in sewer systems; they can build up in sewers, coating walls and allowing grease balls up to 1 m in diameter to form. The increasing use of dishwashers is encouraging solids and grease to be more efficiently removed from the household and to be conveyed further downstream before precipitating. Surprisingly there appears to have been no consistent research studies published about these problems, although they are referred to in a number of proprietary guidance documents. See for example, Stuth and Guichard (1989), Stall and Gupta (1997) and Thermaco Inc. (1997).

Although many countries require traps to be installed prior to commercial discharges into sewers, these are rarely maintained and most evidence of deposition occurs in areas with a high population of restaurants and similar premises. Fats and greases can bind other solids, particularly gross solids, and if retained, can produce odour problems. These problems are particularly noticeable where flows are slack or intermittent, such as upstream of pump stations and where outfalls connect to the sea and tidal effects reduce velocities.

3.6.2 Odours

Odours from sewer systems are usually attributable to the biodegradation of organic matter (fats, protein and carbohydrate), often under anaerobic conditions, leading to the release of volatile compounds into the atmosphere. Odours may originate from deposits of sewer solids and from the sewer itself. Any differentiation between the importance of the wastewater and the solids in a sewer is typically not possible, however, anaerobic conditions in sewer deposits typically prevail and are a potential source of odours.

A number of volatile compounds may be produced from sewer solids under anaerobic conditions. Odours generated include hydrogen sulfide, organic sulfides (mercaptanes), and volatile organic fatty acids (VFAs), such as acetic, propionic and butyric acids. In addition to hydrogen sulfide as a major nuisance, mercaptanes are typically found when odour problems are identified. Among other organic compounds, indole and skatole are particular problematic due to their persistence.

The nuisance of odour is well known to be related to the management of biosolids from treatment plants. The public has become less tolerant of odours. This progressive change in attitude combined with improved standards for emission has encouraged sewerage system operators to ensure that the nuisance is reduced.

In sewer systems, a change from anaerobic to an anoxic or aerobic state will normally reduce the odour problems considerably. Other means for reducing problems in a sewer system and covered storage of biosolids is to collect the odour-laden air and let it undergo treatment, often including combined physical, chemical and biological processes, e.g. oxidation (in addition to oxygen also hydrogen peroxide, ozone and chlorine), adsorption, absorption, incineration, ionisation, UV photolysis, stabilisation and biofilm processes. A great number of such processes and techniques exist (e.g. CIWEM, 1999).

3.6.3 Explosions and hazards to sewer workers

There is little evidence that sewer solids have resulted in in-sewer explosions caused by methane build-up as a result of sewer solid degradation. Problems have more often occurred as a result of natural occurrences of this gas from surrounding soils and the discharge of hydrocarbons.

Asphyxiation of sewer workers due to hydrogen sulphide and/or low oxygen is an extensively reported problem. As described in Sections 2.2 and 3.2, these problems can occur due to the presence of sewer solids.

3.6.4 Miscellaneous

Of less consequence than the preceding sections, are potential problems caused by rats feeding from the solids conveyed in flows and in areas of deposition. Although rats are generally ubiquitous in sewer systems, this is not believed to be a major problem unless industrial or commercial premises discharge significant amounts of edible solids or fats and greases (Ashley *et al.*, 2000).

3.7 References

Ackers, P. (1961) Hydraulic resistance of drainage conduits. *Proc. Inst. Civil Engineers*. 19 July, 307-336.

Ackers, P. (1991) Sediment aspects of drainage and outfall design. *Proc. Int. Symp. on Environmental Hydraulics*, Hong Kong, edited by J.H.W. Lee and Y.K. Cheung, A.A. Balkema, Rotterdam, 19-29.

Ackers, P., Harrison, A.J.M. and Brewer, A.J. (1968) The hydraulic design of overflows incorporating storage. *J. Inst. Mun. Eng.*, 95, 31–37.

Ackers, J.C., Butler, D. and May, R.W.P. (1996) *Design of Sewers to Control Sediment Problems*. CIRIA project report 141. ISBN 086017 443 3. 181 pp.

Ahyerre, M. (1999) *Bilan et mécanismes de migration de la pollution organique en réseau d'assainissement unitaire*. PhD thesis, University Paris 6.

Ahyerre, M., Chebbo, G. and Saad, M. (2000) Sources and erosion of organic solids in a combined sewer. *Urban Water*, No. 2, 305–315.

Arthur, S. (1996) *Near Bed Solids Transport in Combined Sewers*. PhD thesis, University of Abertay Dundee.

Arthur, S., Ashley, R.M. and Nalluri, C. (1996) Near bed solids transport in sewers. *Wat. Sci. Tech.* 33(9), 69–76.

Arthur, S. and Ashley, R.M. (1998) Near bed solids transport and first foul flush in combined sewers. *Wat. Sci. Tech.*, 37(1), 131–138.

Ashley, R.M. and Bertrand-Krajewski, J.-L. (1993) Sewer sediment origins and transport in small catchments. *Proc. 6th ICUSD*, Niagara Falls, Canada, 12–17 September 1993, Vol. 1, 772–777.

Ashley, R.M., Longair, I.M., Wotherspoon, D.J.J. and Kirby, K. (1993) Flow and sediment movement monitoring in large sewers. *Wat. Sci. Tech.* 28(11–12), 55–62.

Ashley, R.M., Marr, I. and Watson, N. (2000) *Rodent Control in Sewers*. UKWIR Report Ref. No. 00/WM/07/4, 273 pp.

Ashley, R.M., Blanksby, J., Dudley, J., Saul, A.J., Smith, H. and Vollertsen, J. (2002) *Impact of Extended Storage on Sewage Treatment*. Report Reference No. 02/WW/04/3 UKWIR, ISBN 1-84057-266-3.

Ashley, R.M., Crabtree, R.W., Fraser, A. and Hvitved-Jacobsen, T. (2003) Recent European research into the behaviour of sewer sediments and associated pollutants and processes. *ASCE J. Hydr. Eng.*, 129(4), April, 267–275.

Bachoc, A. (1992) *Le transfert des solides en réseaux d'assainissement unitaires*. PhD thesis, INPT, Toulouse, France, 281pp + appendices.

Balmforth, D. (1997) *Aesthetic Pollutant Modelling*. UK Wastewater Planners Users' Group Conference, Blackpool, November.

Bertrand-Krajewski, J.L., Lefebvre, M. and Audic, J.M (1995) Flow and pollutant measurements in a combined sewer system to operate a waste water treatment plant and its storage tank during storm events. *Wat. Sci. Tech.*, 31(7), 1–12.

Bertrand-Krajewski, J.-L., Chebbo, G. and Saget, A. (1998) Distribution of pollutant mass versus volume in stormwater discharges and the first flush phenomenon. *Water Research*, 32(8), 2341-2356.

Boon, A.G. and A.R. Lister (1975) Formation of sulphide in rising main sewers and its prevention by injection of oxygen. *Prog. Wat. Tech.* 7(2), 289–300.

Butler, D., May, R.W.P. and Ackers, J.C. (1996) Sediment transport in sewers: Part 1. Background. *Proc. Instn. Civ. Engrs Wat., Marit. and Energy*, 118, 103–112.

CEN (1997) *Drain and Sewer Systems outside Buildings: Part 6. Pumping installations*. European Draft Standard prEN 752-6, CEN, Brussels, Belgium, November, 29 pp.

Chebbo, G., Ashley, R M. and Gromaire, M.C. (2002) The nature and pollutant role of solids at the water-sediment interface in combined sewer networks. *Proc. 3rd Int. Conf. on Sewer Systems and Processes*, Paris, April.

CIRIA (1986) *Sediment movement in combined sewerage and storm-water drainage systems: Phase 1. Project report*. CIRIA research project No. 366, Binnie and Partners and Hydraulics Research.

CIWEM (1999) Proceedings from CIWEM (The Chartered Institution of Water and Environmental Management) and IAWQ Joint International Conference on Control and Prevention of Odours in the Water Industry, London, UK, 22–24 September 1999.

Dauber, L. and Novak, B. (1983) *Quellen und Mengen der Schmutszstoffe in Regenabflussen einer stadtischen Mischkanalisation*. Eidgenossische Technische Hochschulen (ETH), Zurich, Separatum Nr. 927.

Davies, J.W. (1987) Laboratory studies related to the first foul flush in combined sewers. *Proceedings of 4th International Conference on Urban Storm Drainage*, Lausanne, Switzerland, 47–48.

Davies, J.W. (1990a) Laboratory study relating to the modelling of stormwater quality in combined sewers. *Proceedings of 5th ICUSD*, Osaka, Japan, 215–219.

Davies, J.W. (1990b) *Modelling of Storm Water Quality in Combined Sewers*. PhD thesis, Polytechnic of Central London.

Davies, J.W., Xu, Y. and Butler, D. (1996) Laboratory study and computer model development related to gross solids movement in combined sewers. *Wat. Sci. Tech.*, 33(9), 39–47.

Degrémont (1989) *Mémento technique de l'eau.* Editions Lavoisier, Paris, 1989, 1500 pp.

Deletic, A. (1998) The first flush load of urban surface runoff. *Water Research*, 32(8), 2462–2470.

Ellis, J.B. and Hvitved-Jacobsen T. (1996) Urban drainage impacts on receiving waters. *J. Hydr. Res.*, 34(6), 771–783.

Fayoux, C. and Herremans, L. (1995) Impacts of storm weather discharges on the sludge treatment process of waste water treatment plant: state of the art. *Proceedings of SHF/AGHTM conference 'Control of urban storm drainage: from theory to practice'*, Créteil, France, 11–12 October, 245-258 (in French).

FNDAE (1986) *Stormwater tanks in sewer systems*. FNDAE, Paris, France, Technical Report No. 6, 61 pp. (in French).

Geiger, W.F. (1984) Characteristics of combined sewer runoff. *Proceedings of 3rd International Conference on Urban Storm Drainage*, Göteborg, Sweden, 4–8 June, 851–860.

Geiger, W.F. (1987) Flushing effects in combined sewer systems. *Proceedings of 4th International Conference on Urban Storm Drainage*, Lausanne, Switzerland, 31 Aug.–4 Sept., 40–46.

Gent, R., Crabtree R.W. and Ashley, R.M. (1996) A review of model developmennt based on sewer sediments research in the UK. *Wat. Sci. Tech.*, 33(9), 1–7.

Göttle, A. (1978) *Ursachen und Mechanismen der Regenwasserverschmutzung. Berichte aus Wassergütewirtschaft und Gesundheitsingenieurwesen*, Heft nr 23, Technische Universität München, Germany, 313 pp. + appendices.

Gromaire, M.-C., (1998) *La pollution des eaux pluviales urbaines en réseau d'assainissement unitaire, caractéristiques et origines*. PhD thesis, ENPC, Paris, France, 503pp. + appendices.

Gromaire-Mertz, M.-C., Chebbo, G. and Saad, M. (1998) Origins and characteristics of urban wet weather pollution in combined sewer systems: the experimental urban catchment 'le Marais' in Paris. *Wat. Sci. Tech.*, 37(1), 35–43.

Gromaire, M.-C., Garnaud, S., Saad, M. and Chebbo, G. (2001a) Contribution of different sources to the pollution of wet weather flows in combined sewers. *Wat. Res.*, 35(2), 521–533.

Gromaire, M.-C., Garnaud, S., Moilleron, R. and Chebbo, G. (2001b) Contribution des différentes sources à la pollution des effluents unitaires de temps de pluie. *TSM*, n°5, 72–80.

Gupta, K. and Saul, A.J. (1996) Specific relationships for the first flush load in combined sewer systems. *Wat. Res.*, 30(5), 1244–1252.

Harrison, A.J.M. and Holmes, D.W. (1967) *The Movement of Stormwater in Combined Sewers*. Report No. INT 73, Hydraulics Research Station, Wallingford, UK.

Henderson, R.J. (1984a) *A guide to hydraulic roughness in sewers*. WRc external report No. 131E, March.

Henderson, R.J. (1984b) The hydraulic roughness of used sewers. *Proc. Int. Conf. Planning, Construction, Maintenance and Operation of Sewerage Systems*, Reading, BHRA, September.

Hoffman, R. and Meighan, R. (1984) The impact of combined sewer overflows from San Francisco on the western shore of Central San Francisco Bay. *Journal Water Pollution Control Federation*, 56(12), 1277–1285.

Horton, T. (1901) Flow in the sewers of the North Metropolitan System of Massachusetts. *Trans. ASCE*, 56.

Hvitved-Jacobsen, T. (2002) *Sewer Processes*. CRC Press, ISBN 1-56676-926-4.

Hvitved-Jacobsen, T., Keiding, K. and Yousef, Y.A. (1987) Urban runoff pollutant removal in wet detention ponds. *Proceedings of 4th International Conference on Urban Storm Drainage*, Lausanne, Switzerland, 137–142.

Hvitved-Jacobsen, T., Jütte, B., Nielsen, P.H. and Jensen, N.Aa. (1988) Hydrogen sulphide control in municipal sewers. *Pretreatment in Chemical Water and Wastewater Treatment: Proceedings of 3rd International Gothenburg Symposium*, edited by H.H. Hahn and R. Klute, Gothenburg, Sweden, 1–3 June, Springer-Verlag, 239–247.

Hvitved-Jacobsen, T., Raunkjær, K. and Nielsen, P.H. (1995) Volatile fatty acids and sulphide in pressure mains. *Wat. Sci. and Tech.*, 31(7), 169–179.

Hydraulic Models (1995) *Modélisation hydraulique de la station des Quatre Ecluses à Coudekerque. Rapport Hydraulic Models*. Lyonnaise des Eaux, Dunkerque, France, 36 pp.

Institution of Sanitary Engineers (1954) *Field Research into Drainage Problems.* Report of the Joint Committee, London.

Jewell, T.K. and Adrian, D.D. (1982) Statistical analysis to derive improved stormwater quality models. *JWPCF*, 54(5), 489–499.

Johansen, N.B. (1985) *Discharge to Receiving Waters from Sewer Systems during Rain*. PhD thesis, Technical University of Denmark, Lyngby, Denmark, 287 pp.

Kassner, W. (1987) *Regenbecken und Klärbetrieb. Planung, Bau und Betrieb von Regenentlastungen. Fachgespräch Kanalnetzsteuerung und Regenentlastungen.* Berichte aus Wassergütewirtschaft und Gesundheitsingenieurwesen, Technische Universität München.

Kolsky, P. (1998) *Storm Drainage: An Engineering Guide to the Low-Cost Evaluation of System Performance.* IT pub., London.

Krauth, K.H. (1970) *Der Abfluss und die Verschmutzung des Abflusses in Mischwasserkanalisationen bei Regen*. Stuttgarter Berichte zur Siedlungswasserwirtschaft, Heft 45, München, Germany, Oldenburg, 251 pp.

Krejci, V., Dauber, L., Novak, B. and Gujer, W. (1987) Contribution of different sources to pollutant loads in combined sewers. *Proc. 4th Int. Conf. Urban Storm Drainage*, Lausanne, Switzerland, 34-39.

Laursen, E.M. (1956) *The Hydraulics of a Storm-Drain System for Sediment-Transport Flow*. Bulletin No. 5, Iowa Highway Research Board, USA.

Lijklema, L., Tyson, J.M. and Lesouëf, A. (1993) Interactions between sewers, treatment plants and receiving waters in urban areas: a summary of the Interurba'92 workshop conclusions. *Wat. Sci. Tech.*, 27(12), 1–29.

Lillywhite, M.S. (1976) *Current Trends and Research in Building Drainage System Design.* IoP Plumbing Symposium, June, Bristol, UK.

Lillywhite, M.S. and Webster, C.J.D. (1976) Investigations of drain blockages and their implications for design. *Journ. Instn. of Pub. Hlth. Engrs.*, 7(2), 53–60.

Lindholm, O. and Aaby, L. (1989) In-pipe flushing and its implication for overflow quality. *Urban Discharges and Receiving Water Quality Impacts*, edited by J.B. Ellis, Oxford, UK, Pergamon Press, 17–25.

Londong, J. (1990) Sandfanggut in kommunalen Kläranlagen. *Korrespondenz Abwasser*, 12, 1473-1475.

Marine Conservation Society. Annual nationwide beach-clean and survey reports, available from 9 Gloucester Road, Ross-on-Wye, Herefordshire, HR9 5BU, UK.

May, R.W.P. (1982) *Sediment Transport in Sewers*. Report No. IT 222, Hydraulics Research Station, February.

May, R.W.P. (1993) *Sediment Transport in Pipes and Sewers with Deposited Beds*. Hydraulics Research Report SR320.

May, R.W.P., Brown, P.M., Hare, G.R. and Jones, K.D. (1989) *Self-Clearsing Conditions for Sewers Carrying Sediment*. Hydraulics Research Report SR221.

May, R.P.W., Martin, P. and Price, N.J. (1998) *Low-Cost Options for Prevention of Flooding from Sewers.* CIRIA Report C506.

McGregor, I., Ashley, R.M. and Oduyemi, K.O.K. (1993) Pollutant release in sewer systems and their release into receiving waters. *Wat. Sci. Tech.*, 28(8–9), 161–169.

Melbourne and Metropolitan Board of Works (1989) *Hydrogen Sulphide Control Manual: Septicity, Corrosion and Odour Control in Sewerage Systems.* Technological Standing Committee on Hydrogen Sulphide Corrosion in Sewerage Works, Vols 1 and 2.

Michelbach, S. (1995) Origin, resuspension and settling characteristics of solids transported in combined sewage. *Wat. Sci. Tech.*, 31(7), 69–76.

Michelbach, S. and Wöhrle, C. (1992) Settleable solids in a combined sewer system: measurement, quantity, characteristics. *Wat. Sci. Tech.*, 25(8), 181–188.

Nielsen, P.H. and Hvitved-Jacobsen, T. (1988) Effect of sulfate and organic matter on the hydrogen sulphide formation in biofilms of filled sanitary sewers. *J. WPCF*, 60, 627–634.

Pearson, L.G., Thornton, R.C., Saul, A.J. and Howard, K. (1986) An introductory analysis of the factors affecting the concentration of pollutants in the first foul flush of a combined storm sewer system. *Proceedings of International Symposium on Urban Stormwater Quality and Effects upon Receiving Waters*, Wageningen, The Netherlands, 93–102.

Perkins, J.A. and Gardiner, I.M. (1982) *Measurements of the Hydraulic Roughness of Slimed Sewer Pipes*. Report No. IT 237, Hydraulics Research Station, July.

Perrusquía, G., Lyngfelt, S. and Sjöberg, A. (1987) Flow capacity of sewers with a sediment bed. *4th International Conference on Urban Storm Drainage*, Lausanne, Switzerland, 331–336

Philippe, J.-P. and Ranchet, J. (1987) *Pollution of Runoff Water in Urban Areas: Synthesis of data from ten catchments around Paris*. LCPC Research Report n° 142, Paris, France, Laboratoire Central des Ponts et Chaussées (in French).

Pomeroy, R.D. and Parkhurst, J.D. (1977) The forecasting of sulphide build up rates in sewers. *Prog. Water Technol.*, 9(3), 621–628.

Reiff, H. (1992) *Wachstum und Abtrag der Sielhaut in Mischwasserkanälen*. Heft 9 der Schriftenreihe Wasser – Abwasser – Abfall, Verein zur Förderung des Fachgebiet-siedlungswasserwirtschaft an der Universität Kassel.

Riding, E., Balmforth, D.J. and Green, M. (1994) *Monitoring the Performance of Screens at Combined Sewer Overflows*. WRc Report UM 1433.

Saget, A., Chebbo, G. and Bertrand-Krajewski, J.-L. (1996) The first flush in sewer systems. *Wat. Sci. Tech.*, 33(9), 101–108.

Saul, A.J., Gupta, K. and Stovin, V.R. (1998) Gross solids characteristics and an event-based CSO efficiency prediction methodology. *Proceedings of 4th International Conference on Developments in Urban Drainage Modelling*, 21–24 September 1998, London, UK, 373–380.

Saul, A.J., Skipworth, P.J., Tait, S. and Rushforth, P.J. (2003) Movement of total suspended solids in combined sewers. *J. Hydr. Engg. ASCE*, April, 298–307.

Sidwick, J.M. (1984) *Screenings and Grit in Sewage: Removal, Treatment and Disposal. Preliminary report.* CIRIA/WRc Technical Note No. 119, CIRIA, London, UK.

Skipworth, P.J., Tait S.J. and Saul A.J. (1996) Laboratory investigations into cohesive sediment transport in pipes. *Wat. Sci. Tech.*, 33(9), 187–193.

Skipworth, P., Tait, S. and Saul, A.J. (2000) The first foul flush in combined sewers: an investigation of the causes. *Urban Water*, 2(4), 317–326.

Stahre, P. and Urbonas, B. (1990) *Stormwater Detention for Drainage, Water Quality and CSO Management*, Englewood Cliff, N.J., USA, Prentice Hall, 338 pp.

Stall, U. and Gupta, H. (1997) Management strategies for oil and grease residues. *Waste Management and Research*, 15, 23–32.

Stanley, D.W. (1996) Pollutant removal by a stormwater dry detention pond. *Water Environment Research*, 68(6), 1076–1083.

Stokes, R.F. and Morris, J. (1996) Planning and rehabilitation of water distribution and sewerage systems. *CIWEM Yearbook*, UK, 17–23.

Stotz, G. and Krauth, K.H. (1984) Factors affecting first flushes in combined sewers. *Proceedings of 3rd International Conference on Urban Storm Drainage*, Göteborg, Sweden, 4–8 June, 869–878.

Stotz, G. and Krauth, K.H. (1986) Depositions in combined sewers and their flushing behaviour. *Urban Drainage Modelling*, edited by C. Maksimovic, London, UK, Pergamon Press, 443–453.

Stuth, W.L. and Guichard, M. (1989) Managing grease and oil in restaurant waste. *Proc. of 6th Northwest On-Site Wastewater Treatment Short Course*, University of Washington, Seattle, USA, 98–113.

Swaffield, J.A. and Gallowin, L.S. (1992) *The Engineered Design of Building Drainage Systems*. Asgate Pub., ISBN 1-85742-0489.

Tait, S.J., Ashley, R.M., Verhoeven, R., Clemens, F. and Aanen, L. (2002) Sewer sediment transport studies using an environmentally controlled annular flume. *Water Sci.Tech.*, 47(4), 51–60.

Tait, S.J., Chebbo, G., Skipworth, P.J., Ahyerre, M. and Saul, A.J. (2003) Modelling in-sewer deposit erosion to predict sewer flow quality. *J. Hydr. Engg. ASCE*, April, 316–324.

Tait, S.J., Rushforth, P.J. and Saul, A.J. (1998) A laboratory study of the erosion and transport of cohesive like sediment mixtures in sewers. *Wat. Sci. Tech.*, 37(1), 163–170.

Tchobanoglous, G. (ed.) (1981) *Wastewater Engineering: Collection and pumping of wastewater*, Metcalf and Eddy, Inc., McGraw Hill Book Co., New York.

Thermaco Inc. (1997) *Understanding Kitchen Flows and Using Automatic Grease Removal Systems*. Asheboro, N. Carolina, USA.

Thistlethwayte, D.K.B. (ed.) (1972) *The Control of Sulphides in Sewerage Systems*. Butterworth, Sydney.

Thornton, R.C. and Saul, A.J. (1987) Temporal variation of pollutants in two combined sewer systems. *Proceedings of 4th International Conference on Urban Storm Drainage*, Lausanne, Switzerland, 51–52.

Tomlinson, R., Bebee, B., Heyward, A., Munger, S. and Swartz, R. (1980) *Fate and Effects of Particulates Discharged by Combined Sewers and Storm Drains*. EPA-Report: EPA 600/2-80-111, USA, August.

USEPA (1974), *Process Design Manual for Sulphide Control in Sanitary Sewerage Systems.* USEPA, Technology transfer. Washington, D.C.

Verbanck, M.A. (1995a) Capturing and releasing settleable solids: the significance of dense undercurrents in combined sewer flows. *Wat. Sci. Tech.*, 31(7), 85–93.

Verbanck, M. (1995b) *Transferts de la charge particulaire dans l'égout principal de la ville de Bruxelles*. PhD thesis, Université Libre de Bruxelles, 1995, 193 pp.

Vollertsen, J., Hvitved-Jacobsen, T., McGregor, I. and Ashley, R.M. (1998) Aerobic microbial transformations of pipe and silt trap sediments from combined sewers. *Wat. Sci. Tech.*, 39(2), 233–242.

Water Research centre (2001) *Sewerage Rehabilitation Manual*. 4th edition, CD ROM.

Wilmot, P.D., Cadee, K., Katinic, J.J. and Kavanagh, B.V. (1988) Kinetics of sulphide oxidation by dissolved oxygen. *J. WPCF*, 60, 1264–1270.

Wise, A.F.E. and Swaffield, J.A. (1995) *Water, Sanitary and Waste Services for Buildings*, 4th edition, Longman Scientific and Technical, Harlow, UK.

Wotherspoon, D.J.J. (1994) *The Movement of Cohesive Sediment in a Large Combined Sewer*. PhD thesis, University of Abertay Dundee.

Yen, B.C. (2002) Open channel flow resistance. *Journal of Hydraulic Engineering*, ASCE, January, 20-39.

4

Solids management for sewerage systems

Devising the most effective and efficient means for management of sewer solids is not straightforward, as the major aspects of sewer solids behaviour related to movement, transformations and effects in receiving waters are still poorly understood. It is probable that much of what is disposed into sewers could be better utilised as a resource. With moves to deal with urban drainage nearer to source, it may be that many 'sewer solids' will become 'extinct' within the current century. Nonetheless, currently, the management of in-sewer solids requires a diversity of approaches, comprising controls:

- at origin, and prior to entry into the drainage system;
- at the entry point into the drainage system;
- within the sewer system.

A recent review of approaches in use is given by Ashley *et al.* (2000). As part of the extensive French studies over the past two decades, the opportunities for recycling of the grit recovered from sewers has been considered in terms of a cost-effective framework (Delattre *et al.*, 1998). Recently, the application of Life Cycle Analysis to the management of sewer solids has been proposed in the UK as helpful to determine the least impacting approach (Gouda *et al.*, 2002). In developing countries only limited studies for sediment management have so far been reported, most notably from India and Malaysia (Kolsky and Butler, 2002; Ab Ghani *et al.*, 2000), where the solids sizes are typically much larger in the former, but more similar to, European solids, in the latter. With an increasing interest in sustainability and integrated urban water management, there are now signs that sewer solids are being more effectively included as part of the wider purview contributing to the development of fully integrated approaches, as illustrated, for example, by Beurden and Geldof (2001).

Many recent publications consider elements of how best to make decisions about managing sewer solids within this broader context. However, little information currently exists on which to determine cost-effective or whole life solutions. Nonetheless significant advances have been made in enhancing the traditional approaches to sewer solids management that have been in use for more than a century. There is still a long way to go to the inclusion of solids origins, behaviour and effects in the full 'integrated environmental assessment' (Harremoës, 2001), that is essential if the most effective, efficient and sustainable ways of dealing with these solids are to be found.

This Chapter deals with the various approaches being used to manage sewer sediments. These are illustrated in Figure 4.1.

Section 4.1 considers the controls at sources and at the input interface (above to below ground). Section 4.2 reviews knowledge for the design and operation of sewers to control and mitigate the effects of solids. Section 4.3 deals with the management of solids in tanks, and overflow structures.

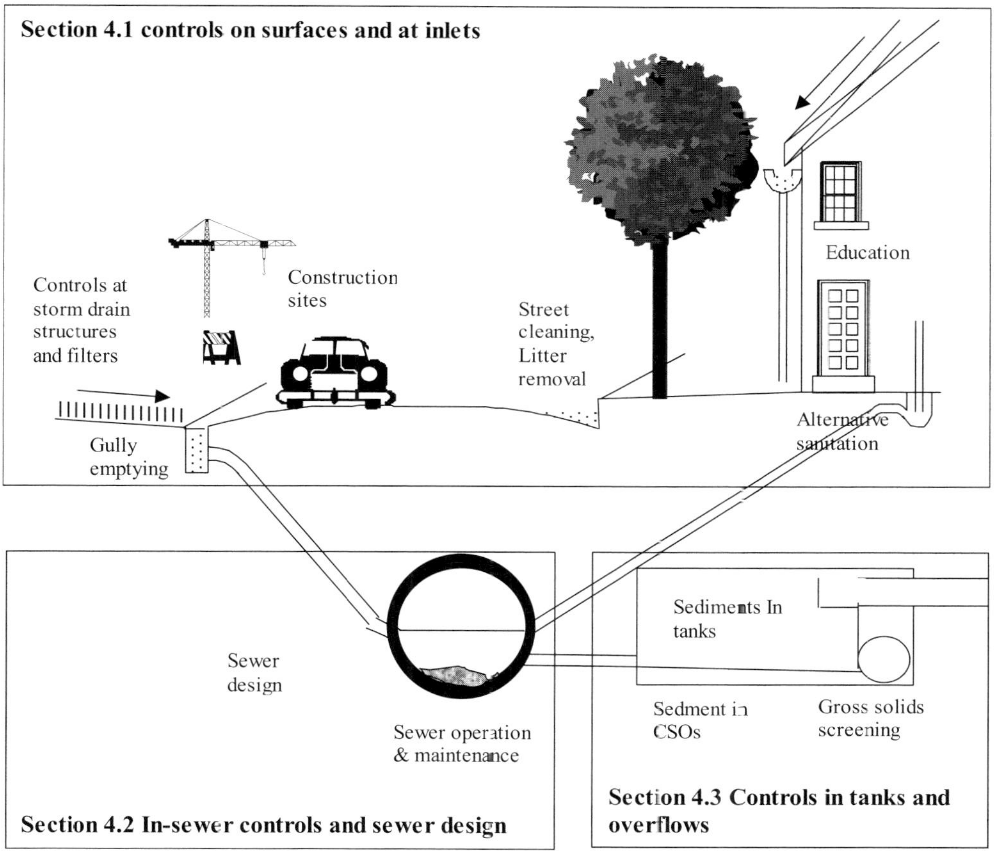

Figure 4.1. Sewer solids controls – introduction to Section 4.

4.1 SOLIDS CONTROL AT SOURCES AND AT INPUTS TO SEWERAGE SYSTEMS

4.1.1 Introduction

The various sources of solids entering sewer systems are described in Section 2.1. They fall into three general categories: urban runoff, domestic (sanitary and kitchen wastes), and commercial and industrial. Specific problems arise from other sources such as construction activities, wind blown sand in coastal regions and corrosion of the fabric of the system itself.

A variety of urban runoff controls can be used at source locations where sediment is generated (eroded), and at inlets to sewerage systems. As an example, Laplace *et al.* (1992) recommends that all particles greater than about 1 to 2 mm in diameter be restricted from entry into drainage systems to prevent deposition in downstream sewerage. The increasing use of 'Best Management Practice' (BMP) type systems to manage sediments and other solids is advocated by Masters-Williams *et al.* (2001) and assessed by D'Arcy *et al.* (2001). Unfortunately, unless the BMP system is specifically designed to collect sediments (as for a trapping device), the retention of sediments and associated pollutants may in itself compromise the performance of the BMP system, particularly where infiltration is involved (e.g. Ellis, 2000). The evolution of stormwater controls has been rapid in recent years, but evaluation of their performance has been irregular and inconsistent and the results not very accessible. Local studies have considered individual systems. For example, the ASCE BMP database project (Strecker, *et al.* 1999) is an on-going effort to collect and conduct quality control evaluations and to make these accessible through a website.

Sanitary inputs are culturally dependent and also depend upon the attitudes of householders. In the UK and USA, for example, a variety of non-faecal solids, including plastics, are routinely introduced via the water closet (WC), whereas in most other countries, these solids are disposed of via the solid waste systems (Ashley *et al.*, 1998). Commercial and industrial inputs to wastewater systems are now generally closely regulated worldwide (although this is not usual in developing countries), and the solids arising from these sources may be considered intrinsically in terms of the other components. Of significance, however, are two particular problems: fats and greases from catering establishments, for which grease traps are typically poorly regulated and maintained, leading to sewerage problems; and runoff from construction sites, which is generally poorly controlled. In developing countries the major problems usually relate to misuse of drainage systems as solid waste dumps, due to the inadequacy of municipal waste collection services (Kolsky, 1998). Even in the USA, inappropriate discharges of sanitary and industrial wastes to separate storm drainage systems are relatively common, with some 10 to 25 % of all stormwater systems affected (Pitt *et al.*, 1993; Field *et al.*, 2000).

The requirements on solids management at source are principally dependent on the transport energy available in the drainage system. The treatment objectives can be implemented via a number of cost-effective source area and inlet treatment practices.

Section 4.1 describes the levels of performance achievable using various control options. The stormwater controls may be conveniently grouped into general categories: regulation and public education; public works and related practices; sedimentation; infiltration; filtration; and construction site erosion controls. Sanitary and kitchen wastes input control may be achieved via public education, or by constraint, which may necessitate changes in domestic plumbing systems. The growing practice of fitting kitchen grinders in new properties to shred food scraps and the increasing use of dishwashers requires new ideas for consequent sewer solids control.

4.1.2 Public education

Public awareness of the consumer role in contributing to the effectiveness of wastewater systems is generally poor. An extensive study in Scotland (Ashley *et al.*, 1998), revealed that few people in four urban test catchments were aware of what happened to their sewage, and as a consequence, they were very relaxed about what items they disposed of via the WC. However, once made aware of the possible environmental effects of their habits, the majority agreed to adopt a more responsible approach. The campaign poster in shown in Figure 4.1.1. Even more than 20 years ago, Poertner (1980) had identified public education as perhaps the most neglected phase of solving stormwater problems in the USA. It was found that very few areas had effective public information or community relations programmes related to stormwater. Those that had programmes, found them to be helpful in making the public aware of the needs of the community and in obtaining their support, especially when promoting public investment in drainage. Public environmental education in the USA has dramatically increased over the past decade as part of the stormwater discharge permit programme of the National Pollutant Discharge Elimination System (NPDES) and renewed interest in the value of urban receiving waters (Field, *et al.* 1994) and schemes for community 'adoption' of watercourses.

Figure 4.1.1. Poster used to promote public behaviour change in relation to WC use in Scotland.

Even if no measurable water quality improvements can be detected in a catchment where public environmental education is being promoted, public support and active participation is still an essential element in ensuring the success of any sewer management programme. Ideally such participation should be from the outset of any proposals to change systems, in order to encourage public awareness of their role as stakeholders in water/wastewater systems (Ashley *et al.*, 2001).

Some recent examples show that public and stakeholder education should be developed as a main approach to improve stormwater and sewer management, e.g. in Australia (Smith, 2001).

Sources connected incorrectly to stormwater sewers include sanitary sewer mis-connections (Section 2.1.5). Wrong connections are often a result of inefficiencies of the service providers (plumbers etc. in, for example, South America) or due to local stakeholder self-action (e.g. in developing countries). Even in developed countries wrong connections occur, often compromising wastewater reuse efforts. The control of the solids from these sources can only be effected by removing the wrong connection. Further prevention of future wrong connections can be encouraged by education.

In summary, it is apparent that public and professional education is essential, for both the medium and longer term, in order to improve the understanding, use, operation and, hence, performance of wastewater systems This may require increased responsibility on the part of householders and others for individual behaviour in relation to these systems.

4.1.3 Alternative sanitary sewage collection and disposal methods and solids implications

The continuing use of waterborne waste disposal systems as developed in the world's first industrial revolution, and employed by most developed countries, is questionable for long-term sustainability in the use of water and disposal of waste (Tyson, 1995; Henze, 1997), and a number of studies are now attempting to formally quantify the environmental and other burdens of contemporary wastewater systems (e.g. Chalmers, 2001). The 'mix-first-and-separate-later' (MIFSLA) (Gunther, 1995) approach is wasteful of both water and key nutrients, as illustrated in Figure 4.1.2. The energy required to treat water to potable standards for subsequent disposal via WC flushes is substantial; given that typically some 30 % of household water in the developed world is used in this way, this raises questions about the current way in which sewers are used. Extension of the technology elsewhere in the world under the 'equity' principle (Agenda 21) is also unlikely to be viable due to economic limitations and environmental impacts, and alternative approaches need to be sought (Otterpohl, 2001).

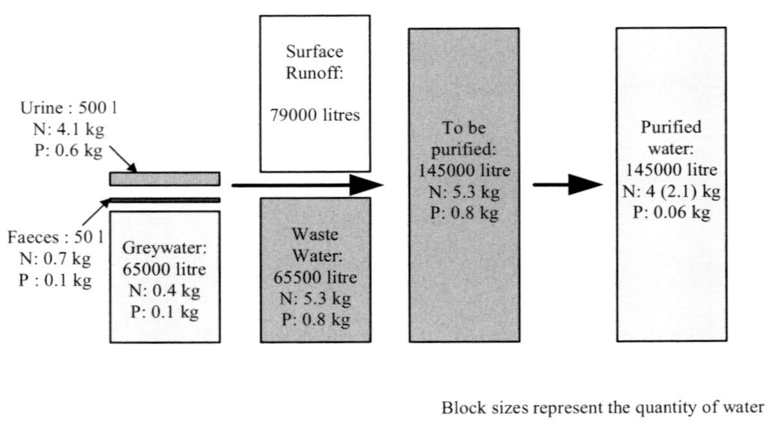

Figure 4.1.2. Current flows and nutrient masses involved in the WC flush-based combined sanitary and stormwater sewer system (per person, per year) (after Gunther, 1995).

Typical sewered disposal systems of the type shown in Figure 4.1.2 require a continuous supply of water of about 75 l/day per person to prevent blockage, compared with less than 20 l/day per person for simpler alternatives (Cairncross and Feacham, 1993). The transitional countries and developing nations are seeking to solve their sanitary waste problems with a variety of lower-cost and more sustainable approaches, which will, in some cases, make conventional sewerage systems unnecessary. For example, Niemczynowicz (1994) proposes 'new solutions', which utilise experience gained in using alternative urban drainage ideas for 'municipal wastewaters' in developing countries and areas where water is scarce, as shown in Table 4.1.1.

Table 4.1.1. Overview of sanitary and commercial/industrial sewage disposal in urban areas (from Niemczynowicz 1994)

	Large city centre	Large city suburbs and small city centre	Small, existing urban areas	Large and small city new developments	New small settlements
Traditional sewerage with end-of-pipe treatment	TRADITIONAL APPROACH IN DEVELOPED COUNTRIES				
Ocean outfalls; Infiltration with soil-aquifer treatment; Local sewers	WHERE APPROPRIATE - BUT NOT WITHOUT TREATMENT			UNLIKELY	
Decentralised - traditional small scale treatment; Local sewers		COMMON		COMMON	
Treatment/recycling at level of local community area: Recycling gray water and nutrients; Sewers minimised				SHOULD BE CONSIDERED	
Treatment/recycling at single house level: Composting toilets, or separating toilets; Virtually no sewers				SHOULD BE CONSIDERED	

Encouragement to move at both institutional and technical levels to the lower part of Table 4.1.1, as the starting point for the formulation of options for sewage disposal has significant implications when considering wastewater solids and their effect on sewerage systems. At the simplest level, the individual 'household' (whatever that comprises in developed or developing countries) may or may not be connected to the sewer system. In the most developed countries, the majority of the population utilise the water closet (WC), and in the USA, some 26.7 % (1990 data: Salvato,1992) are reliant on local septic tanks or other treatment systems that eventually discharge to groundwater (see also Field *et al.*, 2000). There are signs, however, that even more of the population in the USA are moving toward on-site sanitation of some type, with 37 % of new systems in all areas being local (USEPA, 2000), with design and operation standardised (USEPA, 2002). Local systems, comprising septic tanks or other options, necessitate a much higher degree of awareness and commitment to maintenance by the householder; for these and new emerging technologies to work, they require a shift away from the 'out-of-sight-out-of-mind' perspective current amongst WC users in the developed world. In some countries in Europe a change in people's attitudes is proving successful, with eco-villages in Sweden and Germany, for example, utilising alternative sanitary sewage disposal options (Hedberg, 1999). In the USA, on-site systems are now becoming more managed by local collective specialist organisations serving a group of properties.

There are, therefore, two perspectives: the first encompassing the continuing utilisation of existing (and high investment, high impacting) large-scale sewers, and the second, for the future expected shift away from large-scale sewered systems to local, possibly unsewered, systems. In practice, it is probable that for the foreseeable future – several centuries at least – existing and any new systems will have to co-exist. Tyson (1995), however, considers the continuing construction of large-scale sewered systems as an 'admission of failure' to be used only when all other options have been exhausted. Recently, a review of water and sanitation practices worldwide has concluded that a wide range of approaches will need to be adopted in the future to suit local circumstances (e.g. Tejada-Guibert and Maksimovic, 2001).

In an effort to improve the operation of existing large-scale (traditional) sewered systems, there are a number of European initiatives to discourage the use of the flushing toilet (commode or WC) as a means of disposal of solid waste material, which is increasing, particularly in terms of intractable plastics. These initiatives are to some extent complemented by reductions in WC flush volumes (e.g. Swaffield and Gallowin, 1992). Most new housing construction in the USA must now use low-flush commodes (about 7 litre per flush, instead of the older units that used about 26 litres per flush). Unfortunately, the initial models of low-flush commodes (especially when retro-fitted into older homes that did not have specific plumbing modifications) were not very effective and multiple flushes were often required, significantly reducing the benefits of the low flush volumes. In Australia, however, where water is scarce, the introduction in the early 1990s of dual 3/6 litre flush toilets has proven to be very successful (e.g. Cummings and Bonollo, 1999).

Public willingness to adopt new lifestyles will vary culturally, and while the Greek population may accept that faeces are the only solids that can be disposed into the WC, public attitude surveys in the UK and USA, for example, demonstrate a resistance to any change in the *right* to put all kinds of solids into the WC.

In a number of countries, there is an increasing use of in-sink units to shred, grind or pulverise kitchen scraps, making them suitable for disposal via the waterborne rather than the solid waste route. In the USA these are termed 'garbage disposals', in the UK, kitchen sink 'grinders' and in Australia, food waste processor (FWP) units. In Australia, their use is limited by legislation to domestic and existing hospital use only, with the highest per capita installation being in apartment blocks. In the USA there are signs that there may be a shift away from garbage disposals. As an example, New York City has long prohibited their use and in most areas of the USA, homes using on-site septic tanks and infiltration fields are generally required to install septic tanks twice as large as would be required if no garbage disposals are used (in order to contain the increased amounts of solids discharged from the home between septic tank cleaning). Unfortunately, elsewhere in the world, these systems are perceived to be evidence of higher living standards and are therefore becoming more common. For example, in a recent Australian study (CRC, 2000) for In-Sink-Erator, the leading supplier of residential, sewer-based food waste disposal systems, the relative environmental, technical, economic and social impacts of FWP use in apartments was investigated, based on ISO 14040 standards using Life Cycle Assessment (LCA). It was expected that current FWP usage (3–5% of households) could increase to at least 15 % and possibly more. The investigation showed increased solids-based impacts on the wastewater system if more than 15 % of households were to use FWPs. With a 10 times increase in daily solids loads entering the system with 50 % of households using FWPs compared with the current 5 %. There was also an unspecified expected increase in hydrogen sulphide problems in the sewerage.

Jackson (1995) reported that there is a reluctance to spend money on sewered systems in the developing world, where the perception (based on experience) is that latrine systems are more reliable than sewerage systems (ignoring public health implications). Table 4.1.2 shows how potential future changes in the use of waterborne waste systems may or may not be beneficial in terms of the problems and management of sewer solids. Whether some changes are beneficial depends on whether systems exist or not, and any changes are retrofit, or whether they are parts of new developments. Other potential changes relate to the nature and volume of industrial and commercial waste input to sewers. Increasingly stringent controls and costs for industrial (trade) effluents make recycling and/or on-site treatment much more imperative than in the past, and the exclusion of solids and other solids accumulating contaminants such as fats and greases, will become more common.

Table 4.1.2. Changes in domestic sanitary sewage and possible implications for sewer solids and associated pollutants

Decrease problems		Increase problems	
Change	Implication	Change	Implication
WCs and local changes			
Discourage use of WC for solids disposal	Remove gross solids, reduce blockages and solids loads and sewer 'sediments'	Increasing use of intractable plastics in sanitary products	Solids do not degrade, more gross solids in flows and deposits
Macerating toilets	Organics may be more easily transported and/or degrade in-sewer	Macerating toilets	Increase SS loads, finer particles may deposit easier under some conditions
Separate collection of faeces and urine: - vacuum toilets - composting toilets - drying - ANS* to agriculture	Reduces solids entering sewers, smaller sewerage system (may be local) or not needed at all	Reduction in WC flush volumes	Reduction in sediment transport capacity/increase in sewage strength
Storm water use directly for WC flushing	Less potable water use	More 'stormwater' in sanitary or combined sewers	Depends whether stormwater originally discharged directly to sewer or not
Sewers and changes thereto			
Reduction in infiltration	Reduces solids entering system	Reduction in infiltration	Reduction in sediment transport capacity/ increase in sewage strength
Introduction of *new type* sewers: small diameter gravity sewers. settled sewerage. vacuum sewerage	Lower solids concentration, smaller sewers better solids transport	Reduction in water consumption	As above
Alternative grey water disposal systems	Less kitchen solids, fats and detergents, slimes	Alternative grey water disposal systems	Reduction in sediment transport capacity and in-sewer 'treatment'

* Anthropogenic nutrient solution (urine).

In summary, it appears that there are moves to introduce more types of wastewater systems in a number of countries in order to improve locally the overall system performance within the wider context required by sustainability considerations. Traditional and alternative approaches need to be considered in context, with new or retrofit developments being properly integrated with existing infrastructure. However, even where individual behaviour and regulations are changed and new technical innovations introduced, the continuing co-existence of traditional systems and new alternatives needs to be properly managed.

4.1.4 Public works (utility) practices

Institutional arrangements for the various aspects of wastewater systems influence the opportunities for solids management. Locations and facilities where solids originate are owned/operated by a variety of public and private individuals and organisations, and the interfaces between these and wastewater systems may also be public or private. Communication between the various owners and responsible bodies, and the sources of revenue for the construction, operation and maintenance of wastewater systems influence the systems in use to control the solids conveyed in the wastewater. This Section is concerned with sewer solids management in the larger systems (more than one household, for example), which are usually operated as 'utility' systems, although in a number of countries, these systems may be public or owned/operated by private organisations. Numerous utility practices affect stormwater quality and quantity, including street cleaning, storm drainage inlet cleaning, landscaping maintenance on public rights-of-ways, roadway and utility construction erosion controls, erosion controls at landfills, runoff control at garages and maintenance of sewerage systems (see also Section 4.3).

4.1.4.1 Street cleaning

4.1.4.1.1 Introduction

In the late 1960s and early 1970s, street cleaning was presumed to be a potentially effective practice in reducing solids conveyed in runoff and associated pollutants because streets were assumed to contribute most of the urban runoff flows and associated pollutants. Unfortunately, street cleaning has since been shown to be of only limited effectiveness in most areas because other urban non-point pollutant sources may be much more significant than the pollutant contributions from streets for many areas and rainfall events. In addition, conventional street cleaning typically removes only the relatively larger particles ($d > 500$ µm), whilst rainfall preferentially removes the smaller particles ($d < 100$ µm). Furthermore, in many areas, rainfall is relatively frequent and intense and keeps the streets cleaner than can be accomplished using typical street cleaning equipment (Section 2.1.3.2). However, in arid areas where rain is infrequent, streets become quite dirty during the inter-storm dry periods. Intensive street cleaning during this dry period has been shown to result in moderate suspended solids and heavy metal reductions in the runoff (by up to about 50 %) (Pitt, 1979). Street cleaning does, however, play an important role as an aesthetic and traffic safety control measure. Street cleaning is also important in reducing the substantial loadings of street dirt material which may occur after snowmelt in northern countries. Leaf cleanup, frequently augmented by street cleaning, is also necessary in many temperate areas to avoid blocking drainage systems.

4.1.4.1.2 Street cleaning efficiency related to particle size

Typical street dirt total solids loadings show a 'saw-tooth' pattern with time between street cleaning and rain washoff events (Figure 4.1.3). Rain removes very little of the large particles, but can remove large amounts (about 50 %) of the finest particles whose diameter is less than 100 µm (Bannerman *et al.*, 1983; Pitt, 1985) which contribute most significantly to stormwater pollution (see Section 2.1.3.2).

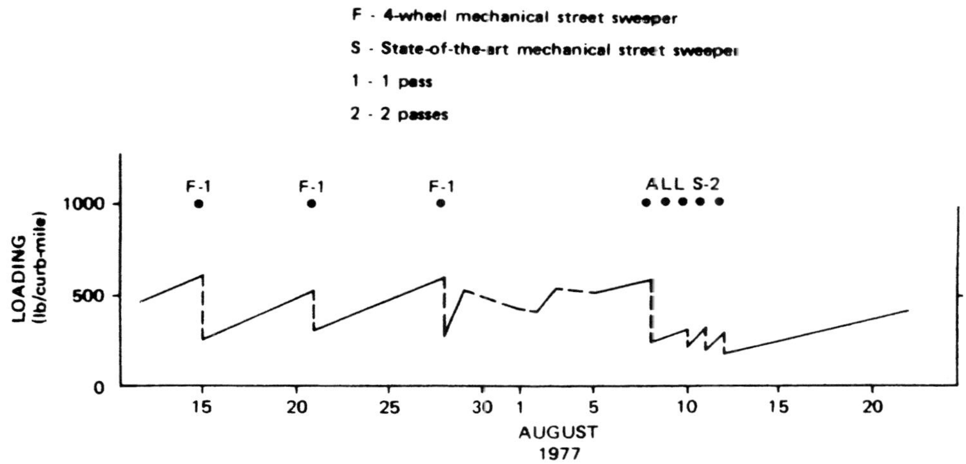

Figure 4.1.3. Saw-tooth pattern for accumulation and removal of street dirt by street cleaning, smooth asphalt street test area in San Jose, California, USA (Pitt 1979).

Unfortunately, typical mechanical street cleaners remove much of the coarser particles in their path, but very little of the finer particles (Sartor and Boyd 1972; Pitt, 1979, 1985). Valiron and Tabuchi (1992) reported the removal rates shown in Table 4.1.3 for street dirt of different particle sizes, obtained with traditional cleaning machines.

Table 4.1.3. Removal rates for street cleaning for various particle sizes

Particle size (µm)	Removal efficiency (%)
0–40	16
40–100	0
100–250	48
250–850	60
850–2 000	67
>2 000	79

These removal rates are also highly dependent on the dirt loading on the street, with much smaller removal rates with lower loadings. As the amount of solids on the surface increases, so does the removal efficiency. With loadings of about 10 kg/curb-km, less than 25 % removal can be expected, whilst removals of up to about 50 % can be expected if the initial loadings are as high as 40 kg/curb-km for the 500–1000 µm particle size range. Note, however, that while removals are higher for higher initial loadings, the residual solids may well be greater than for the low initial solids load condition.

New street cleaners, e.g. using pressurised water and more efficient brushing and sweeping tools, are becoming available which appear to be more effective at removing large fractions of most of the street dirt particles, including the smaller particles that are the most heavily contaminated. For example, Sutherland (1996) showed that by using an *Enviro Whirl I* (now *Schwarze EV series:* Schwarze Industries, Inc.) with monthly cleaning in residential areas, the suspended solids discharges in stormwater could be reduced by about 50 %, compared with only about 15 % when using conventional mechanical street cleaners. Enhanced street cleaner performance was obtained with a modified regenerative-air street cleaner, even at low loadings, during tests in Bellevue, Washington USA (Pitt 1985). The improved performance was much greater for the finest particle sizes, although the mechanical street cleaner did not remove any significant quantities of material. Other tests of mechanical and vacuum street cleaners and regenerative-air street cleaners in areas having much higher street dirt loadings, found no differences in performance between the different types of street cleaners (Pitt 1979). In areas having high loadings, it may be best to use a tandem operation, where the streets are first cleaned with a mechanical street cleaner to remove the large particles, followed by a regenerative-air street cleaner to remove the finer particles.

Factors significantly affecting street cleaning performance include (Pitt, 1979):

- particle loadings;
- street texture;
- moisture;
- parked car conditions;
- equipment operating conditions
- frequency of cleaning.

The pollutant removal benefits of street cleaning are a function of the relative contributions of pollutants from the streets. In the Pacific northwest region of the USA, for example, the large number of light rainfall events results in much of the runoff pollutants originating from the streets. In the southeast region of the USA, in contrast, where the rainfall is much more intense and more frequent, the streets contribute a much smaller fraction of the total annual pollutant loads. However, for heavily paved areas, such as large parking areas or paved storage areas, street cleaning may still result in significant runoff quality improvements, even in areas of frequent and intense rainfall.

In Paris, intensive studies of the Le Marais catchment have included detailed investigations of the solids and metals found from road surface inputs (Rocher *et al.*, 2003). The daily suspended solids pollutant load removed was found to be similar to the amount removed during one rainfall event. It was also shown that the total mass of pollutants stored on the street surface is significant, even with street cleaning, and the effects of street cleaning may therefore be limited (Gromaire, *et al.* 2000).

4.1.4.1.3 Litter control

The primary and historical role of street cleaning is for litter control, and is encapsulated in a number of countries' environmental legislation. For example, the City of New York investigated the role of enhanced street cleaning using intensive manual street sweeping to reduce litter entering the City's storm drainage (Newman *et al.*, 1996, 1999). During the summer of 1993, the City manually swept near-curb street areas and sidewalks in a pilot area, supplementing the twice per week mechanical street cleaning the area normally received. Adding four manual sweepings each week to the two mechanical cleanings reduced the litter loadings by about 64 % by weight. In other areas where almost continuous manual sweeping was undertaken (8 to 12 daily sweeps, 7 days per week), the litter loadings were about 80 % cleaner than comparable areas that only received twice weekly mechanical cleaning. Thus, manual sweeping can be an important tool in reducing floatable solids (debris) pollution, especially in heavily congested areas.

In many areas, street cleaning is not frequent enough to maintain low street dirt loadings. A reported study in nine municipalities in the south-west of France has shown that street cleaning frequencies were generally low, with a mean frequency of about once per month depending upon the nature of the local site (city centre, pedestrianised street, etc.). However, frequencies varied widely from every day to four times a year. Other studies have shown that a frequency of about 6 to 7 cleanings per week was required to remove about 50 to 55 % of the particles from the streets (Valiron and Tabuchi, 1992; Vignoles and Herremans, 1992). Ellis (1986) also found that sweeping in London is more efficient if it is associated with vacuuming and if it is done three times per week.

More than 780,000 tonnes of solids per year are washed into the drainage system in South Africa. The Water Research Commission of South Africa and the Cape Metropolitan Council have funded a 4-year investigation into the reduction of urban litter in drainage systems through the development of catchment specific litter management plans (Armitage, *et al.* 2001). The results of the litter audits from eight catchments will measure the effectiveness of the various litter management strategies.

The California Department of Transportation (Caltrans) conducted a 2-year litter management pilot study in the Los Angeles, CA, area to investigate the characteristics of highway litter and the effectiveness of BMPs for removing the litter (Lippner, *et al.* 2001). Half the catchments were treated with one of five BMPs while the others were controls. The BMPs tested included increased street cleaning frequency, increased frequency of manual litter pickup, a modified drain inlet, a bicycle grate inlet and a litter inlet deflector. Roughly half the freeway stormwater litter was paper, plastic and Styrofoam. Except for cigarette butts, the origins of most litter could not be identified because of its small size. Of the five BMPs tested, only increased litter pickup and the modified drain inlet demonstrated some apparent reduction of litter in stormwater runoff, although the data were highly variable.

Butler and Clark (1993) examined the benefits and costs of street cleaning and gully pot use to prevent sediments from entering combined sewerage in the UK. They compared these costs with those associated with removal of the sediment from the sewerage and removal at the sewage treatment facility. For example (Figure 4.1.4), it was found that the minimum total cost would be achieved with a street-cleaning interval of about once every 6 to 8 weeks, but the total costs of sediment removal would not be significantly increased if there were no street cleaning whatever. For this combined sewerage system example, it was concluded that it was more cost-effective to remove the sediment at the end-of-pipe treatment facility, provided the sediment can be conveyed there. It is important to note that the main requirements for street cleaning in the UK are determined by the Environmental Protection Act which stresses litter removal.

This section may be summarised:

- Street cleaning may be effective at removing the larger particles, responsible for reducing the capacity inside sewers.
- Street cleaning techniques used for aesthetic and hygiene purposes preferentially remove the largest particles, but are very inefficient for the smaller particles (to which most of the pollutant loads are attached). Where street dirt loadings are initially low, when the street surface texture is coarse, or where parked cars interfere, cleaning is inefficient. Hence cleaning is not very effective at controlling the fine and highly polluted particles likely to be conveyed in rainfall washoff into sewers.
- New machines for general street cleaning are emerging which appear to be more efficient at removing the finer particles. Additionally, more efficient street cleaning machines are now available to clean porous pavements and infiltration structures with recycling of the cleaning water (see Section 4.1.8).

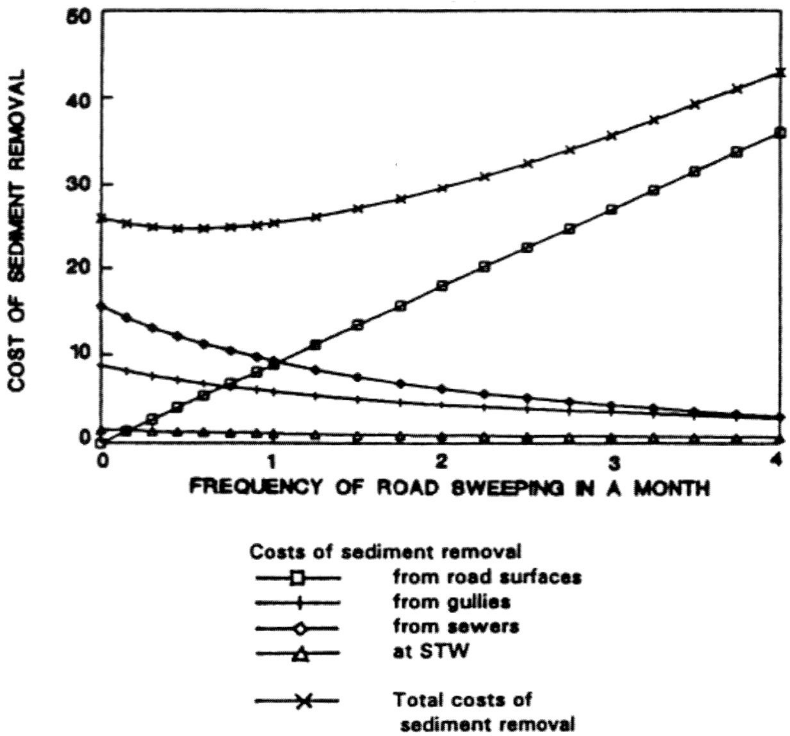

Figure 4.1.4. Example of costs of sewer related sediment removal in UK (Butler and Clark, 1993; original figure reproduced by kind permission of CIRIA).

4.1.5 Storm drainage system inlet structures

Storm drainage system inlet structures from small areas (up to a few hundred m^2) can be considered in four general categories:

(a) *Simple inlet* (Figure 4.1.5a). An inlet comprising a grating at the curb and a box with the discharge located at the bottom of the box, which connects directly to the main storm drainage or combined sewerage system. This inlet simply directs the runoff to the drainage system and contains no attributes that would improve water quality and solids removal. However, large debris (several cm in size) may accumulate in the inlet box if present in the stormwater.

(b) *Inlet with a sump* (Figure 4.1.5b). Similar to the simple inlet above, but incorporating a sump that typically extends about 0.5 to 1 m below the bottom of the outlet. This is termed a catchbasin in the USA, or a gully pot in Europe, and has been shown to trap moderate amounts of the coarse sediment. A hooded outlet is important to improve trapping of debris and floatable material.

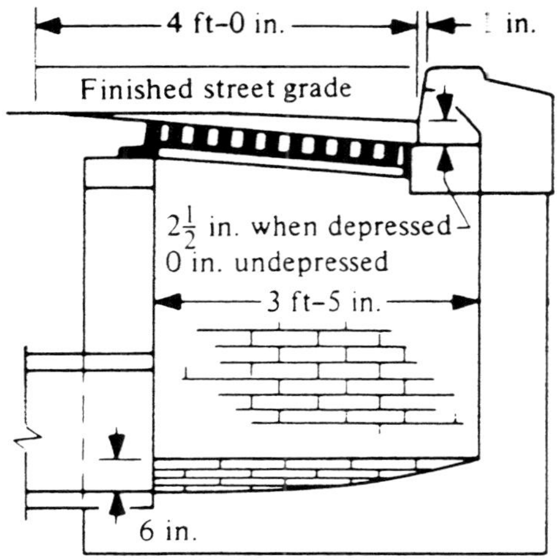

Figure 4.1.5a. Conventional storm sewer inlet (Baltimore County Dept. of Public Works, by permission).

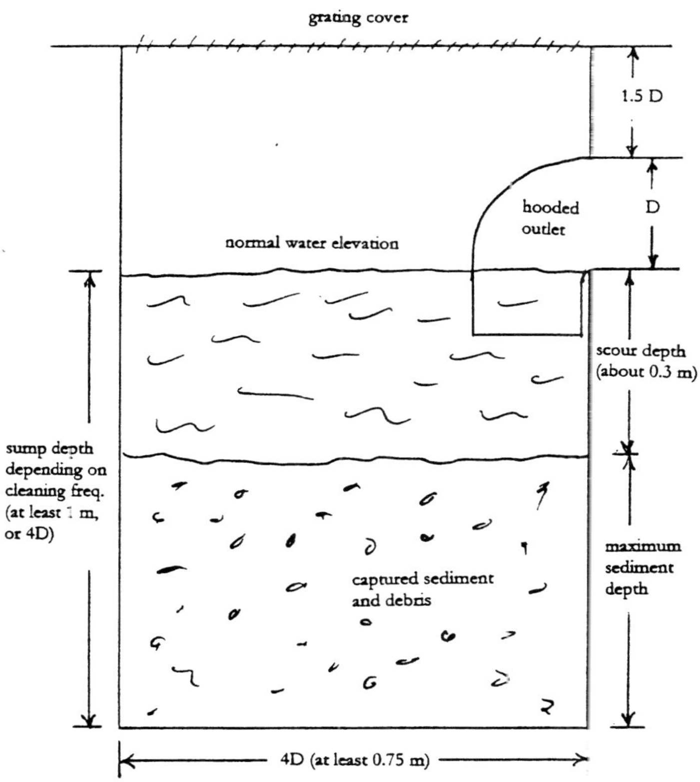

Figure 4.1.5b. Conventional catchbasin with a sump, with effective dimensions (adapted from Lager *et al.*, 1977).

(c) *Screened inlet* (Figure 4.1.5c). Similar to the simple inlet, but including a screen to trap debris. This type typically includes small perforated metal buckets placed under the street grating (common in Germany), large perforated stainless steel plates placed under the inlet grating (used in Austin, Texas, USA), and a number of proprietary devices incorporating filter fabric or other types of screening placed to intercept the stormwater flow. This may trap large debris and litter, depending on the overflow arrangement, but has not been shown to produce significant water quality improvements and typically has a very limited useful life before clogging and requiring maintenance.

(d) *Slot drain*, also named Sajuto drain (Figure 4.1.5d). These devices, invented in 1954 by the French company Sajuto to drain NATO airports, are used worldwide in airport taxiways, along highways, and also in other urban environments. The runoff and the solids directly enter into the drain. As the opened length is equal to the drain length, there is no risk of inlet blockage. However, all solids smaller than the slot width may enter the sewer system. Cleaning of this type of drain has to be carried out with specific equipment. Some slot drains are not fully open, but are covered with a top grid to prevent the entry of large solids.

On a larger scale, there are guidelines for fluvial settling basin design (Garde *et al.*, 1990) that could be applied to the design of structures on larger inlets to wastewater systems. However, such traps are not normally used in urban drainage systems. Figure 4.1.6 shows a large trap at an inlet to the storm drainage system into Bogotà (Colombia). This is located where the rural stream flow enters the urban area. Figure 4.1.7 shows a large-scale inlet vortex grit trap for a stormwater inlet.

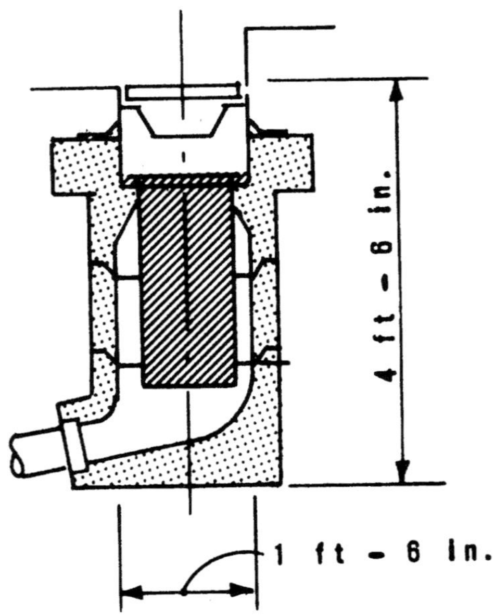

Figure 4.1.5c. Inlet strainer basket as used in Germany (Lager *et al.*, 1977).

Figure 4.1.5d. Slot inlet drain (Sajuto drain) (from the website for Sajuto.com, 2001).

Figure 4.1.6. Sediment trap at the inlet to part of the urban drainage system in Bogotà, Colombia (photograph Richard Ashley).

4.1.5.1 Catchbasins and gully pots

Catchbasin performance has been extensively investigated in the USA. For example, Lager *et al.* (1977) conducted an EPA-funded study to investigate sedimentation in catchbasins, and to develop effective designs. Pitt (1979) undertook long-duration sediment mobility tests using a retro-fitted 'idealised' catchbasin (as shown in Figure 4.1.5b) based on Lager *et al.*'s 1977 design. These studies showed that the sediment that accumulated in catchbasins was large (in size) in comparison with the solids found in storm runoff and the sediments were not very mobile once trapped in the catchbasin sumps.

Figure 4.1.7. Vortex grit trap inlet to a storm drain Madison WI, USA (photograph Bob Pitt).

Catchbasins, simple inlets, manholes, and sewer sediment accumulations were monitored at more than 200 locations for 3 years in Bellevue, Washington, at two mixed residential and commercial study areas (Pitt, 1985). A few atypical locations were affected by erosion of the sediment from steep hillsides adjacent to the storm sewer inlets. The sewer and catchbasin sediments had a much smaller median particle size than the street dirt, but the particles were much larger than those generally found in stormwater. Hence, catchbasins are effective at removing the largest particles that are washed from surfaces during rainfall, preventing them from being deposited in downstream sewerage and in the receiving waters. If the catchbasins are full, they cannot remove the particles from the runoff. Cleaning catchbasins twice a year was found to be most effective in the Washington study. This cleaning schedule was found to reduce the annual discharges of total solids and lead by between 10 and 25 %, and COD, total Kjeldahl nitrogen, total phosphorus and zinc by between 5 and 10 % (Pitt, 1985).

Butler and Karunaratne (1995) reported particle sizes trapped in gully pot sumps using UK and German data (see Section 2.1). The median particle size of the sump particles was shown to be between about 300 and 3000 μm, with less than 10 % of the particles smaller than 100 μm, a typical near-upper limit of particles found in stormwater. Relatively few pollutants are associated with these coarser solids found in the sumps, compared with the finer particles. Memon and Butler (2002) investigated the transformation processes in gully pots in simulated storm-drainage conditions, and concluded that conditions approximated to those assumed in conventional batch reactors at treatment plants.

Valiron and Tabuchi (1992) quote the results of a study carried out in the North of France, which showed that most of the solids trapped in gully pots were made of sand with a mean diameter greater than 200 μm. The organic content was very low, ranging from 4 to 8 %. Table 4.1.4 gives the mean grain size distributions measured for five gully pots.

Table 4.1.4 Mean grain size distributions of solids collected in five gully pots (measurements have been carried out after preliminary drying of the solids) (from Valiron and Tabuchi 1992)

Grain size (µm)	< 50	50–100	100–200	200–500	500–1000	1000–2000	> 2000
Gully no. 1	24	3	6	11	6	8	42
Gully no. 2	24	6	8	18	14	17	13
Gully no. 3	5	2	5	16	13	15	44
Gully no. 4	15	4	14	29	11	10	17
Gully no. 5	56	6	8	12	7	8	4

Butler and Memon (1999) present sediment trapping equations for sediment in gully pots based on simulated wet weather events (see Section 2.1 for details). Earlier studies (Butler and Karunaratne, 1995) had shown that the sediment trapping performance was dependent on the flow rate passing through the gully pot, and on the particle sizes of the sediment. The depth of sediment in the gully pot had less effect on the capture performance than these other factors. In all cases, decreased flow rates substantially increased the trapping efficiencies and larger particles had substantially greater trapping efficiencies than smaller particles, as expected.

Butler and Clark (1993) studied the build-up rate of sediment in UK roadside gully pots. Figure 4.1.8 shows the build-up of sediments in nineteen gully pots in one area, illustrating the temporal variations in sediment depth and the effects of cleaning. They found that for most gullies examined (average drainage area for each of the gully pots was 228 m^2), the build-up rate was fairly constant, at about 18 mm per month.

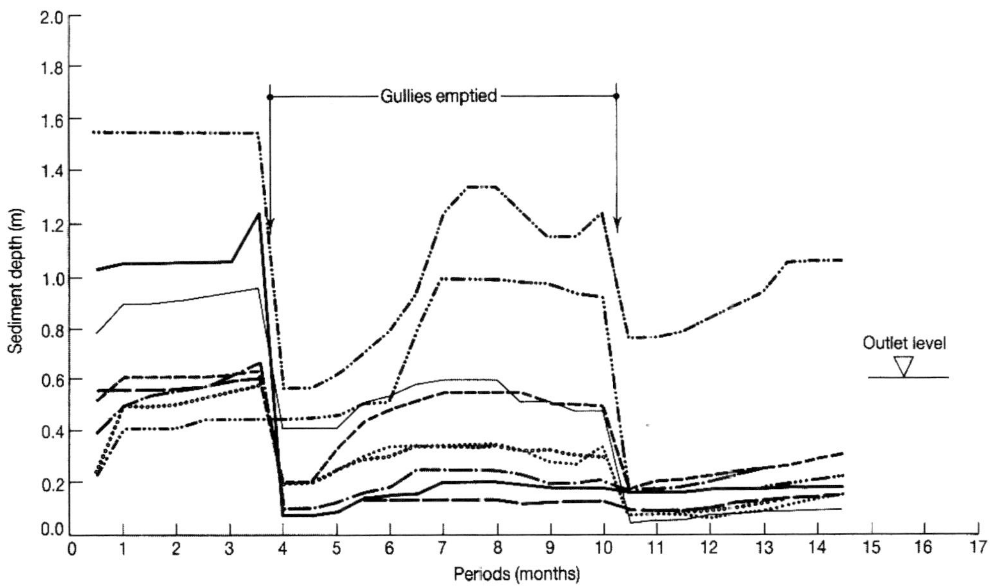

Figure 4.1.8. Gully pot sediment build-up and removal by cleaning for nineteen pots in the UK (Butler and Clark, 1993; original figure reproduced by kind permission of CIRIA).

A study carried out on four gully pots in Toulouse, France (Dastugue *et al.*, 1990) has shown that the amount of solids trapped was highly variable and dependent on the type of urbanisation and intensity of traffic. The annual mass of solids ranged from 300 kg in the residential or the highway areas, up to 700 kg in the dense residential and market place areas (see Table 4.1.5). However, expressed as specific mass per m^2 of upstream catchment, the mass of solids was found to be twice as large in urbanised and highway areas (758 g/m^2/year) as in moderately urbanised areas (351 g/m^2/year).

Table 4.1.5. Mass of solids entering 4 gully pots in Toulouse, France (from Dastugue *et al.*, 1990)

	Type of site			
	Highway	Low-density residential area	Multistorey houses	Buildings and market place
Impervious catchment surface (m^2)	630	600	365	875
Annual volume of runoff flowing through the gully pot (m^3/year)	504	484	292	700
Annual mass of solids entering the gully pot (kg/year)	288	313	350	686

All types of gully pots should be cleaned by means of conventional dry or wet vacuum lorries, which are also used for sewer cleansing. In many cities, responsibilities and tasks are not always appropriately distributed and co-ordinated between those responsible for sewers or drains and those managing street cleaning. Hence there are no attempts to situate and operate gully pots optimally. Frequently, the utility responsible for street cleaning supports most of the gully maintenance costs, whilst the sewer operator has to accept whatever is (or is not) done in this regard. If there is to be any improvement in efficiency and effectiveness, then an integrated approach is required, possibly necessitating revisions to institutional arrangements.

A New York City study indicated that street litter is a major contributor of floatables to the New York harbour (Grey and Oliveri, 1998). Field studies found that hooded catchbasins were 80 to 90 % effective in retaining floatables. A city-wide programme was therefore conducted to inspect, map and hood all catchbasins in the city.

Valiron and Tabuchi (1992) indicate that the cost of gully pot cleaning ranged from €3.5 to €7 (1992 value) per pot, with a cost of €150 to €450 (mean value = 260) (1992 value) per m^3 of solids extracted from the gully pots. This may be compared with the cost of sewer cleaning, which ranges from €150 to €900 (mean value = 450) (1992 value) per m^3 of solids extracted and also with the costs for street cleaning and end-of-pipe management (Section 4.1.4.1).

4.1.5.2 Storm drain inlets with filters

Various inserts into gullies or catchbasins are available. These comprise inserts that filter or trap particles. Figure 4.1.9a illustrates an example as recommended by King County. Figure 4.1.9b shows alternatives for construction sites. While it is likely that the filter material would clog rapidly, causing ponding, material will still be excluded from the drainage system.

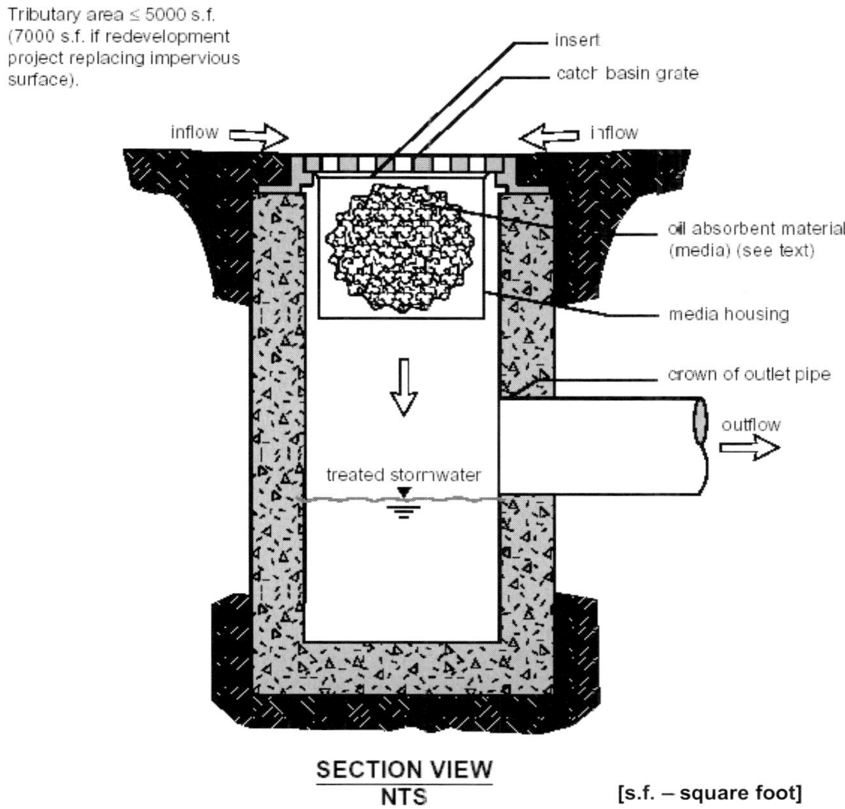

Figure 4.1.9a. Catchbasin with sorbent insert (*King County Surface Water Design Manual*, 1998; reproduced with permission).

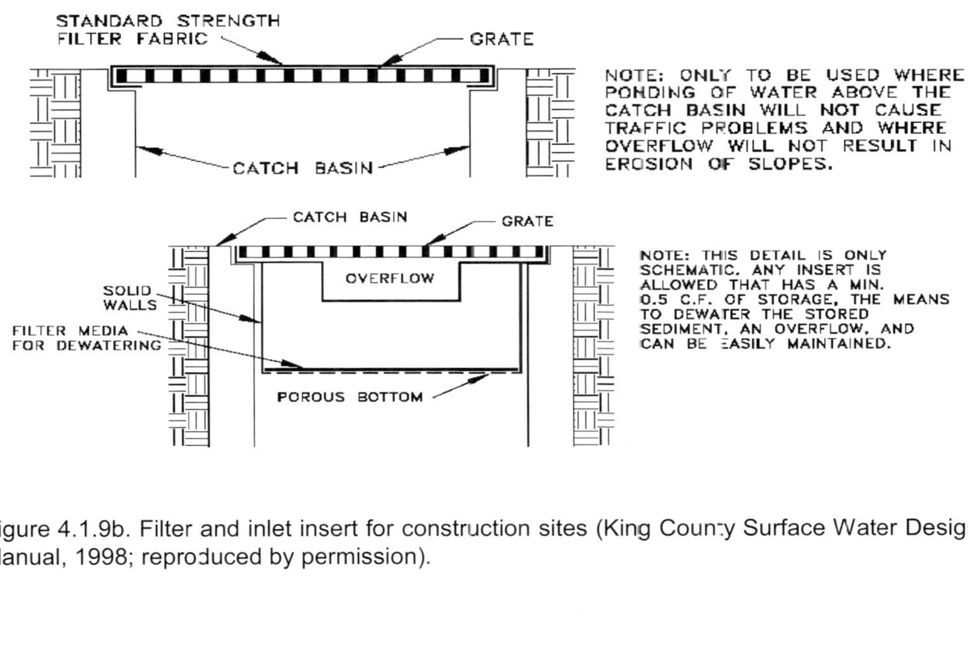

Figure 4.1.9b. Filter and inlet insert for construction sites (King County Surface Water Design Manual, 1998; reproduced by permission).

A typical fine filter fabric installation is illustrated in Figure 4.1.10. This clogged very rapidly after installation (clogging occurs after filter fabrics capture about 2 or 3 mm of fine grained sediment). Limited information is available concerning the performance of filter fabrics in removing stormwater pollutants. During controlled laboratory tests using a wide variety of available woven and non-woven filter fabrics, Clark and Pitt (1999) found that the filters provided moderate reductions (about 50 %) in suspended solids and COD. However, the filter fabrics can only withstand very thin accumulations of sediment before they clog. The maximum sediment thickness on the fabrics before clogging was only some 1 to 2 mm (the maximum sediment loading was about 3.8 kg sediment per m^2 of fabric). A series of tests were carried out to evaluate the performance of inserts in both dry periods (at controlling litter) and wet periods in Los Angeles California (Lau *et al.*, 2001). In some areas the gully inlets have been closed during dry periods to prevent the accumulation of litter and dust, which was prone to being swept into receiving water when the first storm occurred. A key objective in evaluating inserts is to ensure that they would not increase flood risks, and hence may need to be removed during wet periods. It was found that screens with 25 mm mesh openings prevented some 95 % of the solids from entering the basins during the dry season. In addition to looking at mesh type inserts, sorbents were also considered as a means of controlling oils and greases (some associated with the sediments). The most effective inserts were polypropylene geotextile that was easy to retrofit into existing catchbasins coupled with sorbents for oil and grease removal.

Figure 4.1.10. Filter fabric inlet screen (photograph Bob Pitt).

Figure 4.1.11 shows a coarse mesh screen at a storm drain inlet for the capture of large debris, such as leaves and litter. It worked quite well for the capture of this material, requiring maintenance about twice a year. Unfortunately, the decomposition of the captured coarse material caused an increase in downstream suspended solids. This problem would have been minimised if a sump was used to contain the material below the screen, preventing water from flowing through the captured material.

Figure 4.1.11. Illustration of an inlet coarse screen (photograph: Bob Pitt).

Pitt and Field (1998) evaluated three storm drain inlet designs in a residential area in New Jersey, USA: a conventional catchbasin with a sump, and two representative designs that used filter materials. Twelve storms were evaluated for each of the three types of inlet units, with influent and effluent samples being analysed. The catchbasin with the sump was the only device that showed statistically significant removals for solids and several other pollutants. Although the units, which incorporated screen fabric, did not indicate any significant water quality benefits, one type did capture an appreciable amount of leaves and litter. Unfortunately, the coarse screening of the unit trapped leaves in the flow path, where they then decomposed and increased the concentrations of some of the pollutants in the runoff. This adverse impact could be reduced if the filters were cleaned frequently. The basket inserts used in Germany and the large stainless steel perforated trays used in Texas, could provide a similar benefit: direct water quality improvements would be minimal, but litter capture would be substantial. The Texan devices are cleaned frequently, typically after every storm, as they are located in an area of major litter production, and prevention of litter fouling to the receiving water bodies is a major local environmental and aesthetic issue. Similarly, the German inserts are also emptied frequently because they are relatively small (about 5 litres capacity) and would otherwise clog or create a nuisance.

4.1.5.3 Optimisation of storm drain inlet structures

The objective should be to use storm drainage inlet devices which have the following performance characteristics:

- be not prone to flooding due to clogging with debris;
- control the ingress of settleable solids into the downstream system
- do not encourage stormwater to release any previously captured solids;
- do not have unacceptable hydraulic head loss properties;
- require only inexpensive and infrequent maintenance.

The following guidelines meet these criteria (Pitt and Field, 1998) and are suitable for retro-fitting inserts into existing simple storm drainage inlets. The materials used should be concrete, plastic, aluminium or stainless steel. It is especially important to avoid galvanised metal or treated wood due to the associated micropollutants. Catchbasins in newly developing areas could be more optimally designed than the suggestions below, especially by enlarging the sumps and by providing large and separate offset litter traps. Tokyo's Experimental Sewer System, as described by Fujita (1993, 1994, 1998), is an excellent example of the large-scale use of modified catchbasins to increase solids capture and to allow water infiltration.

1 The 'standard' catchbasin (having an appropriately sized sump) and a hooded outlet should be used in most areas. Figure 4.1.5(b) shows diagrammatically a standard recommended configuration (from Lager *et al.*, 1977). This is the most robust configuration. In almost all full-scale field investigations, this design has been shown to withstand extreme flows with little scouring losses, no significant differences between supernatant water quality and runoff quality, and minimum insect problems in warmer climates. It will trap the transported heavier bed-load material from the stormwater (especially important in areas using sand for vehicle traction control) and will trap a moderate amount of suspended solids (about 30 to 45 % of the annual loading). The largest-sized particles will be trapped in preference to the finer material that has proportionately higher concentrations of associated pollutants. The hydraulic capacity should be designed using conventional procedures (grating and outlet dimensions), whilst the sump design should be based on the desired cleaning frequency, maintaining a minimum depth as shown previously.

2 A relatively safe (in terms of the required performance objectives listed) add-on to the standard recommended configuration is an adversely sloped inclined screen covering the outlet side of the catchbasin. The inclined screen should be relatively coarse (12mm spacing) such that it would trap practically all litter and large debris. The screen would tilt outwards at the top to cover the hooded outlet. This design would allow the trapped coarse material to settle into the sump instead of being constantly held against the screen by out-flowing water.

International experience, such as that in France (Valiron and Tabuchi, 1992) suggests that the type of gully pot should be adapted to local conditions, as shown in Table 4.1.6.

Table 4.1.6. Most appropriate locations and types of gully pots (adapted from Valiron and Tabuchi, 1992)

Type of gully pot	Recommended implementation	Remarks
Gully pots with top bar screens and also inlet strainer basket (type (c))	Market places, locations with high density of trees	Coordination between sewer operator and street cleaner is required
Gully pot with sump (type (b))	Areas with high TSS concentration in runoff	Should be used more frequently than currently
Gully pot with direct inlet (type (a))	Should be used only for roof runoff	-
Gully pot with a hooded outlet or a siphon against odours	Where odours are a problem	A ventilation shaft should be utilised remotely. A better solution would be to identify and remove the source(s) of odour

In summary, gully pots with sumps (and including strainer inlets in some specific locations) are recommended as they trap coarse (and some floatable) solids before these enter the sewer or other drainage system. The cost of extracting solids from gullies is typically approximately half of the cost of extracting solids from sewers. The gully pot cleaning frequency should be adapted to local conditions and TSS loads, particularly where strainers are used (and whose blockage could lead to more frequent street flooding). Improved coordination between street cleaning and sewer operating organisations is strongly recommended (for example, Butler and Clark, 1993).

4.1.6 Oil and grease (light liquid) separators

The historical use of oil and grease separators to treat stormwater from petrol station forecourts and other areas has been shown to be ineffective for various reasons, particularly lack of maintenance and poor design for the relatively low levels of free floating oils present in most stormwater (Schueler, 1994). These devices have also been used in an attempt to capture sediment from runoff.

Many proprietary manufacturers promote the use of small prefabricated separators to remove solids transported by runoff. Until the late 1990s, these separators were rarely designed and sized specifically for stormwater discharges, but usually consisted of modified grease and oil separators. The solids are supposed to settle within the separators, either by free decantation or by counter-current or cross-current lamellar decantation. Many of these units have been installed in France alongside highways (Ruperd, 1993). Despite the number of installations, few studies have measured their actual efficiency (Aires and Tabuchi, 1995; Ruperd *et al.*, 1998). Stormwater treatment studies by Ruperd (1993) and Legrand *et al.* (1994) investigating prefabricated oil and grease separators in France have shown that:

- Prefabricated oil and grease separators are usually significantly undersized. They would work reasonably well only at flow rates typically between 20 and 30 % of their published hydraulic design capacity. For higher flow rates, the flow is very turbulent (Reynolds numbers can exceed 6000), the lamellar decantation efficiency is very poor, and the scouring of previous deposits frequently occurs.
- The separators need to be cleaned very frequently. If they are not cleaned, the deposits are scoured out during storm events, resulting in negative efficiencies (see Figure 4.1.12). However, cleaning is usually manual, and expensive, and not straightforward because the separators are so small. Some of the most frequently used devices are equipped with automatic clean-out pumps that can improve cleaning efficiency and facilitate the extraction of deposited sludge and solids. Typical cleaning frequencies advocated of only once or twice a year are inadequate in most locations. As for gully pots, the cleaning frequency of such separators should be adapted to local conditions. Regular inspections are needed to define appropriate cleaning frequencies to ensure satisfactory pollutant removal.

Manufacturers, utilities, operators and researchers are now working to improve the design and operation of separators. New specialised devices, specifically designed for stormwater have recently been developed (e.g. Daligaud *et al.,* 1998) and can be used in areas where conventional oil and grease separators have historically been used. Their mean efficiency for TSS removal is between 30 to 50 %, but may range from 0 to 60–80 % for individual rainfall events. The removal efficiency is lowest for the lower initial TSS concentrations in the runoff. Further development of these new style separators is required to ensure consistent performance and to improve maintenance and sludge extraction.

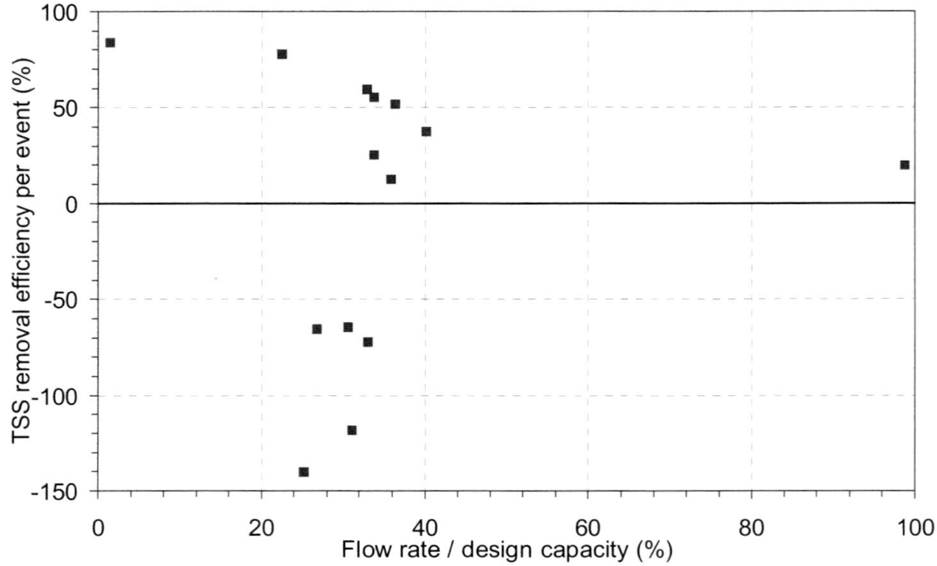

Figure 4.1.12. TSS removal efficiencies in an oil/grease separator for fourteen storm events at different flow rates (after Ruperd 1993).

A number of new proprietary devices have been developed during the past decade for the treatment of stormwater from critical areas. These are typically used in areas having high oil and grease concentrations in the runoff (such as vehicle maintenance and storage yards and automobile salvage yards). The Wisconsin Department of Natural Resources (WDNR) and the US Geological Service (USGS) recently tested several of these devices. In one example, Washbusch (1999) reported on an evaluation of a Stormceptor unit in Madison, Wisconsin. The evaluation was conducted at a 1.7 ha (4.3 acre) city maintenance yard and included forty-five runoff events, plus snowmelt. About 90 % of the runoff was treated by the unit, and the remaining flow bypassed the unit during periods of high flows. The overall effectiveness of the unit was only 33 % for suspended solids. In general, dissolved constituents were unaffected by the unit, but large quantities of oil were captured in the unit. Performance during early spring events was degraded by high saline snowmelt water, which had elevated conductivity from on-site de-icer storage that significantly slowed particle settling. In another series of tests, Ball *et al.* (2000) evaluated the effectiveness of stormwater treatment devices installed on a roadway in Australia, including detention tanks, a Continuous Deflective Separation (CDS) GPT and a sand filter. A CDS unit was expected to reduce suspended solids, bed loads, floating solids, free oil and grease from stormwater. Field studies verified the removal efficiencies for gross solids to be greater than 95 % and TSS removals greater than 70 % (Kohzad 2000). These devices appear to operate like grit chambers and over-sized catchbasins/gully pots, but data has shown them to be quite effective in trapping significant quantities of oil and grease, if present in the stormwater.

In summary, small prefabricated separators may be used to trap some of the solids transported in stormwater runoff. Unfortunately, traditional separators have been found not to be reliable, with highly variable solids removal efficiency, and with difficulties for cleaning and maintenance. A new generation of more efficient separators, designed specifically for stormwater, is now emerging. However, these require further development to be reliable under a wide range of flow conditions.

4.1.7. Sediment control by detention ponds and wetlands

Detention ponds (Figure 4.1.13) are one of the most common management practices for the control of stormwater sediment. If properly designed, constructed, and maintained, they can be very effective in controlling a wide range of pollutants, in addition to solids loads and peak runoff flow rates. In most cases, these relatively large stormwater controls are located at drainage outfalls and treat all the runoff before discharge to receiving waters. However, it is appropriate to consider locating these devices in critical upland areas to capture sediment and other pollutants before entry to the drainage system, as part of the stormwater treatment train concept.

Figure 4.1.13. Detention pond/wetland shopping centre Birmingham Alabama, USA (photograph Bob Pitt).

There is more information available concerning the design and performance of detention ponds than for any other stormwater control device. However, the use of detention ponds for both water quality and quantity benefits is relatively new. Detention ponds typically show significant pollutant reductions as long as certain design-related attributes are met (most important being size and permanent pool depth). The most appropriate conditions for providing the necessary biological and chemical processes for stormwater treatment can be provided by ensuring optimal hydrologic effectiveness and hydraulic efficiency. Persson *et al.* (1999) found that stormwater pond design (length/width ratio and location of inlets and outlets) had a significant effect on these attributes. This was to avoid short-circuiting. Suitably sized wet detention ponds are quite robust and show good to excellent levels of pollutant control under a wide range of shape and outlet conditions.

Table 4.1.7 is a brief summary of the pollutant removal capabilities of some of the better-documented detention ponds, retention ponds, and wetlands that have been used for stormwater management. Note that these results cover a wide range of residence times, hydraulic conditions and relative pond to catchment size. Detention ponds are seen to be the most effective option under the widest variety of conditions. Retention ponds tend to have a high variability in terms of pollutant removal, probably due to scour and resuspension of solids and associated pollutants. Wetlands also show substantial potential, but questions remain concerning their long-term performance in cold climates and in the absence of careful plant management.

Whilst this section is concerned with the control of solids in stormwater, it must be appreciated that the most important wet detention pond design guidelines are to maintain public safety. Studies of pond safety have been made by Marcy and Flack (1981), and Eccher (1991) lists the important pond attributes to ensure maximum safety, while retaining good ecological control, including prevention of shoreline erosion.

Table 4.1.7. Reported performance of ponds and wetlands used for stormwater quality management

Sites	Sources	% removals						
		SS	BOD_5	COD	TP	TKN	Heavy metals	Bacteria
Detention ponds								
US EPA's Nationwide Urban Runoff Program (NURP)	EPA, 1983	90	70	70	60–70	60–70	60–95	
Lansing, MI	Luzkow *et al.*, 1981	70			>50	30–50		
Toronto, Ontario	Brydges and Robinson, 1986	70–90			80	25–60		
Rolla, MO	Oliver and Grigoropoulis, 1981	88		54–60	54–60	22		
Ottawa, Ontario	Gietz, 1983	80–95	35–45		70–85	45–50		50–95
Aalborg, Denmark	Hvitved-Jacobsen *et al.*, 1994	80–90			60–70		40–90	
Retention ponds								
France	Bertrand-Krajewski *et al.*, 1994	80–85		60–70			75–80	
Greenville, NC	Stanley, 1996	42–83			-5–36	-66–43	2–79	
Wetlands								
UK	Ellis, 1993	50–90					<90	
Virginia	Yu *et al.*, 1996	>50		30	>80			
Orlando, FL	Schueler, 1996	81			14		21–78	78

4.1.7.1 Maintenance of detention ponds and safety of disposal of removed sediments

Detention ponds must be regularly maintained. Poor operation and maintenance not only reduce the pollutant and flow rate reduction effectiveness, but can also cause detention facilities to become eyesores, nuisances and health hazards (Poertner, 1974). If a pond does not 'need' maintenance, particularly sediment removal, then it is not providing significant water quality benefits. SEMCOG (1981) for example, found that major detention pond maintenance (dredging) is usually needed within about 10 years after pond construction. More frequent (routine) maintenance may include: bank stabilisation; removal of debris and litter from the water and surrounding land; grass cutting; and algal control.

The US Milwaukee Nationwide Urban Runoff Program (NURP) project (Bannerman *et al.*, 1983) estimated an annual sediment delivery of about 500 kg per hectare for medium density residential land uses and about 2500 kg per hectare for commercial areas. Other land uses contribute sediment generally between these values. Assuming a solids bulk density of about 1930 kg/m^3, about 320 and 1600 m^3 of sediment would be deposited in a well designed detention pond for each hectare of medium density residential or commercial area per year. With a typical pond sized about 0.6 % of the contributing area, this would only result in the deposition of between 5 and 25 mm of sediment per year (considered across a pond as a whole) based on deposition rates commonly found during monitoring of wet detention ponds. Most of the sedimentation would, however, occur near the inlet and the resulting sediment accumulation would be very unevenly distributed throughout the pond. As an example. Marsalek and Marsalek (1997) observed sediment sorting in ponds draining a commercial complex, with the coarsest particles settling nearest the inlet.

During a study of a wet detention pond in Sydney, Australia, Ball (2000) reported removal rates of 87 % for suspended solids and 50 % for phosphorus. Caltrans recently began detailed studies in Los Angeles and San Diego, California, to examine the benefits, technical feasibility, costs and operation and maintenance requirements of retrofitting extended detention facilities at existing highway and associated locations (Taylor *et al.*, 2001). Initial sampling has indicated average suspended solids removal rates of 73 %. Concrete lined basins showed generally lower removal rates. Major removal of sediment from the pond is estimated to be required every 10 years.

Concerns about the safety of disposal of sediments from wet detention ponds have been expressed by, for example, Dewberry and Davis (1990), who analysed sediments from twenty-one ponds in northern Virginia, USA. They found trace metals in many of the sediments, but the available forms of the metals were significantly less than applicable toxic thresholds. They concluded that the dredged materials could be safely disposed either on-site or at landfills without danger of health problems. However, they recommend that sediment samples from specific ponds be analysed before dredging and selecting the most appropriate disposal option (see also Section 4.2.3 for sediment treatment and reuse).

4.1.7.2 Enhancing pond performance during severe winter conditions

Oberts (1990, 1994) monitored four urban wet detention ponds during both warm and cold weather in Minnesota, USA. The ponds performed as expected during warm weather, providing typical removals of suspended solids (80 %), lead (68 %) and total phosphorus (52 %). However, the ponds were less efficient at removing fewer suspended solids (39 %), organic matter (12 % for COD), nutrients (4 % for TKN, 17 % for TP) and lead (20 %) in the winter. Thick ice effectively eliminated much of the detention volume for incoming snowmelt water. In addition, the first melting water was forced under the ice, causing scour of the previously deposited sediments. Later snowmelt water flowed across the surface of the ice, with very little sedimentation opportunities. Any sediment that was accumulated on top of the underlying ice was later discharged out of the pond when the ice melted. Similar research in Minnesota wetlands also showed similar poor performance during winter conditions, for much the same reasons.

Oberts (1990, 1994) proposed several improvements in stormwater management practices during winter conditions. Initial recommendations are to utilise infiltration and grass filters in drainage pathways prior to the detention facilities. The design of the detention pond should be modified for winter operations. A low flow channel leading to and through the pond will discourage the formation of ice. The pond can also be aerated to prevent ice formation; however, if it gets extremely cold, ice formation could then be very thick and rapid.

The most important suggestion is to use a special riser for the outlet of the pond that can be used to draw down the water elevation during the winter. Ice would then form near the bottom of the pond and seal off the sediments. As the snowmelt occurs, the bottom outlets on the riser should be closed, forming a deeper pond for better sedimentation.

4.1.7.3 Detention pond design guidelines for suspended solids reductions

The most important design criterion for wet detention ponds is to provide a large enough water surface area. Table 4.1.8 was extrapolated from extensive wet detention pond monitoring data, mainly the EPA's NURP (EPA, 1983) studies.

Table 4.1.8. Percent of drainage area required as pond surface area for control of suspended solids

Land use	Approximate suspended solids reduction efficiency	
	90%	65%
	Required proportion of drainage area for pond (%)	
Freeways	2.8	1.0
Industrial areas	2.0	0.8
Commercial areas	1.7	0.6
Institutional areas	1.7	0.6
Residential areas	0.8	0.3
Open space areas	0.6	0.2

In France, design rules or ratios for both detention and retention ponds are based on different approaches, using hydrologic and/or hydraulic calculations (e.g. Bergue and Ruperd, 1994). However, a specific volume of about 100 m^3 of storage/active ha of catchment is required to intercept some 80 % of the annual TSS load, for either combined or separate wastewater systems. More detail may be found elsewhere (Bergue and Ruperd, 1994; Bertrand-Krajewski *et al.*, 2000). These values are based on expected runoff volumes for typical development conditions and would therefore vary considerably for different developments, especially if infiltration systems are used in the catchment. Other important design attributes include providing sufficient freeboard storage (the storage volume above the normal wet pond surface and below the invert of the emergency spillway). This storage volume should be adequate to provide for the resulting runoff from the watershed associated with a rainfall volume of about 30 mm. These values should be verified for local conditions, using a risk assessment procedure and continuous modelling for a long period of time. The pond outfall structure must also be designed to provide capture of necessary particles and for peak runoff rate objectives. Detailed detention pond design guidelines for different conditions in various countries have been published (e.g. Stahre and Urbonas, 1990; Livingston *et al.*, 1988; NVPDC, 1992; Azzout *et al.*, 1994; Bergue and Ruperd, 1994; Taylor *et al.*, 2001).

In summary, it is apparent that the use of ponds and wetlands can be very effective to prevent solids from passing downstream into sewer systems. These systems are normally used to manage stormwater flows rather than water quality, and restrictions to pond use, due to land area needs, safety concerns etc., may prevent their utilisation.

4.1.8 Infiltration systems to control solids

4.1.8.1 Benefits and problems associated with stormwater infiltration

Infiltration techniques have been used for many years to control stormwater quality and local drainage problems. Infiltration systems are designed to infiltrate runoff water near source areas, thus reducing energy available to promote erosion or washoff of solids (and associated pollutants) to the drainage system. Infiltration in swales or in grass areas also effectively traps sediment as the flowing water is infiltrated. Infiltration devices offer many advantages when integrated into conventional drainage systems (Azzout *et al.*, 1994; Novatech 1992, 1995, 1998, 2001):

- lower costs of associated sewerage systems;
- limited maintenance;
- effective integration into urban environments;
- better preservation of the hydrological balance in the environment.

Upland infiltration devices are located in urban source areas and can significantly reduce stormwater runoff volumes and peak flow rates. All infiltration devices redirect runoff waters from the surface to the sub-surface environment. They must, therefore, be carefully designed using site-specific information to protect groundwater resources and to achieve the desired management goals.

Numerous publications describe the successful use of stormwater infiltration throughout the world: Musiake *et al.* (1990) the use of shallow infiltration facilities in Tokyo, and Stenmark (1990) the use of infiltration facilities in cold climates. Other stormwater infiltration experience has been described by Wada and Miura (1990), Harada and Ichikawa (1993), Yamada (1993) and Duchene *et al.* (1993). The Technical University of Denmark has conducted numerous research projects concerning the benefits of infiltration as a source area control to reduce combined sewer overflows (Geldof *et al.*, 1993, 1994; Mikkelsen *et al.*, 1994; Petersen *et al.*, 1994; Jacobsen and Mikkelsen, 1996). Petersen *et al.* (1994), for example, found that the optimum solution for reducing CSO volumes by 40 % required infiltration from 65 % of the paved areas using trenches having total storage volumes for 3.6 mm of the total rainfall. This corresponded to a return period of 0.04 years (about 2 weeks), in contrast to the commonly applied design return periods of between 2 and 10 years.

One of the major concerns of stormwater infiltration is the question of adversely impacting groundwater quality. Pitt *et al.* (1996; 1999b) reviewed many studies that investigated groundwater contamination from stormwater infiltration. They developed a methodology to evaluate the contamination potential of stormwater nutrients, pesticides, other organic compounds, pathogens, metals, salts and other dissolved minerals, suspended solids, and gases, based on the concentrations of the contaminant in stormwater, the treatability of the contaminant and the mobility of the contaminant through the vadose zone. The stormwater pollutants of most concern (those that may have the greatest adverse impacts on groundwaters) include pollutants that associate with the solids (see Section 2.1).

4.1.8.2 Grass swales

Grass swales can be used in place of concrete curb and gutter systems in many areas, except strip commercial, manufacturing industrial, and high-density residential areas. Infiltration of the runoff causes trapping of the associated particulate pollutants near the soil surface, while many of the 'dissolved' pollutants are sorbed to organic soils Shallow flows slow dramatically due to increased roughness of the grass, enhancing particulate settling (Deletic, 2000; 2001). Shallow flows also almost entirely infiltrate in most swales, significantly reducing the sediment-carrying capacity of the water.

A project to study specifically the effects of grass swale drainage was conducted in Florida, USA by Kercher *et al.* (1983). This study examined two adjacent low-density residential areas, of about 5.6 ha, having about fifty homes. One area had conventional concrete curbs and gutters, while the other had grass swales for roadside drainage. The two areas had very similar characteristics (soil, percentage imperviousness, slope, vegetation, etc.). Thirteen rainfall events were monitored in the areas for flow and selected pollutants. The curb and gutter area produced runoff flows during all thirteen events, while the grass swale area produced runoff during only three of the events. Estimated annual pollutant yields from the curb and gutter area were much greater than for the grass swale area. BOD_5 annual discharges from the guttered area were estimated to be about 130 times the discharges from the swale drained area. Other yield increases from the guttered area as compared with the swale area included 90 times for suspended solids. The grass swale system also cost about one-half the cost of the 'conventional' curb and gutter system to install. Yu *et al.* (2001) field tested grass swales in Virginia and Taiwan. Average measured pollutant removal efficiencies varied from 14 to 99 % for total suspended solids (TSS).

The efficiency of particle trapping by grass swales has been investigated by Bäckström (2001) in two laboratory and seven field sites in Lulea, Sweden. The main conclusions were:

- Grass swales have a potential to capture considerable amounts of solids.
- Observed tss removal efficiency ranged from 79 to 98 %.
- Sedimentation processes were more significant than grass filtration.
- Highest particle trapping efficiency was observed in the field swales with dense and fully developed turf.
- High infiltration rates in the grass swale increased the particle trapping efficiency.
- Increasing the swale length led to a better trapping efficiency for the smaller particles.
- Some erosion of solids, leading to negative 'removal' efficiency can occur if the flow over the swale is too rapid.

Figure 4.1.14 gives the particle trapping efficiency for the two laboratory (L1 and L2) and the seven field (F1 to F7) grass swales, for three classes of particles. For each class, the experimentally measured particle settling velocity corresponds to a given particle diameter: $d = 13.5$ μm and $w = 0.1$ m/h, $d = 24.9$ μm and $w = 1$ m/h, $d = 45.9$ μm and $w = 10$ m/h. Swales F3 and F7 with lower vegetation density, had less trapping efficiency. Recently completed field and laboratory tests in Tuscaloosa, Alabama, USA (Johnson *et al.*, 2002) resulted in very similar conclusions as were found in these Swedish tests.

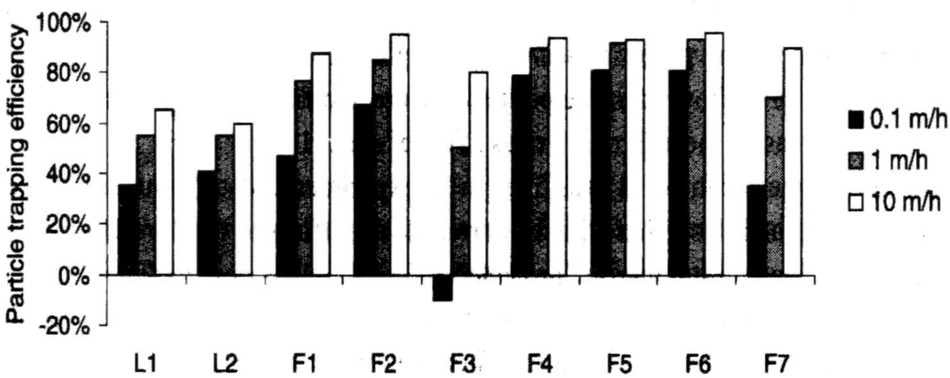

Key: L1, L2 laboratory swales

F1 – F7 field swales with a range of slopes, grass types, heights and spacing

Figure 4.1.14. Particle trapping efficiency in nine different grass swales, for three classes of particles (adapted from Bäckström, 2001).

In summary, it is apparent that grass swales can be used to efficiently trap the solids transported by surface runoff. Swales may be used effectively in housing estates and low density urbanised areas. After some years of working, the topsoil layer infiltration capacity should be checked. This layer may eventually need to be replaced to recover the design infiltration capacity and to remove the deposited solids and pollutants for further disposal and/or treatment.

4.1.8.3 Porous pavements

4.1.8.3.1 Efficiency of porous pavements

Porous pavements are generally used in areas of low traffic, such as service roads, storage and parking areas. These surfaces can be designed to eliminate all the runoff from paved areas for all but the most intense rainfall. Numerous successful applications of porous pavements throughout the world have been described: e.g. Niemczynowicz and Hogland, (1987); Hogland (1990) in Sweden; an overview of porous pavement use in the USA and Sweden; Pratt (1999) design and maintenance issues; Nawang and Saad (1993) and Sztruhar and Wheater (1993) experimental field trials; Fujita (1993) the Tokyo Experimental Sewer System.
French experiments in Nantes, Bordeaux and Paris have shown that porous and reservoir pavements were very efficient in reducing the solids and pollutant loads discharged into drainage systems, with removal rates ranging from 50 to more than 90 % depending on the pollutant and the site conditions (Baladès *et al.*, 1995a, 1995b). The results are similar to those observed by others (e.g. Hogland *et al.*, 1987; Pratt *et al.*, 1989, 1995). As an example, the removal rates at the outlets of one Swedish and two French porous pavement test sites are given in Table 4.1.9. Analyses of samples taken at the outlet of porous pavement structures have shown that the discharged water met the French national standards for raw waters to be used for drinking water production, and that there were no characteristics that would restrict infiltration directly into the ground.

Table 4.1.9. Examples of pollutant removal efficiencies for three porous pavement test sites

Site	Street structure	Source	TSS (%)	COD (%)	Pb (%)
Parking area Bordeaux (France)	Porous surface (5 cm) above a porous concrete layer (17 cm)	Baladès *et al.*, 1992	50	89	93
Large avenue in Paris (France)	Single porous concrete layer (16 cm) above concrete	Ranchet *et al.*, 1993	70	54	78
Parking area Gothenburg (Sweden)	Porous asphalt surface layer above a drained aggregate layer with a geotextile membrane in the bottom	Hogland *et al.*, 1987	95	40–60	-

4.1.8.3.2 Maintenance of porous pavements

Soil density studies using gamma rays and scanner auscultation have verified that clogging of porous pavements occurs typically only in the near-surface depths of the pavement, with the particles causing clogging only extending to a depth of about 1 to 2 cm below the surface. Progressive clogging with time is caused by an increase of accumulated solids in the first few centimetres of the pavement and not by the transport of the particles further down into the pavement structure (Baladès *et al.*, 1995b). The loss in infiltration capacity is the consequence of progressive clogging with time, as shown in Figure 4.1.15. This is an example showing the infiltration rate decreasing to about one-half of the initial rate after about 3 years, after which time it then generally remained constant. Similar results have been observed in many other experimental studies (Baladès *et al.*, 1995b).

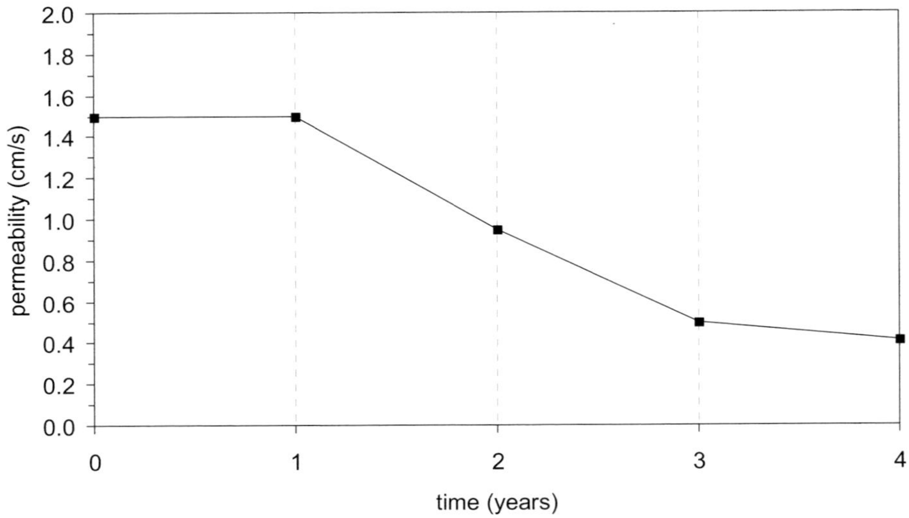

Figure 4.1.15. Example of temporal decrease in permeability of a porous pavement in an urban area (traffic: about 50 vehicles/day) (data from Baladès, 1997).

The mean diameter of the particles responsible for the clogging of porous pavements is about 300 μm (see Figure 4.1.16). For sites where there is only a thin porous pavement layer above an impervious structure layer, it has been observed that the mean diameter of the clogging particles is finer, with about 30 % of particles by mass, finer than 100 μm. Complementary analyses have shown that these solids were highly polluted, especially by heavy metals (e.g. Colandini *et al.*, 1995).

The masses of particles extracted from porous pavements depend on the use of the pavement, on the traffic intensity, on the cleaning machine and on the cleaning frequency. However, it appears that the amount of extracted particles is always very high: 0.2 to 1.5 kg/m^2. The highest values have been measured several times in residential household streets that have not been cleaned during the last 2 or 3 years.

Remediation of severe porous pavement clogging requires very powerful cleaning techniques, whereas regular cleaning with normal techniques should be satisfactory to keep the porous pavement surface functional. For porous pavements with traffic volumes below 100 lorries/day, and especially for parking lots, monthly cleaning by aspiration is sufficient to keep an almost constant infiltration capacity. If clogging has already become a problem, a stronger cleaning technique using both high pressure jetting (a pressure of 80 to 85 MPa is required) and aspiration is necessary (Baladès *et al.*, 1995b). Techniques which recycle the cleaning water are preferable to avoid transfer of the extracted solids to a wastewater system or receiving water. Extraction was found to be very efficient, effectively restoring the porous pavement infiltration capacity (Figure 4.1.17). In France, porous pavement cleaning costs between €0.4 to €1.3 (1996 value) per m^2. More information and case studies about the design and the maintenance of porous pavements may also be found in CERTU (1999).

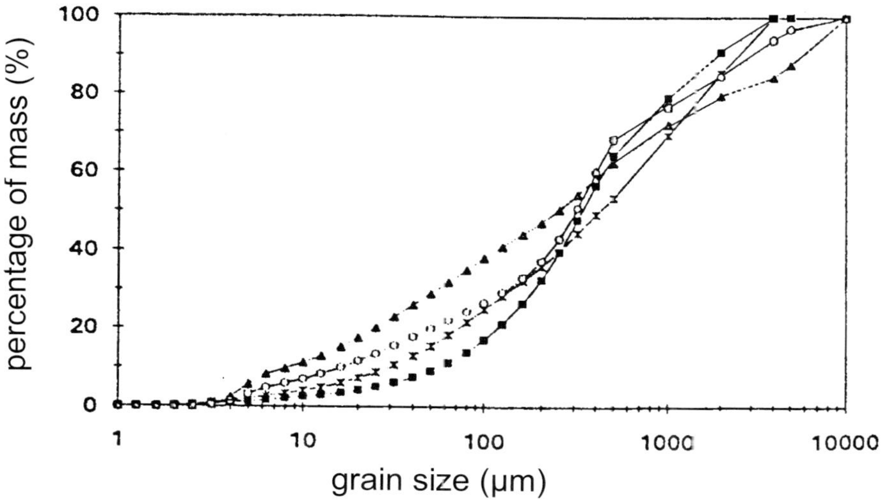

Figure 4.1.16. Grain size distribution of solids responsible for porous pavement clogging in Bordeaux

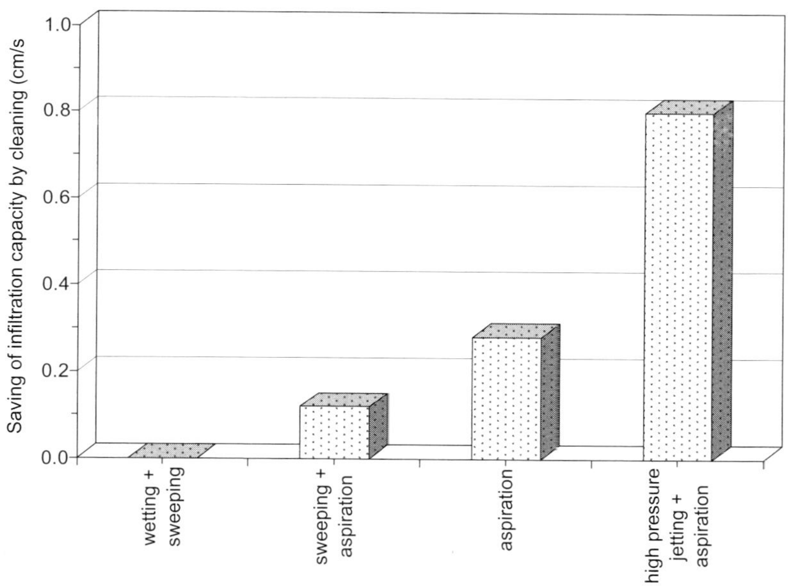

Figure 4.1.17. Recovery of infiltration capacity of porous pavements after cleaning with different techniques.

In summary, it has been found that porous pavements can efficiently trap solids transported by surface runoff. Depending primarily on the traffic type and intensity, the 1-3 cm deep top layer of a porous pavement will eventually become clogged. However, original infiltration capacity may be recovered if very high-pressure jetting and aspiration machines are used to clean the pavement. The solids removed are highly polluted (especially with heavy metals) and further disposal and/or treatment should be carefully considered.

4.1.8.4 Protection of groundwater quality during stormwater infiltration

Infiltration devices (like grass swales, soakaways, porous pavements) have hydrologic/hydraulic effects: they reduce the downstream flow rates, allow local infiltration and groundwater recharge, etc. However, surface runoff contains pollutants that may constitute a risk to groundwater quality, necessitating detailed site-specific evaluation. In soils, there are complex physical, chemical and biological processes and interactions that may affect the particulate and soluble fractions of these pollutants. The risk of significant groundwater contamination is likely to be limited when infiltrated stormwater runoff comes from only residential areas, pedestrian pathways or roofs. More detailed information and regulation concerning the use of such infiltration systems in different countries may be found elsewhere, e.g. Pitt *et al.* (1996) for the USA, Bettes *et al.* (1996) for the UK, Azzout *et al.* (1994) for France, Krejci *et al.* (1993) for Switzerland, Petersen *et al.* (1993) for Scandinavia and other countries.

As many of the pollutants found in runoff are attached to the transported solids, these pollutants are usually trapped by both sedimentation and infiltration, usually in the first layer of soil downstream from any inlet structures (Baladès *et al.*, 1995b; Lind and Karro, 1995). Periodic removal of these contaminated solids is necessary to recover the infiltration capacity and for further disposal and/or treatment of the attached pollutants.

4.1.9 Filtration and combination practices

Filtration systems are an effective means of removing solids from stormwater. In the USA, various municipalities and other organisations have produced manuals that include designs for sand filters to manage solids (e.g. Claytor and Schueler, 1996).

Large-scale use of sand filtration of stormwater has been used for many years in Texas, USA, where sand filters have been used for sites up to 20 ha in size. The filters are designed to hold and treat the first 13 mm of runoff and the pollutant removal ability of the sand filters has been found to be good, as shown in Table 4.1.10, compared with filtration of other media. Studies of a sand filter in a heavily used parking area in Maryland, USA showed that it began to clog only after 6 years in operation. Inspection of the sand below the surface of the filter indicated that oil, grease and finer sediments had migrated into the filter, but only to a depth of approximately 50 to 75 mm.

Table 4.1.10. Reported filtration media performance

Pollutant/media	Sand filter[1]	Leaf compost[2]	Peat moss[3]
Suspended solids	70	95	90
Turbidity	n/a	84	n/a
Total nitrogen	21	n/a	50
Total Kjeldahl nitrogen	46	56	n/a
Nitrate – nitrogen	0	n/a	n/a
Total phosphorus	33	40	70
BOD	70	n/a	90
Fecal coliform bacteria	76	n/a	90
COD	n/a	67	n/a
Total organic carbon	48	n/a	n/a
Iron	45	89	n/a
Copper	n/a	67	80
Lead	45	n/a	80
Zinc	45	88	80
Petroleum hydrocarbons	n/a	87	n/a

[1] City of Austin (1988); [2] W&H Pacific (1992); [3] Galli (1990).

Clogging of filters is assumed to occur when the filtration rate becomes less than about 1 m/day. The most restrictive materials are very fibrous and show evidence of compaction. In all tests, Clark and Pitt (1999) found that the media clogged, even with extensive pretreatment (SS about 5 mg/l), before chemical breakthrough occurred. A three-media mixed filter (peat, sand, and activated carbon) provided the most effective filtration operation (Pitt *et al.,* 1999a). This design is in operation as the polishing chamber of a multi-chamber treatment train (MCTT) in Milwaukee, USA. The MCTT includes a catchbasin with a sump as the inlet, with litter and grit trapping capability, a sedimentation chamber having inclined tube settlers and sorption pads, and a final filtration/sorption chamber. The MCTT is most suitable for use for relatively small and isolated paved critical source areas, from about 0.1 to 1 ha in area, where surface land is not available for traditional stormwater controls.

Typical locations include fuel stations, junk yards, bus depots, utilities yards, car washes, fast food restaurants, convenience stores, and other areas where the stormwater has a high probability of containing high concentrations of oils and filterable toxic pollutants that are difficult to treat by other means. A typical MCTT requires between 0.5 and 1.5 % of the paved drainage area, which is about one-third of the area required for a well-designed wet detention pond, and is generally installed below ground. After monitoring some thirty-five storms at a pilot-scale MCTT and two full-scale MCTTs, the average suspended solids removal was found to be greater than 95 %, and the average particulate and filterable organic and metallic toxicant removals were 90 %, or greater. Detailed design guidance is provided in a USEPA research report (Pitt *et al.*, 1999a).

4.1.10 Construction site erosion control

Construction site erosion can be temporarily the most significant local source of sediment over a limited timescale (Section 2.1). In the USA, erosion losses from uncontrolled construction sites have been measured to be in the range of 30 to more than 100 kg/ha/yr. Although very localised, typically these discharges substantially exceed those from urban stormwater. In addition, high sediment discharges can occur for several years after development has been completed as landscaping matures. Numerous states in the USA have published construction guidance for erosion controls (such as Livingston *et al.*, 1988 for Florida; Smoot *et al.*, 1992 for Tennessee; Association of Bay Area Governments, 1995 for California), but these generally lack suitable guidance on selecting the most appropriate controls. In the UK, CIRIA (Murnane *et al.*, 2002) has published a guide to good practice for the control of water pollution from construction sites.

Construction site erosion control requires various elements to protect the sewerage system and receiving waters:

- Diversion of water from the higher ground, where the area is undisturbed, so that the runoff does not flow across the disturbed part of the site. This preventive measure can reduce the volume of water and energy available to transport soil exposed by construction activity.
- Mulching of disturbed ground at time intervals that permits necessary grading but also reduces erosion losses during intense rains. Stabilisation of these inactive but disturbed areas is needed to prevent site erosion, to eliminate the cost of regrading severely eroded areas, and to protect off-site areas from erosion products. In many cases, better timing of grading operations could also reduce the time over which an area is left disturbed.
- Down-slope controls to minimise the quantity of erosion products that leave the site from the areas which have been disturbed. This is necessary because plants

(the longer term solution) normally require at least several weeks to become established and thence more effective at reducing erosion.

- The active construction area has to be left exposed. For small sites (less than 4 ha) with no channelled flow, filter fences or other perimeter controls may be adequate. These controls are fragile, however, and suitable only for sheetflows at low velocities. When larger flows are expected, sedimentation ponds are needed because high water flows can quickly destroy filter fences.
- On-site 'good house-keeping' practices. These practices include proper disposal of waste materials and wastes from equipment maintenance. Stabilised roads, entranceways, and storage areas are also needed, along with inlet protection.

The effectiveness of erosion controls in field observations are highly variable, but generally rather low (Barrett, *et al.* (1995), for example). Reported sediment removal rates for common filter fabric fences are about 30 to 50 % during laboratory tests, but typically reduce to close to zero during actual field applications. Sediment ponds on construction sites also provide typically low levels of control, especially in areas having fine grain (clayey) soils. Erosion prevention is much more effective than sediment control.

Probably the most common reason for failure of construction site erosion control devices is inadequate maintenance. These devices are often reluctantly installed and then ignored. If control devices are properly constructed, but not properly or frequently maintained, very little benefit can be expected.

Newly installed devices will perform as initially expected until their 'capacity' is exceeded. Filter fences, for example, should be maintained before the material that accumulates behind them becomes excessive. More importantly, the integrity of the fence also needs to be checked frequently. Many filter fences at construction sites are undermined or bypassed because of large flows or large sediment accumulation. Sedimentation ponds, silt traps, catchbasins, etc., also need to be cleaned frequently. The necessary cleaning frequency of these devices located in areas undergoing construction can be quite high. Rill or gully erosion must be corrected immediately when first observed. Similarly, mulched or planted areas need frequent inspections and maintenance to avoid large amounts of material being lost. Notwithstanding the effectiveness of these good practices, significant sediment and associated pollutants may still leave a construction site, as illustrated in a study by the University of Texas (Tenney *et al.*, 1995).

New practices for the delivery of sand and aggregates to construction sites in many countries in Europe are emerging, requiring, for example, bagging rather than bulk storage in heaps on site. In general, construction site practices are now being better controlled in developed countries to avoid environmental problems, with commensurate benefits for sewered discharges (Murnane *et al.*, 2002).

4.1.11 References

Ab Ghani, A., Zakaria, N.A., Kassim, M. and Nasir, B.A. (2000) Sediment size characteristics of urban drains in Malaysia. *Urban Water*, 2(4), 335–341.

Andral, M.C., Roger, S., Montrejaud-Vignoles, M. and Herremans, L. (1999) Particle size distribution and hydrodynamic characteristics of solid matter carried by runoff from motorways, *Water Environ. Res.*, 71(4), 398.

Aires, N. and Tabuchi, J.P. (1995) Hydrocarbons separators and stormwater treatment. (in French) *TSM, Special Issue 'Stormwater'*, 11, 862–864.

Armitage, N., Marais, M. and Pithey, S. (2001) *Reducing Urban Litter in South Africa through Catchment Based Litter Management Plans*. Models and Applications to Urban Water Systems, Monograph 9. 37. CHI, Guelph, Ontario.

Ashley, R.M., Souter, N., Butler, D., Davies, J., Dunkerley, J. and Hendry, S. (1998) Assessment of the sustainability of alternatives for the disposal of domestic sanitary waste. Conference on 'options for closed water systems: sustainable management' Wageningen, March. *Wat. Sci. Tech.*, 39(5), 251–258.

Ashley, R.M., Fraser, A., Burrows, R. and Blanksby J. (2000) The management of sediment in combined sewers. *Urban Water*, 2, 263–275.

Ashley, R. M., Souter, N. and Hendry, S. (2001) The importance of public attitudes and behaviour for the management of urban water systems. *Proc. International symposium 'Frontiers of Urban Water Management: Deadlock or Hope?'* June. IHP-V Technical documents in hydrology No. 45, edited by J.A.Tejada-Guibert and C. Maksimovic, UNESCO, Paris, 318–326.

Association of Bay Area Governments (1995) *Erosion and Sediment Control Measures*, 2nd edition, May, Oakland, California.

Azzout, Y., Barraud, S., Crès, F.N. and Alfakih, E. (1994) *Alternative Techniques for Urban Storm Drainage.* (in French) Paris, France, Tech et Doc, 372pp. ISBN 2-85206-998-9.

Bäckström, M. (2001) Particle trapping in grass swales. *Wat. Sci. Tech.*, 45(7), 41–49.

Baladès, J.-D. (1997) *Clogging and cleaning of porous pavements.* Strasbourg, France, ENGEES, Professional Training Course, 14 pp. (in French).

Baladès, J.-D., Bourgogne, P. and Madiec H. (1992) Assessment of pollutant removal through a compensating technique structure. (in French) *International Conference Novatech 92*, Lyon, 65–75.

Baladès, J.-D., Bourgogne, P. and Gourioux, J. (1995a) A reservoir structure pavement for the urban storm drainage of a supermarket area of 120.000 m^2 in Bordeaux. (in French) *Proceedings of Novatech 95*, Lyon, 329–336.

Baladès, J.-D., Legret, M and Madiec, H. (1995b) Porous pavements as tools to control the pollutant loads on urban streets. (in French) *Proceedings of Novatech 95*, Lyon, 345–352.

Ball, J.E. (2000) *Stormwater Quality at Centennial Park, Sydney, Australia.* Research Report 205, Water Research Laboratory, School of Civil and Environmental Engineering, The University of New South Wales.

Ball, J.E., Wojcik, A. and Tilley, J. (2000) *Stormwater Quality from Road Surfaces: Monitoring of the Hume Highway at South Strathfield.* Research Report 204, Water Research Laboratory, School of Civil and Environmental Engineering, The University of New South Wales, 130 pp.

Bannerman, R., Baun, K., Bohn, M., Hughes, P.E. and Graczyk, D.A. (1983) *Evaluation of Urban Nonpoint Source Pollution Management in Milwaukee County, Wisconsin*, Vol. I. Grant No. P005432-01-5, PB 84-114164. US Environmental Protection Agency, Water Planning Division, November.

Barrett, M.E., Kearney, J.E., Malina, J.F., Charbeneau, R.J. and Ward G.H. (1995) *An Evaluation of the Use and Effectiveness of Temporary Sediment Controls.* Technical Report CRWR 261, Center for Research in Water Resources, The University of Texas at Austin, August 1995.

Beurden van, E. and Geldof, G.D. (2001) The integrated urban water plan: a useful instrument? *Proc. 4th International Conference Innovative Technologies in Urban Drainage (Novatech)*, June, GRAIE ISBN 2-9509337-4-2.

Bergue, J.-M. and Ruperd, Y. (eds) (1994) *Guide technique des bassins de retenue d'eaux pluviales.* Paris, France, Tec et Doc, 275 pp. ISBN 2-85206-934-2.

Bertrand-Krajewski, J.-L., Breuil, B., Herremans, L. and Renard, D. (1994) How to reduce stormwater discharges and their impact on receiving water? (in French) *La Houille Blanche*, 1/2, 155–169.

Bertrand-Krajewski, J.-L., El-Jawhari, A. and Chebbo, G. (2000) Sizing of wet weather urban discharges treatment facilities. Villeurbanne (France): INSA de Lyon – Laboratoire URGC Hydrologie Urbaine et Marne-la-Vallée (France): Cereve, rapport de recherche pour l'Agence de l'Eau Seine-Normandie (commande n° 99-97044), September 2000, 108 pp. (in French).

Bettes, R., Davies, A. and Watkins, D. (1996) *Infiltration Drainage – Hydraulic Design.* Project report 23, HR Wallingford, UK, January.

Brydges, T. and Robinson, G. (1986) Two examples of urban stormwater impoundment for aesthetics and for protection of receiving waters. *Restoration of Lakes and Inland Waters: International Symposium on Inland Waters and Lake Restoration*, Portland, Main, September. Proceedings published by the USA Environmental Protection Agency, Washington, D.C.

Burton, G.A. Jr., and Pitt, R. (2002) *Stormwater Effects Handbook: A tool box for watershed managers, scientists, and engineers.* CRC Press, Inc., Boca Raton, FL., 1085 pp.

Butler, D. and Clark, P. (1993) *Sediment Management in Urban Drainage Catchments.* Report No. RP416, Construction Industry Research and Information Association (CIRIA), London

Butler, D. and Memon, F. (1999) Dynamic modelling of roadside gully pots during wet weather. *Water Research*, 33(15), 3364–3372.

Butler, D. and Karunaratne, S.H.P.G. (1995) The suspended solids trap efficiency of the roadside gully pot. *Wat. Res.*, 29(2), 719–729.

Butler, D., Xiao, Y., Karunaratne, S.H.P.G. and Thedchanamoorthy, S. (1995) The gully pot as a physical and biological reactor. *Water Science and Technology*, 31(7), 219–228.

Cairncross, S. and Feacham, R.G. (1993) *Environmental Health Engineering in the Tropics*. Wiley, NY.

CERTU (1999) Guide technique: chaussées poreuses urbaines. CERTU, collection Références 7, Lyon, France, 150 pp. ISSN 1263–3313.

Chalmers tekniska hogskola (2001) *Urban Water – Progress Report*. ISSN 1650–3791

City of Austin, Texas (1988) *Design Guidelines for Water Quality Control Basins*. Environmental DCM, City of Austin Transportation and Public Services Department.

Clark, S. and Pitt, R. (1999) *Stormwater Treatment at Critical Areas: Vol. 3. Evaluation of Filtration Media for Stormwater Treatment*. U.S. Environmental Protection Agency, Water Supply and Water Resources Division, National Risk Management Research Laboratory, Cincinnati, Ohio, 442 pp.

Claytor, R. and Schueler, T. (1996) *Design of Stormwater Filtering Systems*. Center for Watershed Protection. Silver Spring, MD, 200 pp., September 199.

Colandini, V., Legret, M., Brosseaud, Y. and Baladès, J.-D. (1995) Pollution métallique contenues dans les produits de décolmatage de chaussées poreuses urbaines. *Proceedings of Novatech 1995*, Lyon, France, 30 May–1 June, 353–360.

Corsi, R., Greb, S.R., Bannerman, R.T. and Pitt, R.E. (1999) *Evaluation of the Multi-Chambered Treatment Train, a Retrofit Water-Quality Management Device*. U.S. Geological Survey Open-File Report 99-270, Middleton, WI.

CRC (2000) *Assessment of Food Disposal Options in Multi-Unit Dwellings in Sydney*. Document 2883R, Report for In-Sink-Erator, CRC For Waste Management and Pollution Control Ltd, December.

Cummings, S. and Bonollo, E. (1999) Experience with dual flush technology in Australian WC design. *Proc. CIB W62 Conference Water Supply and Drainage for Buildings*, Heriot Watt University, UK.

Daligaud, A., Meaudre, D., Arnault, D., Duc, V., Bardin, N., Aires, N., Biau, D., Schmid, J., Clement, P. and Viau, J.-Y. (1998) Stormwater and lamellar settlers: efficiency and reality. *Proceedings of Novatech 98*, Lyon, France, 4–6 May 1998, 1, 471–479.

D'Arcy, B., Ellis, J.B., Ferrier, R.C., Jenkins, A. and Dils, R. (eds) (2001) *Diffuse Pollution Impacts*. CIWEM/IWA, Terence Dalton. ISBN 1 870752 46 5.

Dastugue, S., Vignoles, M., Heughebaert, J.-C., and Vignoles, C. (1990) Matières en suspension contenues dans les eaux de ruissellement de la ville de Toulouse: quantification, corrélation au type d'urbanisation et répartition des rejets. *TSM*, 3, 131–143.

Delattre, C., Bertrand-Krajewski, J-L., Picard, V., Berga, P. and Balades, J.-D. (1998) Characteristics and treatability of deposits from urban catchments, sewer systems and storage tanks. *Proc 3rd Int. Conf on Innovative Technologies in Urban Storm Drainage*, Lyons, France, May.

Deletic, A. (2000) *Sediment Behaviour in Overland Flow Over Grassed Areas*. PhD thesis, University of Aberdeen, UK.

Deletic, A. (2001) Modelling of water and sediment transport over grassed areas. *Journal of Hydrology*, 248, 168–182.

Dewberry and Davis, Inc. (1990) *Investigation of Potential Sediment Toxicity from BMP Ponds*. Prepared for Northern Virginia Planning District Commission, Occoquan Policy Board, and Virginia State Water Control Board, May.

Duchene, M., McBean, E. and Thomson, N. (1993) Infiltration characteristics associated with the use of trenches in stormwater management. *Proceedings of 6th International Conference on Urban Storm Drainage*, edited by J. Marsalek and H.C. Torno, Niagara Falls, Ontario, 12–17 September 1993, IAWQ, London. Published by Seapoint Publishers, Victoria, B.C., 1115–1120.

Eccher, C.J. (1991) Thoughtful design is prime factor in water safety. *Lake Line*, 4–8 May.

Ellis, J.B. (1986) Pollutional aspects of urban runoff. *Urban Runoff Pollution*, NATO ASI series, G10, Springer Verlag, Berlin, 1–38.

Ellis, J.B. (1993) Wetland BMP design for urban runoff pollution control in Europe and Australia. *Proceedings of 6th International Conference on Urban Storm Drainage*, edited by J. Marsalek and H.C. Torno, Niagara Falls, Ontario, 12–17 September 1993, IAWQ, London. Published by Seapoint Publishers, Victoria, B.C., 957–962.

Ellis, J.B. (2000) Infiltration systems: a sustainable source-control option for urban storm water quality management? *J CIWEM*, 14, 27–34.

EPA (USA Environmental Protection Agency) (1983) *Final Report for the Nationwide Urban Runoff Programme*. Water Planning Division, Washington, D.C., December.

Field, R., Brown, M.P. and Vilkelis, W.V. (1994) *Stormwater Pollution Abatement Technologies*. EPA/600/R-94/129, U.S. Environmental Protection Agency, Office of Research and Development, Cincinnati, OH.

Field, R., Heaney, J.P. and Pitt, R. (2000) Innovative urban wet-weather flow management systems. *Technomic*. ISBN 1-56676-914-0.

Fujita, S. (1993) Infiltration in congested urban areas of Tokyo. *Proceedings of 6th International Conference on Urban Storm Drainage*, edited by J. Marsalek and H.C. Torno, Niagara Falls, Ontario, 12–17 September 1993, IAWQ, London. Published by Seapoint Publishers, Victoria, B.C., 993–998.

Fujita, S. (1994) Infiltration structures in Tokyo. *Water Science and Technology*, 30(1), 33–41.

Fujita, S. (1998) Restoration of polluted urban watercourses in Tokyo for community use. *Sustaining Urban Water Resources in the 21st Century: Proceedings of an Engineering Foundation Conference*, 7–12 September 1997, Malmo, Sweden. ASCE/Engineering Foundation, New York.

Galli, J. (1990) *Peat-Sand Filters: A proposed stormwater management practice for urbanized areas*. Prepared for the Coordinated Anacostia Retrofit Programme and Office Of Policy and Planning, D.C. Department of Public Works, December.

Garde, R.J., Ranga Raju, K.G. and Sujudi, A.W.R. (1990) Design of settling basins. *J. Hydr. Res.*, 28(1), 81–91.

Geldof, G., Jacobsen, P. and Fujita. S. (1993) Urban stormwater infiltration perspectives. *Proceedings of 6th International Conference on Urban Storm Drainage*, edited by J. Marsalek and H.C. Torno, Niagara Falls, Ontario, 12–17 September 1993, IAWQ, London. Published by Seapoint Publishers, Victoria, B.C., 963–968.

Geldof, G., Jacobsen, P. and Fujita, S. (1994) Urban stormwater infiltration perspectives. *Water Science and Technology*, 29(1–2), 245–254.

Gietz, R.J. (1983) *Urban Runoff Treatment in the Kennedy-Burnett Settling Pond*. For the Rideau River Stormwater Management Study, Pollution Control Division, Works Department, Regional Municipality of Ottawa-Carleton, Ottawa, Ontario, March.

Gouda, H., Ashley, R.M., Gilmour, D. and Smith, H. (2002) Life cycle analysis and sewer solids. *Wat. Sci. Tech.*, 47(4), 185–192.

Grey, G. and Oliveri, F. (1998) Catch basins: effective floatables control devices. *Proc. Adv. in Urban Wet Weather Pollut. Reduction*, Cleveland, Ohio, WEF (CP3805), 267, Arlington, VA.

Gromaire, M.C., Garnaud, S., Ahyerre, M. and Chebbo, G. (2000) The quality of street cleaning waters: comparison with dry and wet weather flows in a Parisian combined sewer system. *Urban Water*, 2, 39.

Gunther, F. (1995) Simplifying waste water treatment by source separation: outline for a biologically based grey water purification plant. *Proc. of UNESCO symposium on Integrated Water Management in Urban Areas*, Lund, 26–30 September.

Hall, M.J., Hockin, D.L. and Ellis, J.B. (1993) *The Design of Flood Storage Reservoirs*. CIRIA/Butterworths. ISBN 0-7506-1057-3.

Harada, S. and Ichikawa, A. (1993) A study on the performance of the drainage infiltration strata. *Proceedings of 6th International Conference on Urban Storm Drainage*, edited by J. Marsalek and H.C. Torno, Niagara Falls, Ontario, 12–17 September 1993, IAWQ, London. Published by Seapoint Publishers, Victoria, B.C., 1104–1109.

Harremoës, P. (2001) Integrated urban drainage, status and perspectives. *Wat. Sci. Tech.*, 45(3), 1–10.

Hedberg, T. (1999) Attitudes to traditional and alternative sustainable sanitary systems. *Wat. Sci. Tech.*, 39(5).

Henze, M. (1997) Waste design for households with respect to water, organics and nutrients. *Wat. Sci. Tech.*, 35(9), 113–120.

Hogland, W., Niemczynowicz, J. and Wahlman, T. (1987) The unit superstructure during the construction period. *The Science of the Total Environment*, 59, 411–424.

Hogland, W. (1990) Pervious asphalt construction: an overview of the situation in Sweden and the USA. *Developments in Storm Drainage – A Symposium on Infiltration and Storage of Stormwater in New Developments*, Sheffield City, January 1990, 122–135.

Hvitved-Jacobsen, T., Johansen, N.B. and Yousef, Y.A. (1994) Treatment systems for urban and highway runoff in Denmark. *The Science of the Total Environment*, 146/147, 499–506.

Jackson, H.B. (1995) Global needs and developments in urban sanitation. Paper No. 1, *Proc. Int. Conf. On Low Cost Sewerage*, University of Leeds, UK, July.

Jacobsen, P. and Mikkelsen, P.S. (1996) Reduction of urban runoff pollutant loads using detention ponds and stormwater infiltration techniques. *Proceedings of 7th Conference of Urban Storm Drainage*, Hanover, Germany, 9–13 September.

Johnson, P., Pitt, R., Durrans, R. and Clark, S. (2002) *Innovative Metals Removal Technologies for Urban Stormwater*. Water Environment Research Foundation, Arlington, VA.

Kercher, W.C., Landon, J.C. and Massarelli, R. (1983) Grassy swales prove cost-effective for water pollution control. *Public Works*, April.

King County Surface Water Design Manual (1998) King County Department of Natural Resources and Parks, Land and Water Resources Division, Seattle, Washington.

Kohzad, F. (2000) Continuous deflective separation: an integrated technology for removal of storm-water gross pollutants utilizing special flow mechanics. *WEFTEC2000, 73rd Annual Conference and Exposition*, October 2000, Anaheim, CA. Water Environment Federation, CD-ROM.

Kolsky P. (1998) Surface water drainage: how evaluation can improve performance. Intermediate Technology Development Group Publications, *Waterlines*, 17(1), 15–18.

Kolsky, P. and Butler, D. (2002) Performance indicators for urban storm drainage in developing countries. *Urban Water*, 4(2), 37–144.

Krejci, V., Haldimann, P. and Grottker, M. (1993) Administrative aspects of stormwater infiltration in Switzerland. *Proceedings of 6th International Conference on Urban Storm Drainage*, edited by J. Marsalek and H.C. Torno, Niagara Falls, Ontario, 12–17 September, IAWQ, London. Published by Seapoint Publishers, Victoria, B.C., 999–1004.

Lager, J.A., Smith, W.G. and Tchobanoglous, G. (1977) *Catchbasin Technology Overview and Assessment*. EPA-600/2-77-051, USA EPA, Contract No. 68-03-0274, Cincinnati, May, 129 pp.

Laplace, D., Bachoc, A., Sanchez, Y. and Dartus, D. (1992) Truck sewer clogging development: description and solutions. *Wat. Sci. Tech.*, 25(8), 91–100.

Lau, S.-L., Khan, E. and Stenstrom, M.K. (2001) Catch basin inserts to reduce pollution from stormwater. *Wat. Sci. Tech.*, 44(7), 23–34.

Legrand, J., Maillot, H., Nougarède, F. and Defontaine, S. (1994) A device for stormwater treatment in the urban development zone of Annœullin. (in French) *TSM*, 11, 639–643.

Lind, B.B. and Karro, E. (1995) Stormwater infiltration and accumulation of heavy metals in roadside green areas in Goteborg, Sweden. *Ecological Engineering*, 5(4), 533–529.

Lippner, G., Johnston, J., Combs, S., Walter, K. and Marx, D. (2001) Results of California Department of Transportation Litter Management Pilot Study. *Transportation Res. Record*, 1743, 10.

Livingston, E., McCarron, E., Cox, J. and Sanzone P. (1988) *The Florida Development Manual: A guide to sound land and water management*. Florida Department of Environmental Regulation, Tallahassee, Florida, June, 1231 pp.

Lloyd, S.D., and Wong, T.H.F. (1999) Particulates, associated pollutants and urban stormwater treatment. *Proc. 8th International Conference on Urban Storm Drainage*. 30 August–3 September 1999, Sydney, Australia, edited by I.B. Joliffe and J.E. Ball, The Institution of Engineers of Australia, The International Association for Hydraulic Research, and The International Association on Water Quality, 1833.

Luzkow, S.M., Scherger, D.A. and Davis, J.A. (1981) Effectiveness of two in-line urban stormwater best management practices (BMP's). *1981 International Symposium on Urban Hydrology, Hydraulics, and Sediment Control*, University of Kentucky, Lexington, Kentucky, July.

Marcy, S.J. and Flack, J.E. (1981) Safety considerations in urban storm drainage design. *2nd International Conference on Urban Storm Drainage*, Urbana, Illinois, June.

Marsalek, J. and Marsalek, P.M. (1997) Characteristics of sediments from a stormwater management pond. *Wat. Sci. Tech.*, 36(8–9), 117–122.

Masters-Williams, H., Heap, A., Kitts, H., Greenshaw, L., Davis, S., Fisher, P., Hendrie, M. and Owens D. (2001) *Control of Water Pollution from Construction Sites: Guidance for consultants and contractors*. CIRIA C532. ISBN 0-86017-532-4.

Murnane, E. *et al.* (2002) *Control of Water Pollution from Construction Sites: Guide to good practice*. CIRIA. ISBN 0-86017-807-2.

Memon, F. and Butler, D. (2002) Identification and modelling of dry weather processes in gully pots. *Water Research*, 36, 1351–1359.

Mikkelsen, P.S., Weyer, G., Berry, C., Walden, Y., Colandini, V., Faulsen, S., Grotehusmann, D. and Rohlfing, R. (1994) Pollution from urban stormwater infiltration. *Water Science and Technology*, 29(1–2), 293–302.

Musiake, K., Okamura, J. and Matsushita, J. (1990) The urban hydrological system with shallow infiltration facilities: a case study of an area in Tokyo Metropolis. *Proceedings of 5th International Conference on Urban Storm Drainage*, Osaka, Japan, July, 113–118.

Nawang, W.M. and Saad, S. (1993) Stormwater infiltration investigation using porous pavement. *Proceedings of 6th International Conference on Urban Storm Drainage*, edited by J. Marsalek and H.C. Torno, Niagara Falls, Ontario, 12–17September 1993, IAWQ, London. Published by Seapoint Publishers, Victoria, B.C., 1092–1097.

Newman, T.L., Leo, W.M., Muekker, J.A. and Gaffoglio, R. (1996) Effectiveness of street cleaning for floatables control. *Proceedings Urban Wet Weather Pollution: Controlling Sewer Overflows and Stormwater Pollution*, 16–19 June, Quebec City, Quebec. Water Environment Federation, Alexandria, VA.

Newman, T.L., Leo, W.M. and Gaffoglio, R. (1999) A No Cost, Best Management Practice for Floatables Control in New York City. *Proc. Water Environ. Fed. 72nd Annual Conf. Exposition*, New Orleans, LA. [CD-ROM]

Niemczynowicz, J. and Hogland, W. (1987) Tests of porous pavement in Lund, Sweden. *Proceedings of 4th International Conference on Urban Storm Drainage*, Lausanne, Switzerland, August. Water Resources Publications, 79–80.

Niemczynowicz, J. (1994) New aspects of urban drainage and pollution reduction towards sustainability. *Wat. Sci. Tech.*, 30(5), 269–277.

Novatech (1992) International conference on new technologies for urban storm drainage. (in French and English), *GRAIE/Eurydice 92*, Lyon, France, 3–5 November 1992, 568 pp.

Novatech (1995) 2nd International conference on new technologies for urban storm drainage. (in French and English), *GRAIE/Eurydice 92*, Lyon, France, 30 May–1 June, 670 pp. ISBN 2-9509337-0-X.

Novatech (1998) 3rd International conference on new technologies for urban storm drainage. (in French and English), *GRAIE/Eurydice 92*, Lyon, France, 4–6 May, Vol. 1, 588 pp. ISBN 2-9509337-1-8; Vol. 2, 588 pp. ISBN 2-9509337-2-6.

Novatech (2001) 4th International conference on new technologies for urban storm drainage. (in French and English), *GRAIE/Eurydice 92*, Lyon, France, 25–27 June, 1135 pp. ISBN 2-9509337-3-4; ISBN 2-9509337-4-2.

NVPDC (1992) *Northern Virginia BMP Handbook: A guide to planning and design best management practices in Northern Virginia.* Northern Virginia Planning District Commission, Engineers and Surveyors Institute (NVPDC), Engineers and Surveyors Institute (ESI), Annandale, VA, USA, 6 November 1992, 302 pp. Available as PDF file at http://www.novaregion.org/bmp.htm/.

Oberts, G. (1990) Design considerations for management of urban runoff in wintry conditions. *Proceedings International Conference on Urban Hydrology under Wintry Conditions.* 19–21 March, 1990, Narvik, Norway.

Oberts, G. (1994) Performance of stormwater ponds and wetlands in winter. Technical note 16, *Watershed Protection Techniques*, 1(2), 64–68, Summer.

Oliver, L.J. and Grigoropoulis, S.G. (1981) Control of storm-generated pollution using a small urban lake. *Journal Water Pollution Control Federation*, 53(5), May.

Otterpohl R. (2001) Options for alternative types of sewerage and treatment systems directed to improvement of the overall performance. *Wat. Sci. Tech.*, 45(3), 149–158.

Persson, J., Somes, N.L.G., and Wong, T.H.F. (1999) Hydraulics efficiency of constructed wetlands and ponds. *Wat. Sci. Tech.*, 40(3), 291–300.

Petersen, C.R., Faarbaek, T., Jensen, G.H., Weyer, G., Fujita, S., Ishikawa, K., Geldof, G., Stenmark, C. and Pratt, C.J. (1993) Urban stormwater infiltration design practice and technology: state of the art assessment. *Proceedings of 6th International Conference on Urban Storm Drainage*, edited by J. Marsalek and H.C. Torno, Niagara Falls, Ontario, 12–17 September, IAWQ, London. Published by Seapoint Publishers, Victoria, B.C., 969–974.

Petersen, C.R., Jacobsen, P. and Mikkelsen, P.S. (1994) Design of stormwater infiltration for reduction of combined sewer overflow (CSO). *Wat. Sci. Tech.*, 30(1), 53–61.

Pitt, R. (1979) *Demonstration of Nonpoint Pollution Abatement Through Improved Street Cleaning Practices.* USA EPA, Grant No. S-804432. EPA-600/2-79-161, Cincinnati, August, 270pp.

Pitt, R. (1985) *Characterizing and Controlling Urban Runoff through Street and Sewerage Cleaning*. USA EPA, Contract No. R-805929012. EPA/2-85/038, PB 85-186500/AS, Cincinnati, June, 467 pp.

Pitt, R., Lalor, M., Field, R., Adrian, D.D. and Barbe, D. (1993) *A User's Guide for the Assessment of Non-Stormwater Discharges into Separate Storm Drainage Systems*. U.S. Environmental Protection Agency, Storm and Combined Sewer Program, Risk Reduction Engineering Laboratory, EPA/600/R-92/238, PB93-131472, Cincinnati, Ohio, January, 87 pp.

Pitt, R., Clark, S., Parmer, K. and Field, R. (1996) *Groundwater Contamination from Stormwater.* Ann Arbor Press, Inc. Chelsea, Michigan, 219 pp. ISBN 1-57504-015-8.

Pitt, R. and Field, R. (1998) An Evaluation of Storm Drainage Inlet Devices for Stormwater Quality Treatment. *Water Environment Federation 71st Annual Conference and Exposition, WEFTEC Technology Forum*, Orlando, FL, October.

Pitt, R., Robertson, B., Barron, P., Ayyoubi, A. and Clark S. (1999a) *Stormwater Treatment at Critical Areas: The Multi-Chambered Treatment Train (MCTT)*. U.S. Environmental Protection Agency, Wet Weather Flow Management Program, National Risk Management Research Laboratory, EPA/600/R-99/017, Cincinnati, Ohio, March, 505 pp.

Pitt, R., Clark, S. and Field, R.. (1999b) Groundwater contamination potential from stormwater infiltration practices. *Urban Water*, 1(3), 217–236, September.

Poertner, H.G. (1974) *Practices in Detention of Urban Stormwater Runoff.* American Public Works Association, OWRR Contract No. 14-31-0001-3722, Chicago.

Poertner, H.G. (1980) *Stormwater Management in the United States*. Office of Water Research and Technology, Dept. of the Interior, Project No. C-5239, September.

Pratt C.J., Mantle J.D.G. and Schoffield P.A. (1989) *Urban stormwater reduction and quality improvement through the use of permeable pavements. Urban discharges and receiving water quality impacts.* Pergamon Press, Oxford, 123–132.

Pratt C.J., Mantle J.D.G. and Schoffield P.A. (1995) UK research into the performance of permeable pavement, reservoir structures in controlling stormwater discharge quantity and quality. *Proceedings of Novatech 95*, Lyon, France, 30 May–1 June, 337–344.

Pratt, C.J. (1999) Design guidance for porous/permeable pavements. *Sustaining Urban Water Resources in the 21st Century*, edited by A.C. Rowney, P. Stahre and L.A. Roesner, ASCE, 196-211. ISBN 0 7844 0424 0.

Ranchet, J., Penaud, F., Le Grand, R., Constant, A., Obry, P. and Soudieu. B. (1993) Comparison of an impervious pavement and a porous pavement from hydraulic and stormwater pollutant removal points of view. (in French) *Bulletin de Liaison des Laboratoires des Ponts et Chaussées*, 188, November/December.

Rocher, V., Azinis, S., Moilleron, R. and Chebbo, G. (2003) Hydrocarbons and heavy metals in the different sectors in 'Le Marais' catchment (Paris, France): stocks distributions and origins. *Science of the Total Environment*. In press.

Ruperd, Y. (1993) A lamellar separator for urban street runoff treatment. (in French) *Bulletin de Liaison des Laboratoires des Ponts et Chaussées*, 183, 85–90.

Ruperd, Y., Bourgogne, P. and Blanloeil, C. (1998) Experimental platform for testing compact devices for the treatment of surface runoff. *Proceedings of Novatech 98*, Lyon, France, 4–6 May, 1, 405-412.

Salvato, J.A. (1992) *Environmental Engineering and Sanitation*. 4th edition, John Wiley, NY.

Sansalone, J.J. and Hird, J.P. (1999) A Passive Source Control Prototype For Heavy Metals Transported by Urban Storm Water. *Proc. Water Environ. Fed. 72nd Annual Conf. Exposition*, New Orleans, LA. [CD-ROM]

Sartor, J. and Boyd, G. (1972) *Water Pollution Aspects of Street Surface Contaminants*. USA EPA, Contract No. 14-12-921. EPA-R2-72-081, Washington, D.C., November, 236 pp.

Schueler, T. (ed.) (1994) Hydrocarbon hotspots in the urban landscape: can they be controlled? *Watershed Protection Techniques*, 1(1), 3–5 February.

Schueler, T. (ed.) (1996) Vegetated rock filter treats stormwater pollutants in Florida. *Watershed Protection Techniques*, 2(2), Spring, 372–374.

SEMCOG (Southeast Michigan Council of Governments) (1981) *Detention Basin Maintenance, Programme Needs and Financing Options*. Technical Bulletin No. 3, Detroit, Michigan, May.

Smith, P. (2001) The role of education in stormwater management. *Proceedings of Novatech 2001*, Lyon, France, 1, 443–450.

Smoot, J.L., Moore, T.D., Deatherage, J.H. and Tschantz, B.A. (1992) *Reducing Nonpoint Source Water Pollution by Preventing Soil Erosion and Controlling Sedimentation on Construction Sites. A training manual for construction inspection personnel.* Tennessee Department of Transportation in cooperation with Tennessee Department of Environment and Conservation, Nonpoint Source Program.

Stahre, P. and Urbonas, B. (1990) *Stormwater Detention for Drainage, Water Quality, and CSO Management*. Prentice-Hall, Englewood Cliffs, New Jersey, 338 pp.

Stanley, D.W. (1996) Pollutant removal by a stormwater dry detention pond. *Water Environment Research*, 68(6), Sept./Oct., 1076–1083.

Stenmark, C. (1990) Local infiltration of urban storm water in cold climate. *Proceedings of 5th International Conference on Urban Storm Drainage*, Osaka, Japan, July 1990, 809–814.

Strecker, E.W. (1999) Considerations and Approaches for Monitoring the Effectiveness of Urban BMPs. *Proc. National Conf. on Retrofit Opport. for Water Rec. Protect. in Urban Enviror.*, Chicago, IL, EPA/625/C-99/001, U.S. EPA, Washington, D.C., 65.

Sutherland, R.C. and Jelen, S. (1996) Studies show sweeping has beneficial impact on stormwater quality. *APWA Reporter*, 8–23.

Swaffield, J.A. and Gallowin, L.S. (1992) *The Engineered Design of Building Drainage Systems*. Ashgate Publishing, Aldershot.

Sztruhar, D. and Wheater, H.S. (1993) Experimental and numerical study of stormwater infiltration through pervious parking lots. *Proceedings of 6th International Conference on Urban Storm Drainage*, edited by J. Marsalek and H.C. Torno, Niagara Falls, Ontario, 12–17 September, IAWQ, London. Published by Seapoint Publishers, Victoria, B.C., 1098–1103.

Taylor, S.M., Hansen, L. and Beitia, C. (2001) Assessment of costs and benefits of detention for water quality enhancement. *ASCE EWRI Conference – Bridging the Gap: Meeting the World's Water and Environmental Resources Challenges*, 20–24 May 2001, Orlando, Florida, edited by D. Phelps and G. Shelke, ASCE, Reston, VA. ISBN 0-7844-0569-7. [CD-ROM]

Tejada-Guibert, J.A. and Maksimovic, C. (eds.) (2001) *Frontiers of Urban Water Management: Deadlock or Hope?* IHP-V Technical documents in hydrology, No. 45, June, UNESCO, Paris.

Tenney, S., Barrett, M.E., Malina, J.F., Charbeneau, R.J. and Ward, G.H. (1995) *An Evaluation of Highway Runoff Filtration Systems*. Center for Research in Water Resources, Technical Report CRWR 265, The University of Texas at Austin, September, 133 pp.

Tyson, J.M. (1995) Quo Vadis - sustainability? *Wat. Sci. Tech.*, 32(5–6), 1–5.

USEPA (1993) *Guidance Specifying Management Measures for Sources of Nonpoint Pollution in Coastal Waters*. EPA 840-B-92-002. U.S. Environmental Protection Agency, Office of Water, Washington.

USEPA (2000) *Guidelines for Management of On-site/decentralised wastewater systems*. EPA 832-F-00-038.

USEPA (2002) *Onsite Wastewater Treatment Systems Manual*. EPA/625/R-00/008.

Valiron, F. and Tabuchi, J.-P. (1992) *Maîtrise de la pollution urbaine par temps de pluie - Etat de l'art*. Paris, France, Tec et Doc, 564 pp. ISBN 2-85206-863-X.

Vignoles, M. and Herremans, L. (1992) Street cleaning: an alternative to stormwater treatment? (in French) *Novatech International Conference*, Lyon, 272–283.

Waschbusch, R.J. (1999) *Evaluation of the Effectiveness of an Urban Stormwater Treatment Unit in Madison, Wisconsin, 1996–97*. U.S. Geological Survey Open-File Report 99-4195, Middleton, WI.

W&H Pacific (now Stormwater Management) (1992) *Compost Storm Water Filter (CSF): Answers to Typical Questions. Compost Storm Water Treatment System: Executive Summary. Methods and Results Summary: Compost Storm Water Filter System*. Stormwater Management, Portland, Oregon.

Wada, Y. and Miura, H. (1990) Effect and evaluation of storm runoff control by permeable combined infiltration facilities for controlling storm runoff. *Proceedings of 5th International Conference on Urban Storm Drainage*, Osaka, Japan, July, 789–794.

Yamada, K. (1993) Infiltration pitch for individual houses to mitigate storm water impact. *Proceedings of 6th International Conference on Urban Storm Drainage*, edited by J. Marsalek and H.C. Torno, Niagara Falls, Ontario, 12–17 September, IAWQ, London. Published by Seapoint Publishers, Victoria, B.C., 1110–1114.

Yu, S.L., Liao, S-L., Fitch, G.M. and Long, S. (1996) Constructed wetlands for stormwater management. *69th Annual Water Environment Federation Technical Exposition and Conference (WEFTEC)*, Dallas, TX. Water Environment Federation, Alexandria, VA. 7(I, II and III), 167-178.

Yu, S.L., Kuo, J.T., Fassman, E.A. and Pan, H. (2001) Field test of grassed-swale performance in removing runoff pollution. *J. Water Resour. Plan. Man.*, 127, 168.

4.2 SEWERAGE DESIGN, OPERATION AND MAINTENANCE TO CONTROL SEDIMENT PROBLEMS

4.2.1 Design of sewer systems to control sediment

Sewer network operators have long faced the challenge of how to manage their systems with regard to in-sewer sediments. Sewer systems were primarily designed to remove wastewater to prevent urban flooding and disease. The configuration and use of the inlets to sewer systems meant that wastewater also conveyed high proportions of sediment particles of highly variable characteristics. Traditionally it was expected that sewer systems would cope with these sediments by transporting them through the drainage system to treatment or some point of convenient disposal. However, the mismatch between the sediment transporting capacity and sediment input rates from the catchment has resulted in many systems experiencing periods of sediment deposition and then erosion (see Section 2.3). Sediment deposits in sewers cause two major problems (Section 3). Firstly, the presence of an in-pipe sediment deposit can increase the overall hydraulic roughness of a pipe and reduce the cross-section, significantly reducing the hydraulic capacity and so detrimentally affecting the primary role of the system. Secondly, the in-pipe deposits can have a substantial impact on flow quality; at higher flows they can suddenly release large quantities of sediment into the flow, significantly affecting sewer flow quality. For these two reasons renewed attempts are now being made to operate and design sewer systems to more effectively control the impact of sewer sediments.

4.2.1.1 Design criteria – self-cleansing sewers

Historically the design of systems has been based on attaining a flow capacity to prevent flooding and then achieving a set of conditions that will ensure the transport of all sediment and so prevent deposition within the system. These conditions have been determined based on a minimum velocity of flow or a minimum shear stress that the flow should exert on the walls of the pipe to maintain *self-cleansing* conditions, corresponding to a particular depth of flow, or a particular frequency of occurrence. The range of design criteria currently in use have been reported by CIRIA (1986) and a summary of some examples of these, as outlined by Nalluri and Ab Ghani (1993), together with an update of German practice, is given in Tables 4.2.1 and 4.2.2.

The criteria in Tables 4.2.1 and 4.2.2 take no account of the characteristics of the sediment, the sediment load entering a pipe, or of any cohesion between the sediment particles. Recent work by many researchers, has shown that a single value of minimum velocity or shear stress cannot adequately be used to determine the self-cleansing conditions in pipes of different size, roughness and gradient for the range of sediment characteristics and flow conditions found in sewers (Section 2.3).

Brombach *et al.* (1992), amongst others, showed by field and laboratory experiments that the sediment transport phenomena present in sewer pipes were much more complex than previously thought. It was found that the shear stress, and sediment characteristics such as grain diameter and density of *representative* sand, were not sufficient to describe the sediment transport processes. Based on their observations they concluded that in normal combined sewers about 90 % of the normal sediments can be transported with a minimum shear stress of 1.6 N/m^2.

Table 4.2.1. Minimum velocity criteria for self-cleansing sewer design

Source	Country	Sewer type	Minimum velocity (m/s)	Pipe conditions
American Society of Civil Engineers (1970)	USA	Foul Storm	0.6 0.9	Full/half-full Full/half-full
British Standard BS 8001 (1987)	UK	Storm Combined	0.75 1.0	Full Full
Minister of Interior (1977)	France	Foul Combined or separate storm sewer	0.3 0.6 0.3	Mean daily flow For a flow equal to 1/10 of the full section flow for a flow equal to 1/100 of the full section flow
European Standard EN 752-4 (1997)	Europe	All sewers	0.7 once per day for pipes with D < 300 mm 0.7 or more if necessary in sewers larger than D = 300 mm	N/A
Abwassertechnische Vereinigung ATV, Standard A 110 (1998) (replaced by ATV 110 (2001))	Germany	Foul Storm Combined	Depends on diameter of pipe ranging from 0.48 (D=150 mm) to 2.03 (D=3000 mm)	0.3 to full for 0.1 to 0.3, velocity plus 10 %

Table 4.2.2. Minimum shear stress criteria for sewer design

Reference source	Country	Sewer type	Minimum shear stress (N/m^2)	Pipe conditions
Lysne (1969)	USA		2.0–4.0	
ASCE (1970)	USA		1.3–12.6	
Yao (1974)	USA	Storm Foul	3.0–4.0 1.0–2.0	
Maguire rule (CIRIA, 1986)	UK		6.2	Full/half full
Lindholm (1984)	Norway	Combined Separate	3.0 – 4.0 2.0	-
Scandiaconsult (1974)	Sweden	All	1.0 – 1.5	1.5 if sand is present
Macke (1982)	Germany	Foul Storm Combined	depending on transport capacity and transport concentration	0.1 to full typical combined systems under long-term conditions
Brombach *et al.* (1992)	Germany	Combined	1.6 to transport 90% of all sediments	

4.2.1.2 Transport capacity design approaches

Bertrand-Krajewski *et al.* (1993) presented a summary of the design equations which may be used to predict the deposition, erosion and transport of sewer sediments. Nalluri and Ab Ghani (1993) showed that the presence of a limited depth of sediment deposit in the invert of the pipe reduced the slope required to achieve a fixed sediment transport capacity over the entire range of pipe diameters. They presented a series of design options for self-cleansing storm sewers and for non-cohesive granular sediments.

Many other researchers, for example Nalluri and Alvarez (1992), Ashley *et al.* (1993), Arthur and Ashley (1997), Berlamont and Torfs (1996), Ristenpart (1996), Skipworth *et al.* (1999) and Wotherspoon (1994) have presented work associated with the description and modelling of the transport processes of cohesive or cohesive-like sediments in sewers, but to date, these have not yet been translated into design guidance for practising engineers. A fuller discussion of this work is presented in Section 2.3.

The diurnal pattern of the dry weather flow and the temporal distribution and nature of sediments found in sewer flows (Section 2.1) will result in the deposition of some sediments at times of low flow and the subsequent erosion and transport of these sediments, either as suspended load or bed-load, at times of higher flow. If the peak dry weather flow velocity or bed shear stress is of sufficient magnitude to erode and transport sufficient sediments, the sewer will maintain self-cleansing conditions during periods of dry weather. Ackers *et al.* (1996) presented an alternative definition to describe a self-cleansing sewer:

> An efficient self-cleansing sewer is one having a sediment-transporting capacity that is sufficient to maintain a balance between the amounts of deposition and erosion, with a time-averaged depth of sediment deposit that minimises the combined costs of construction, operation and maintenance.

They promoted the concept that a 'reasonable' depth of deposition could be permitted providing the system was in a long-term equilibrium based on the incoming sediment loads and flows.

Their work was adopted by CIRIA in the UK as an alternative approach to the design of sewer pipes to control sediment problems. The approach recommends that sewers should be designed to satisfy one of the three following criteria to achieve adequate 'self-cleansing' performance:

I Flows equalling or exceeding a limit appropriate to the sewer should have the capacity to transport a minimum concentration of fine-grained or low-density particles in suspension.

II The capacity of flows to transport coarser granular material as bed-load should be sufficient to limit the depth of deposition to a specified proportion of the pipe diameter.

III Flows with a specified frequency of occurrence should have the ability to erode bed particles from a deposited granular bed that may have developed a certain degree of cohesive strength.

The level of the flows required to satisfy the above criteria will be primarily dependent on the incoming sediment load and sediment characteristics as well as the characteristics of the sewer pipe, e.g. slope, diameter and roughness.

Criterion I – Capacity to transport a minimum suspended sediment load

In order to assess whether the above criteria had been met, Ackers *et al.* (1996) recommended a number of empirically calibrated relationships to assess the capacity of flows to transport various types of sediment, in conditions of a clean pipe and with a deposited bed.

For suspended sediments, at the limit of deposition (clean pipe), an equation outlined by Macke (1982) was recommended:

$$C_v = \frac{\lambda^3 V_L^5}{30.4(s-1)W_s^{1.5} A} \tag{4.1}$$

where		
	C_v	is volumetric sediment concentration (m^3/m^3)
	λ	is Darcy friction factor (-) (see Equation 4 5)
	V_L	is limiting flow velocity without deposition (m/s)
	s	is specific gravity of the sediment (-)
	W_s	is settling velocity of the sediment (m/s)
	A	is cross sectional area of the flow (m^2)

For pipes with a deposited bed Ackers *et al.* (1996) recommended that the Ackers-White equations be used with the 1990 revisions to the empirical calibration co-efficients along with the reduced sediment bed width (Ackers, 1991). As might be expected the acceptance of an 'allowable' depth of deposition permits much lower design velocities in the sewer pipes for the same sediment input load and particle characteristics, particularly for large diameter pipes (Figure 4.2.1).

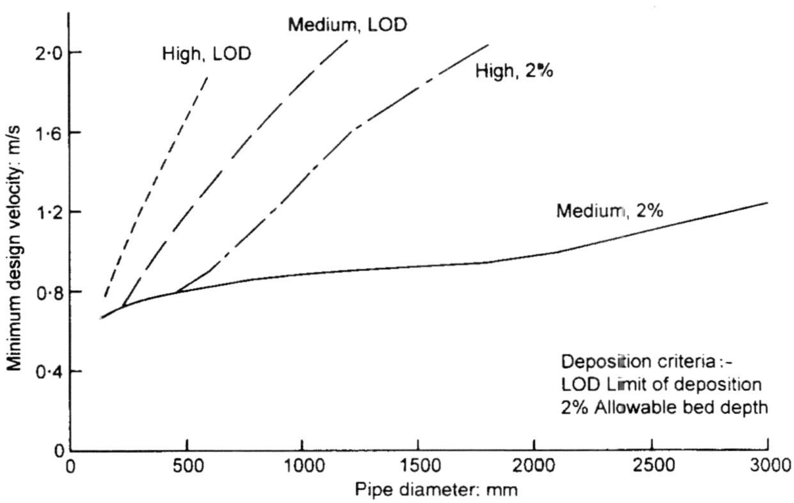

Figure 4.2.1. Diagram illustrating that by allowing a 2 % depth of deposit, a large pipe, with either a medium or high sediment load, requires a lower critical design velocity than with no deposit, at Limit of Deposition (LOD) (adapted from Butler *et al.*, 2003).

Criterion II – Bed load

At the limit of deposition (clean pipe conditions) the empirically calibrated bed load transport equation developed by May (1996) was proposed, with d_{50} being the particle diameter (m) and d (m) being the pipe diameter:

$$C_v = 3.03 \, x \, 10^{-2}\left(\frac{d^2}{A}\right)\left(\frac{d_{50}}{d}\right)^{0.6}\left[1-\frac{V_t}{V_L}\right]^4\left[\frac{V_L^2}{g(s-1)d}\right]^{1.5} \tag{4.2}$$

where the threshold velocity, V_t, required to initiate movement is given by:

$$V_t = 0.125\left[g(s-1)d_{50}\right]^{0.50}\left[y/d\right]_{50}^{0.47} \tag{4.3}$$

and y (m) is the depth of flow in the pipe. Equation 4.2 is valid for bed-load transport in both rough and smooth pipes and the proportion of sediment moving in suspension was considered significant when

$$U_* > 0.75\, W_s \tag{4.4}$$

where U∗ (m/s) is the overall shear velocity of the flow that is related to the mean flow velocity V (m/s) by:

$$\mathrm{U}_* = \mathrm{V}\sqrt{\frac{\lambda}{8}} \tag{4.5}$$

With limited deposition (deposited bed) the presence of the bed will reduce the flow capacity but the corresponding increase in the effective width of the sediment at the invert may significantly increase the transporting capacity. For this case Equation 4.6 was adopted as the design equation, to describe the bedload transport capacity of a sewer with a deposit.

$$C_v = \eta\left(\frac{W_b}{d}\right)\left(\frac{d^2}{A}\right)\left(\frac{\theta\lambda_g V^2}{8g(s-1)d}\right) \tag{4.6}$$

where
- η is transport parameter for continuous bed (-)
- $\mathrm{W_b}$ is sediment bed width (m)
- θ is transition coefficient for particle Reynolds number (-)
- λ_g is friction factor for grain shear stress (-)
- V is mean flow velocity (m/s)
- d is sediment particle diameter (m)
- A is cross sectional flow area (m^2)

This equation was developed based on a series of pipe transport tests using a range of single sized sand. It requires the flow roughness associated with the bed and pipe walls to be estimated so that the effect of the composite roughness of a pipe containing a deposit on the transport capacity can be ascertained. θ is an empirically derived coefficient which can account for the effect of this composite roughness on the grain mobility number (Fg) to produce an effective sediment mobility parameter (Fs), where

$$Fg = [\lambda_g V^2/8g(s-1)d_{50}] \qquad 4.7$$

λ_g is the roughness associated with grains and is determined using a Colebrook-White form of relationship. The transport parameter (η) is based on adjusting the values of Fs and was calibrated using data collected in laboratory pipe tests mainly using single-sized sands (May, 1993).

Criterion III – Cohesion within in-sewer deposits

The third criterion stated that flows with a specified frequency should have the ability to erode particles from a deposited bed that may have demonstrable cohesive properties when subject to fluid shear. Sediment deposits in sewers may demonstrate such properties due to their clay/silt content, and the possible transformation of organic sediments. It was recommended that the design flow condition should produce a minimum value of bed shear stress of 2.0 N/m^2 on a flat bed with Colebrook-White equivalent roughness of 1.2 mm.

The equations and criteria above, along with simplified methods for estimating incoming sediment loads, were used to formulate a new design procedure for practising engineers. For the first time sediment characteristics and loads were acknowledged as primary design inputs, and economic factors were considered in that small amounts of in-pipe deposition, in long-term equilibrium, were permissible; this is important if costs are to be reduced whilst attempting to concurrently manage and control sediments. This is illustrated in Figure 4.2.1 from Butler *et al.* (2003), which shows that even with a small depth of allowable deposit (1 or 2 %) the required design velocity for suspended sediment transport may be reduced in larger pipes by typical factors of 2–3.

Unfortunately, although the conceptual framework behind this design method is significantly better than that behind the 'self-cleansing' velocity or shear stress criteria, its current use by engineers in the UK is limited. It is notable that when the design procedure becomes too complex, either because of the data or computational burden in application, or because of a lack of confidence in the value of application when compared with more simple traditional approaches, the utilisation of new methods can be limited. One aspect of the burden of applying a new, more complex methodology is obviously the larger amount of external data required in application of the various transport capacity equations. In a recent UK survey, it was found that very few designers were utilising the CIRIA approach, believing it to be too complex and not likely to lead to significant savings compared with 'traditional' approaches (Arthur *et al.*, 1999).

4.2.1.3 Alternative practical approach based on energy slope and geometrical considerations

Appropriate sewer design should allow for an adequate value of the slope of the hydraulic (energy) grade line, *S*. This is the primary driver for both water transport and particle transport (see Section 2.3) and hence should be the primary design parameter. In most practical situations, *S* will *differ* from the sewer invert gradient I_f. Pipes with negative invert slopes, pipe lengths with backwater effects caused by downstream controls, together with temporal and spatial variations in flow and temporal and spatial 'bed' gradient changes due to sediment deposition and erosion and structural defects, can all discourage steady uniform flows from occurring.

Whenever *S* reduces, there are increased risks of sediment deposition (Section 2.3.4.5). Hence recent initiatives to utilise in-sewer storage to alleviate the occurrence of flooding (e.g. Schilling and Kollatsch, 1990) may tend to enhance sedimentation.

For the discussion below, which focuses on the design of new sewers, it is considered that gradually varied flow conditions are generally those which tend to occur the most often in sewers and that, at least locally, the energy slope S can be considered in most instances to be approximately equal to the pipe invert slope I_f.

The minimum sewer gradient (denoted S_{min}) is defined here as the limiting hydraulic gradient at which there is no deposition of the transported particulate load. The S_{min} criterion is thus a valid 'self-cleansing' condition, expressed as a minimum gradient. Without considering key factors such as pipe geometry or sediment particle size and density and incoming sediment transport rate it is still useful to attempt to specify a value for S_{min}.

There is no definitive consensus about an appropriate calibrated relationship to estimate the value of S_{min}. The value of the minimum energy slope required to maintain 'clean pipe' conditions is primarily influenced by the sediment particle characteristics (diameter and density), deposit characteristics such as cohesion, the incoming sediment transport rate, hydraulic and pipe characteristics such as the pipe geometry, the hydraulic roughness, and the prevailing flowrate. It is possible to relate the value of S_{min} to the hydraulic radius R_H (m), in order to estimate the influence of pipe size and shape. Dyer (1986) stated that many total-load transport capacity formulae found in sediment transport literature are of the form of Equation 4.8, where C_v is the sediment volumetric concentration (m^3/m^3):

$$C_v \propto u_m^{3} / w_s R_H \tag{4.8}$$

with flow velocity u_m (m/s) normally replaced by shear velocity U* (m/s) if the hydraulic roughness can be estimated (see Section 2.3), and the particle settling velocity w_s (m/s) sometimes replaced by $(g\Delta d_{50})^{1/2}$ when only sedimentation associated with coarse granular particles is relevant.

The dependence of the sediment transport capacity (expressed as C_v) upon the fluid shear velocity cubed is widely accepted. Relationships of this form have been utilised in a range of applications in which fluid transports sediment, such as slurry transport in pipes flowing full (Newitt *et al.*, 1955) and wind blown sand transport in deserts and coastal environments (Bagnold, 1954).

S is used implicitly in Equation 4.8, as u_m is proportional to $\sqrt{R_H S}$ according,for example, to the Chezy formula (this is also true where shear velocity U* is used instead of u_m). In sewer design it is necessary to estimate the delivery rate of sediment from upstream, and this will be equal to C_v, if 'clean pipe' conditions are to be achieved. The upstream conditions will also determine the incoming particle characteristics, i.e. there is no design control on either C_v or the characteristic settling velocity w_s of the incoming sediment in sewers.

If C_v and w_s are fixed by the upstream conditions, then a relationship between the slope S_{min} required to maintain a 'clean' pipe and the hydraulic radius R_h can be developed in general terms, as illustrated in Equation 4.9.

$$S_{min} \propto 1 / R_H^{1/3} \tag{4.9}$$

This expression describes the threshold criteria (for S) below which sediment deposits begin to build up in the conduit.

Consider the case of a circular sewer section, the diameter D of which is selected to accommodate the specified discharge. In terms of self-cleansing slope, Equation 4.9 indicates that the larger the hydraulic radius, the smaller the pipe gradient should be to achieve adequate cleansing. Hence a circular sewer will be most effective in transporting sediment without deposition if it operates at $y/D = 81$ % (the relative depth at which the hydraulic radius is maximum for a given D). This result is particularly relevant for the sizing of smaller conduits, such as upstream sanitary sewers.

Equation 4.9 indicates that situations where there are low values of hydraulic radius (small flow cross-section compared with wetted boundary) are very demanding in terms of required slope (steep). For this reason, large-sized combined sewers should not be designed to have a circular cross-section. Equation 4.9 thus provides a rough theoretical basis to explore which cross-sectional shapes are most appropriate. It is obvious that narrow sections such as the egg-shape provide a comparatively better hydraulic radius for a given discharge. They are, therefore, most appropriate for low y/D conditions, being more efficient in sediment transport than circular conduits. Ma (2003) recently developed these ideas and suggested a corresponding set of guidance rules for sewer design that match the latest ATV 110 (2001) recommendations

When designing new systems progressively from the uppermost pipes in the catchment to the outfall, it is evident that successive flow increases require R_H to be progressively increased pipe by pipe to cope with the increased flow discharge. Where this is accomplished by an increase in sewer size (and not by an increase of y/D ratios), Equation 4.9 demonstrates that (with an assumption of unchanged Cv and w_s) it is possible to accept relatively slacker gradients in the downstream parts of the sewer network. However, caution is required with this approach, as the value of Cv within the system will increase as proportionately more sediment may be washed from catchment surfaces into the sewer system as flows progress downstream. Therefore Equation 4.9 needs to be used with care. The usefulness is in the determination of the optimum shape of sewer cross-section to maximise the transport capacity of the flow. It can be seen that if the hydraulic radius can be maximised for a particular flow condition, by use of a particular pipe shape, then the energy slope required to maintain clean pipe conditions can be minimised and thus excessive excavation depths can be avoided. In a recent study (Ma, 2003), the use of egg-shaped rather than circular sewers is recommended only where local sedimentation may be a problem due, for example, to limited topographical slope and also when the Qmax/Qmin ratio (ratio of design-rain flow to average dry weather flow) is high. Where y/D remains relatively large, retaining the simpler (and lower-cost) circular conduit shape is a viable design option. Constructing two parallel circular conduits can also be more effective where sediment transport may be a problem.

4.2.2 Operation and maintenance to control sediments in sewers

4.2.2.1 Introduction

There are two elements to be considered when assessing the maintenance needs for dealing with solids in sewer systems (Section 3):

- loss in hydraulic capacity caused by the deposits constricting cross-sections and increasing the hydraulic roughness of the conduit;
- pollution consequences of the re-erosion and wash-out of in-sewer deposits, particularly via combined sewer overflows to receiving waters.

Deposition of heavier solids is particularly a problem in combined sewer systems and storm drains, but is also a problem in separate sanitary systems. However, the most intractable problems relating to solids deposition are in combined sewers as these are sized to convey many times the anticipated peak dry weather flow. In some cases, the smaller connecting combined sewers can carry flows larger by many orders of magnitude than those expected under average dry weather flow conditions. In the larger interceptor sewers, ratios of peak flow to average dry weather flow usually range from 2 to 10. Dry weather flow velocities are typically inadequate to maintain most solids in suspension, and therefore substantial amounts of solids accumulate on inverts. During storm events, the accumulated solids can be eroded and then re-suspend into the flow. The flow and suspended solids can then be released via an overflow to receiving waters. Oversized combined sewers thus have the potential to experience substantial sedimentation during dry weather periods. This is particularly a problem where an allowance for future increases in flow has to be built-in to the design of a new sewer, which is thus temporarily oversized.

The way in which solids deposit and build-up in sewers is described in more detail in Sections 2.2.4 and 2.3.4.5.

There are two approaches to in-sewer solids operational management:

- manual or mechanical removal of problematic solids;
- encouragement of the solids to continue to be transported downstream using 'artificial' means (other than the normal operation of the sewer system), with devices permanently or temporarily introduced into the system.

In the first case, the problem arises as to what to do with the extracted solids, whereas in the second, the solids will be ultimately transported to the original 'end-of-pipe' destination, usually a treatment plant.

Solutions for the operational management of deposits in combined sewers can be presented from two perspectives.

Preventative solutions

- Encouraging hydraulic conditions which ensure self-cleansing during dry weather;
- Limiting the size (density) of the solids entering the network;
- Improving the hydraulics of the system generally;
- Improving cross sections in lengths likely to be vulnerable to deposition (e.g. replace circular pipes by egg-shape pipes).

Curative solutions

- Implementation of 'partial removal' visits;
- Specific interception of solids and localisation of deposits (e.g. sediment traps);
- Tuning of the frequency of deposit removal;
- Deposit displacement using dry weather flow (manual flushing, flushing gates, balls, vane wagons, etc.)

Maintenance of larger 'man-entry' sewers has not benefited significantly from major recent developments and many of the approaches developed at the end of the 19th century are still in use for networks which have since become more complex and typically undersized due to contributing population expansion (Bertrand-Krajewski, 2003). Figure 4.2.2 shows the results from a recent survey of the UK sewerage operators and their reported techniques used to clean sewers. Although many of the approaches have not changed, the development of new equipment has helped operators to become more efficient at extracting sediments and encouraging downstream transport, with less labour. However, because maintenance of these systems still requires significant manual labour it is very expensive; consequently many operators undertake only reactive rather than proactive maintenance, when blockages or flooding necessitates. Operators are often unaware of the link between sewer deposits and increased frequency of CSO discharges (e.g. Fraser *et al.*, 1998).

Nowadays it is important to try to plan sewer sediment management in terms of whole life cost of operation of wastewater systems. Unfortunately, little guidance is as yet available for this. Examples of initial attempts are given by Delattre *et al.,* (1998), Pisano *et al.* (1998) and Ashley *et al.* (2003). In each case the relative cost of dealing with sediment has been considered. In the latter example, a risk-based approach is advocated as illustrated in Figure 4.2.3. This may be related to Table 4.2.3. that provides an examples of the risk conceptualisation.

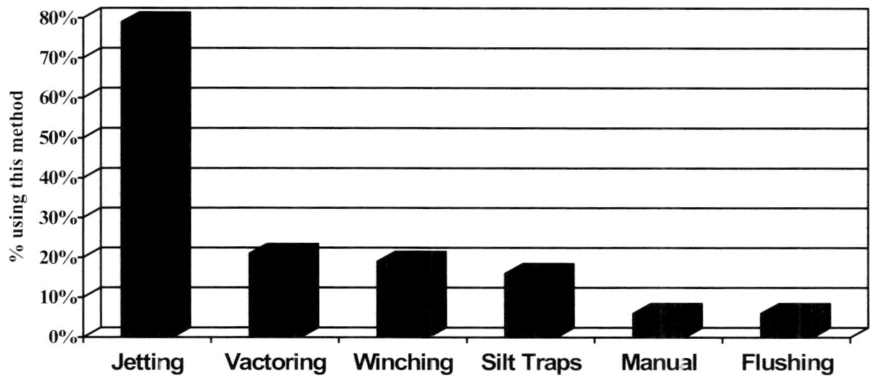

Figure 4.2.2. Methods used for sewer cleaning – survey of UK operators (adapted from Fraser *et al.*, 1993).

Table 4.2.3. Example of conceptualisation for risk assessment for managing sewer solids (Ashley *et al.*, 2003).

Risk no.	Hazard	Transport mechanism	Pathway	Receptor
1	Pathogenic material, organic material and toxic substances	Sewage	Storm water overflows	Adjacent watercourses Humans (water supply and recreational users)
2	Pathogenic material and toxic substances	Sewage	Manhole covers and gulleys	Property basements and foundations Humans (residential and commercial)
3	Pathogenic material and toxic substances	Sewage	Manhole covers and gulleys	Adjacent ground surface Humans (recreational users) Eco-system damage Crop accumulation and damage
4	Grit	Sewage	Sewerage system	Downstream sewers
5	Grit	Sewage	Sewerage system	WTP inlet works, other ancillaries
6	Other organic materials, plastics etc.	Sewage	Storm water overflows	Eco-system damage

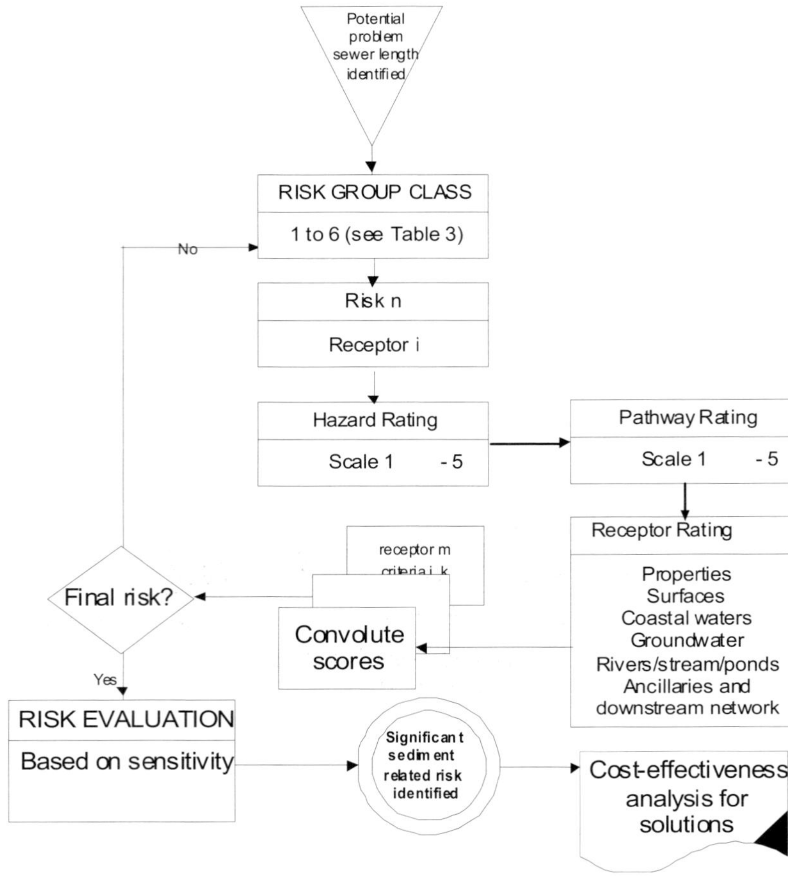

Figure 4.2.3. Risk based approach to sewer sediment management (Ashley *et al.*, 2003).

4.2.2.2 Cleaning

Because of the high costs of maintenance and the difficulties associated with cleaning sewers there has been a number of developments in recent years to make the approaches more efficient. In the USA there is considerable interest in this topic and the USEPA have produced a lot of recent guidance (USEPA, 1999; Fan *et al.*, 2003).

4.2.2.2.1 Cleaning techniques for small sewers

Common smaller sewer cleaning techniques may be manual or automated and include the use of rodding, balling, flushing, poly pigs and bucket machines (Dinkelacker, 1992). These methods are used to clear blockages once they have formed, and as preventative maintenance to minimise future problems. With the exception of flushing, these methods are generally used in a 'reactive' mode to prevent or remove hydraulic restrictions. As a control concept, the flushing of sewers by manual methods or using special automated equipment is a means both to reduce hydraulic restriction problems and to prevent pollution.

Power rodding includes an engine and drive unit, steel rods and a variety of cleaning and driving units. The power equipment applies torque to the rod as it is pushed through the line, rotating the cleaning device attached to the lead end. Power rodders can be used for routine preventative maintenance, cutting roots and breaking up grease deposits and are efficient in pipes up to 300 mm in size.

Balling is a hydraulic cleaning method in which the pressure of a water head creates high velocity water flow around a ball, such as an inflated rubber cleaning ball. The ball has an outside spiral thread and swivel connection that causes it to spin, resulting in a scrubbing action of the water along the pipe (Figure 4.2.4). The balls can remove settled grit and grease and this technique is useful for sewers up to 600 mm in size. For larger sewers, alternatives such as the Göttingen Cleaning Balls System (Lorenzen *et al.*, 1992; 1996) are reportedly effective. The system relies on a suitable method of ball introduction and subsequent capture and the technique is only appropriate for sewers of reasonably large diameter. In Paris (France), large balls (diameters up to more than 2 m) made of wood are also used to clean the inverted siphons below the Seine river.

Figure 4.2.4. Flush ball as used in Gottingen (photograph Lorenzen).

Kites, bags and poly pigs are used in a similar manner as balls. The rigid rims of bags and kites cause the scouring action. Water pressure moves these devices against the tension of restraining lines. The shape of each type of device creates a forward jet of water which removes the sediments. The poly pig is used for large sanitary sewers and is not restrained by a line. The pig moves through the pipe due to a build-up of water pressure behind the pig which provides the motive force.

Jetting is a hydraulic cleaning method that directs high velocities of water against the sewer walls at various angles. The jetting equipment is usually mounted on a truck or trailer and consists of a large water supply tank of at least 4000 litres, a high pressure water pump, an auxiliary engine, a powered drum reel holding at least 150 m of 25 mm hose and a variety of nozzles. Jetting is efficient for routine cleaning of small diameter sewers in which there is low flow. It effectively removes grease build-up and debris.

The power bucket machine is a mechanical cleaning device which is effective in partially removing large deposits of silt, sand, gravel and grit. These machines are used mainly to remove debris from a blockage or an accumulation that cannot be cleared by hydraulic methods. In cases where the pipe is so completely blocked that a cable cannot be threaded between manholes, the bucket machine cannot be used. The bucket machine is usually trailer or truck mounted and consists of a cable storage drum coupled to an engine with controllable drive train, up to 300 m of 13 mm steel cable and various sized buckets and tools ranging up to 900 mm in diameter. The cable drum and engine are mounted on a framework that includes a vertical A-frame that has sufficient height to permit the lifting of the cleaning bucket above ground level. Typically two machines of the same design are required. One machine at the upstream manhole is used to thread the sewer line from manhole to manhole. The other machine at the downstream manhole has a small swing boom or arm attached to the top of the A-frame for emptying the buckets that are pulled through the system. The bucket is cylindrical and at the bottom has two opposing hinged jaws. When the bucket is plugged through the material obstructing the line, these jaws are opened and dig into the material and fill the bucket. When the bucket is pulled in the reverse direction, the jaws are forced closed by a slide action and any material in the bucket is retained. The collected material is removed from the bucket after it has been pulled out of the system at the downstream manhole.

A much simpler example of a manual technique is that employed within the Paris and Brussels sewer system where a (*vane*) wagon incorporating a flushing vane is manually operated along the length of the cunette shaped sewer system, largely via differences in hydrostatic head across the vane, which can be lowered into the flow. The vane disturbs the sediment that is subsequently transported downstream with the sewer flow (Figure 4.2.5).

Figure 4.2.5. Paris vane wagon used for sewer cleaning (photograph Mairie de Paris).

4.2.2.2.2 Cleaning requirements for man-entry combined sewers

The mechanical methods above are often not very efficient for cleaning larger sewers and frequent manual maintenance of large sewers is onerous and expensive. The French technical guidelines relating to urban sewer networks (Minister of Interior, 1977) defines sewers with internal height greater than or equal to 1.6 m as large *man-entry* sewers. Sewers of a height greater than 1.2 m are considered to be semi-man-entry. The proportion of man-entry sewers in a network, as well as their size classes, related to the size of agglomerations are shown in Table 4.2.4.

Table 4.2.4 Distribution of the sizes of sewers in France (from Sanchez, 1987)

Size of built-up area	Percentage of man-entry sewers as a proportion of total network	Distribution of man-entry sewers (%) according to size classes		
		h < 1.20 m	1.20 < h < 1.80 m	h > 1.80 m
< 500 000 inhabitants	20	50	30	20
> 500 000 inhabitant	35	15	45	40

Routine maintenance removal of sewer sediments in man-entry sewers depends on many factors. Sanchez (1987) indicates that the range of sediment volume removed per capita or per hectare was relatively small, as detailed in Table 4.2.5, with the exception of Paris where the density of the population is greater than in the other cities. The volumes of solids removed averaged some 200 l/ha/year. In Seine-Saint-Denis (Paris area) 30 000 m^3 of deposits are extracted annually from the main sewers, which represents a completely full pipe 13 km long and 2 m in diameter.

Table 4.2.5. Sediment volumes removed from combined man-entry sewers in France (from Sanchez, 1987)

City	Population connected to the sewers	Catchment area (ha)	Volume of sediment extracted (l/inhab/year)	Volume of sediment extracted (l/ha/year)
Bordeaux	400 000	35 000	12.5	143
Le Havre	200 000	4 000	6.0	300
Lyon	1 200 000	50 000	8.5	204
Marseille	960 000	60 000	9.6	154
Nancy	260 000	16 000	10.4	169
Paris	2 500 000	10 000	6.0	1500
Hauts de Seine (near Paris)	750 000	-	10.6	-

The cost of the removal of sewer deposits varies widely. For example, in a separate sewer system in Toulouse, France the removal of deposits from 930 km of sewer cost about €1 million in 1979 whereas the cost, using specialist contractors, of cleaning some 5 km of the main (approximately 1500 mm high) Dundee interceptor sewer was approximately €50 000 in 1996; i.e. costs of from €1000–€10 000 per kilometre for the two locations, presumably depending upon scale. In a survey carried out in ten French built-up areas in 1987, the costs for removing 1 m^3 of deposit from man-entry sewers was found to vary between €150 and €900 (Sanchez, 1987).

In many cases, the sewer operator accepts the temporary presence of deposits and decides to remove them when a given deposit depth is reached. For example, in Bordeaux, France, when the sediment depth reaches 20 % of the sewer height (or only 10 % in some particularly sensitive areas), cleaning is deemed necessary. Periodic visits are required to evaluate the amount of deposits. In Marseille, a minimum of 2 visits per year are carried out, to investigate the sediment deposits throughout the sewer network.

Ideally, a move from a reactive to a proactive (preventive) sewer sediment maintenance strategy would be expected to be more efficient and effective. Gérard (1999) recently developed and tested a conceptual model, including parametric expert rules, which forecasts the risk of sediment deposition in man entry sewers based on the previous work of Bachoc (1992). The latter showed the links between the presence of deposits and the sewer system structure (geometry, slope, connections, 'local singularities'). The risk is evaluated by combining the vulnerability (i.e. the propensity of a given sewer pipe to accumulate sediments) and the amount of sediments the pipe receives from upstream inputs. The model was satisfactorily tested in the City of Lyon, France, where operational plans are now being produced to direct maintenance activities more effectively.

4.2.2.3 Sewer flushing

Flushing of sewers has been an option dating back to the first sewers (Bertrand-Krajewski, 2003). Historical uses of this approach are described by Ogden (1898), for cleaning sewers in Syracuse, New York and by Bechmann (1900) for the City of Paris, France. The concept of sewer flushing is to induce an unsteady waveform in the sewer either by rapidly adding external water or creating a 'dambreak' effect by the storage and release of water or effluent. The cleaning efficiency created by the wave is a function of the flush volume, the flush discharge rate and the corresponding flush velocity, the length and diameter of the sewer and the flow conditions in the sewer at the time of flushing. The aim is to re-suspend, scour and transport deposited solids and pollutants out of the sewer length (normally to the sewage treatment facility) during dry weather and/or to displace solids deposited in the upper reaches of large collection systems closer to the system outlet. The objectives are to reduce deposited pollutants that may be at risk of resuspension and overflow to receiving water bodies during wet weather, and/or to speed up the transport of the solids through the system.

Manual methods usually involve discharge from a fire hydrant or a quick opening valve from a tank truck to introduce a high flow of water into the line at a manhole. Flushing removes floatables and some sand and grit, but is not very effective at removing heavier solids. It is most effective when used in combination with mechanical operations, such as rodding or bucket machine cleaning.

Research into sewer flushing has demonstrated the feasibility of reducing pollution from combined sewer overflows by means of periodic flushing of combined sewers during dry weather. US studies considered flushing using a small-scale flushing evaluation facility consisting of 300 and 450 mm sewers (FMC 1967, 1972). This led to flushing of larger sizes of pipe, wave sequencing, and studies of solids build-up over long periods of time. Long-term field studies were undertaken in Boston to investigate the pollution reduction potential of flushing local combined collector sewers (Pisano *et al*., 1979). The results are summarised in Table 4.2.6.

Sewer flushing of large diameter combined sewers in the range 450 to 1380 mm diameter was investigated in the City of Elizabeth, New Jersey (Kaufman and Lai, 1978). It was concluded that daily flushing of troublesome deposition sections within seven sub-areas using twelve automatic flushing systems was likely to reduce about 28 % of the first flush overflow load. Figure 4.2.6 shows a typical automatic flushing module using the concept of 'back-up and release' of upstream dry weather flows into sewers with deposition in sections directly downstream.

Table 4.2.6. Results from US tests of sewer flushing effectiveness (adapted from Pisano *et al.*, 1979)

1. Effectiveness (see Figure 4.2.5 for typical flush wave characteristics) of a single 1200 litre manual flush (tanker) in scouring, entraining and transporting solids and associated pollutants in 300–450 mm pipes			
Distance transported >	75 m	215 m	300m
Pollutant	75–90%	65–75%	35–45%
Organic/Nutrient Deposits (BOD, TP, TN)			
Total suspended solids Deposits	~75%	55–65%	18–25%
2. Metals transported			
Approximately 20–40 % of total heavy metals (cadmium, chromium, copper, lead, nickel and zinc) associated with the flush wave will not settle within a 2 hour period. Figure 4.2.6 shows typical settling column analysis of flush wave samples. Heavy metals associated with the lighter colloidal solids scoured and suspended by the flush wave will be transported long distances.			
3. Effect of rainfall			
Field measurements showed that runoff from slight to moderate rainstorms only partially removed fresh sanitary deposits, whereas 'artificial' flushing immediately following storm events removed significant loads. Flushing frequency intervals need not be determined by slight to moderate rainfall return periods. Load First-flush in runoff during intense storms can be the result of long-term deposition accumulations occurring over both dry and moderately wet periods.			
4. Use of automatic flushes			
An automatic sewer flushing module was installed (Figure 4.2.6) using a simple hydraulic gate powered by an air cylinder and time clock triggered operation for 5 ½ month period to back up sewage, retract and induce a flush wave. Flushed pollutant loads were comparable to removals noted by manual flush tanker means.			

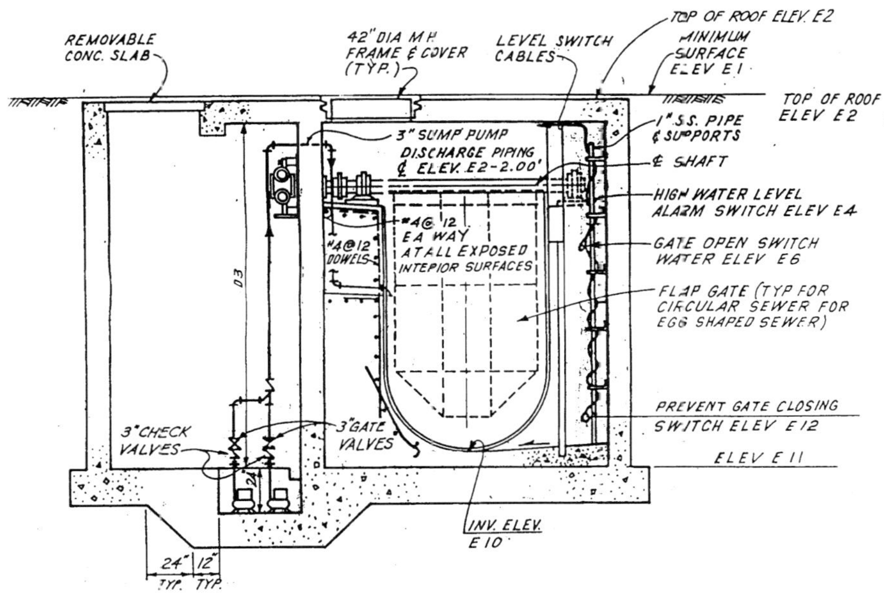

Figure 4.2.6. Typical automatic flushing module using the concept of 'back-up and release' of upstream dry weather flows into sections with deposition directly downstream.

A level sensor connected to a centralised computer controls the system. Since implementation in 1990, no operational problems have been reported. However, one significant disadvantage of this approach is the need for an external power source to activate the flush gate. Typical examples of the pollutant characteristics and heavy metals content of a recorded flush wave are shown in Figures 4.2.7(a) and 4.2.7(b) respectively.

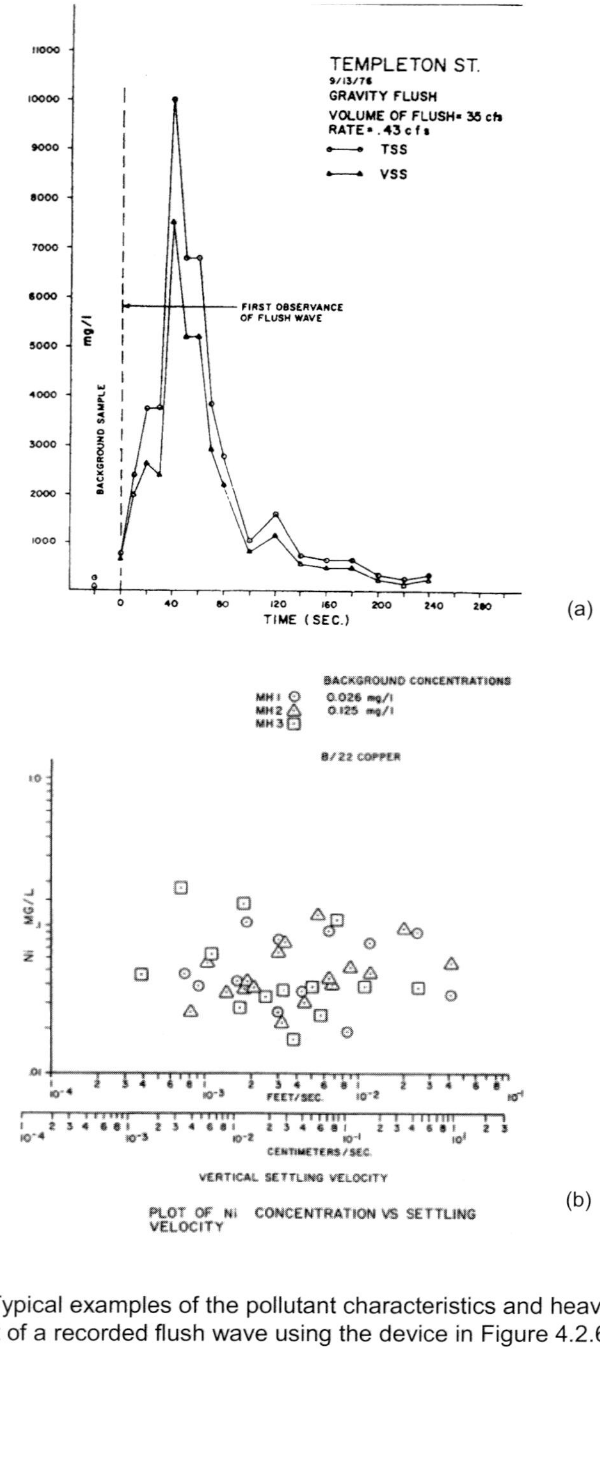

Figure 4.2.7. Typical examples of the pollutant characteristics and heavy metals content of a recorded flush wave using the device in Figure 4.2.6.

Significant recent investment has been made by the Massachusetts Water Resources Authority in the City of Cambridge, USA, in separating the combined sewer system into storm and sanitary components. Part of the project is utilising novel flushing systems with fast-opening gates from off-sewer chambers, to scour downstream sediments with previously retained stormwater (Pisano *et al.*, 2003). The sediments, scoured over distances of 1220 m in 1.4 m and 1.2 m x 2.7 m box culverts, will be collected in downstream grit traps. The sanitary sewers (450–600 mm) will also be scoured using previously collected spent filtrate from the City's new water treatment plant. The designs were based on the necessity to achieve a presumed 'self-cleansing' velocity of 1 m/s, whereas the new sewers were found to achieve only up to 0.6 m/s from model simulations. The potential for sediments to be implicated in the generation of sulphides and thence oxidised to produce corrosive gas, is reviewed by Fan *et al.* (2001). Application of two types of flush device, the tipping flusher and the flushing gate, to help prevent sediment build-up is reviewed by reference to eighteen existing installations in Europe and America.

Two innovations have been developed in Germany and France for cleansing deposits from sewers using the automatic 'dam break' concept. Neither system requires external power for operation. In Germany, there are more than 500 in-line pipe storage tanks, some 1800-2115 mm in diameter with lengths in the range 125–180 m. The outlet discharge from these tanks is throttled to approximately twice the average dry weather flow plus infiltration. As a consequence there is the potential for sediments to settle to the bed of the tank, a feature common to all designs of storage tank. Many different methods for cleaning storage pipes and tanks have been developed.

The HYDROSELF system has been used to clean settled solids in sewers, interceptors, tunnels, retention and detention tanks in Germany and Switzerland (Grande and Novak, 1995). The first installation in North America was for an 8000 m^3 tank in Sardia, Ontario (Parente *et al.*, 1995). The system requires no external water supply, results in negligible waste of storage space, can be easily retrofitted in existing installations and can operate without an external energy source. It comprises a hydraulically operated flap gate, a flush water storage area created by the erection of a concrete wall section, a float or pump to supply hydraulic pressure and valves controlled by either a float system or an electronic control panel as shown in Figure 4.2.8.

The arrangement for a given installation is site dependent. The flushing length, flushway slope and width determine the flush water volume needed for an effective single flush of the system. Flush water is held in reserve in the storage area by an hydraulically operated flap gate and the water level in the sewer or tank is used to activate the release and/or closure of the gate using a permanently sealed, float controlled hydraulic system. The ensuing high celerity wave (2 m/s) scours, resuspends and transports accumulated debris and solids. For extremely long sewers several units may be designed to operate in series.

For large diameter sewers ($\geq$ 2000 mm) the flushing system may be installed directly in the sewer. The required volume for the flush water is created by erecting two walls in the sewer, together with an enhanced internal floor slope (usually 5 %). The requirements for the storage area slope will determine, in most instances, the maximum flushing length possible for a single flush gate. Should the actual flushing length be longer than can be accommodated with one unit, then additional flushing gates must be installed to operate in series. Experience has shown that it is better to start the flushing process at the downstream end of the system. In order to increase the maximum flushing length it is also possible to build additional flush water storage volumes by creating a rectangular chamber in-line or adjacent to the sewer.

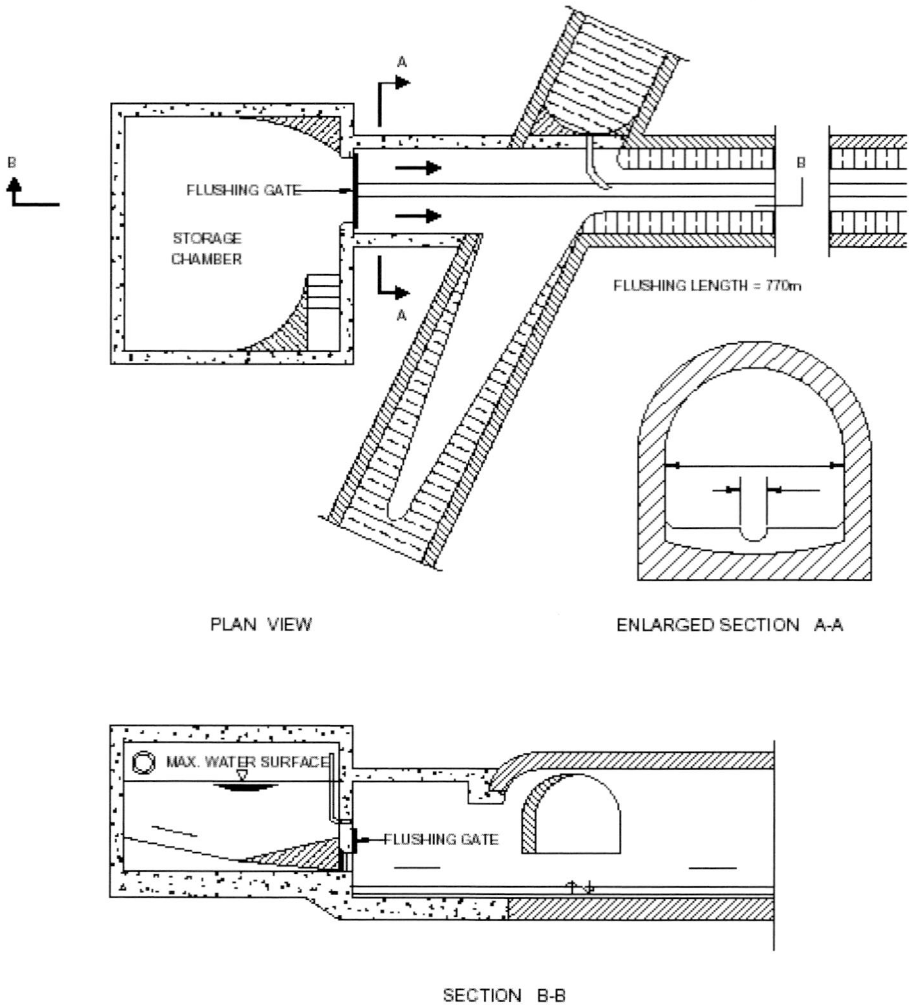

Figure 4.2.8. HYDROSELF system

Comprises hydraulically operated flap gate, flush water storage area created by erection of a concrete wall section, float or pump to supply hydraulic pressure and valves controlled by either a float system or an electronic control panel (Pisano *et al.*, 1998).

For smaller diameter sewers, the requisite flush water volume cannot be built in the pipe itself. A storage chamber either in-line or adjacent to the pipe must be built. In combined or sanitary sewers a 'flow-through-pipe' or underflow is usually installed below the floor of the flush water storage area. This pipe allows the dry weather flow to pass unimpeded through the system. When the inflow to the system exceeds the capacity of the underflow pipe, the water level rises and the storage area is filled with flush water. It is only when the sewer is drained or the peak flow abates that the flushing system commences operation.

This system has also been used extensively in Europe with 209 installations and 436 units in operation since 1985. Approximately 37 % of the projects (77) are to flush sewers, interceptors and tunnels ranging from 250 mm to 4300 mm in diameter and flushing lengths of up to 340 m for large diameter pipes and 1000 m for small diameter pipes. The rest of the installations are for cleaning sediments from CSO tanks. Figure 4.2.9 shows the distribution of flushing length versus number of installations and Figure 4.2.10 illustrates the distribution of pipe diameter versus number of installations (Grande *et al.*, 1995). In an evaluation of ten in-line sewer sediment flushing facilities, Pisano *et al.* (2003) reported that flushing gate technology appeared to be the most cost-effective means of flushing sediments in large diameter flat sewers compared with conventional methods and that a velocity of 1 m/s was adequate for cleansing.

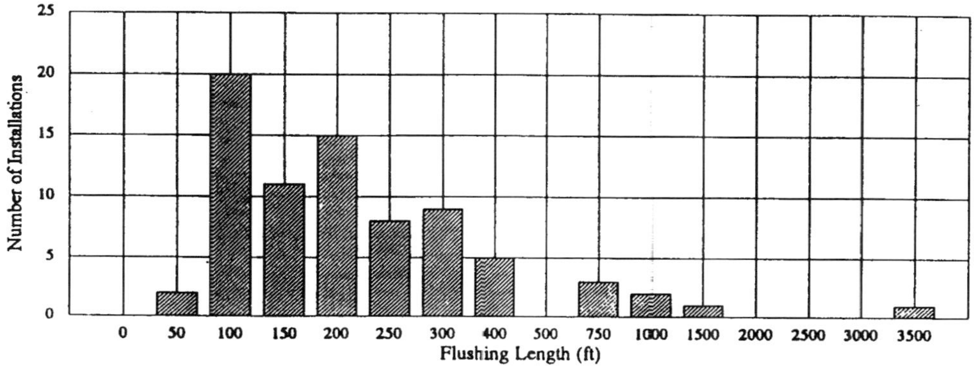

Figure 4.2.9. Distribution of flushing length versus number of projects (adapted from Grande *et al.*, 1995).

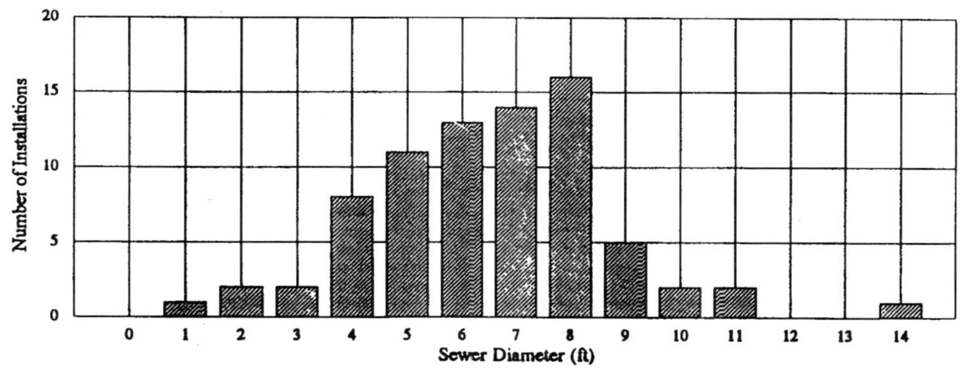

Figure 4.2.10. Distribution of pipe diameter versus number of projects (adapted from Grande *et al.*, 1995).

A similar system to HYDROSELF has been developed and patented in France, called HYDRASS (Sykora, 1989) (Figure 4.2.11a); it operates automatically, as shown in Figure 4.2.11b The system comprises a simple vertical gate pinned along a horizontal axis. The position of the axis causes automatic pivoting of the gate without mechanical assistance due to an imbalance of the hydrostatic head; when the pressure becomes greater on the portion of the gate above the axis (step 1), the gate pivots (step 2) and releases a flushing wave downstream (step 3). When the storage upstream of the gate empties, gravity and water pressures reset the gate into a vertical position to start another cycle (step 4). The filling speed depends on the dry weather flow rate, geometry of the sewer and the height of the gate. This system has been installed on a part of the Marseille No. 13 trunk sewer. It was found that a 100 m length of sewer required 700 flushes to clean an initial deposit of about 100 mm (Figure 4.2.11c). The flushing frequency can be reduced if the upstream head can be increased, for example, the number of flushes required at a head of 0.5 m is 24 times that required for a head of 1.5 m. Laboratory experiments (Gendreau *et al.*, 1993) and 3D modelling (Lainé *et al.*, 1998) have also been carried out to investigate and improve the functioning of the gate.

Many HYDRASS gates have been installed in Lyon, France, where they are used both to cleanse sediments and to maintain clean sewers in a clean state. In some locations, with downstream sewers that have straight alignments, the length over which sewers are efficiently maintained free of sediment may reach 400 m or more.

The HYDRASS gate may be used in various conditions:

- fixed at a given location where there is a continuous risk of sediment deposition;
- mobile, where the upstream water level and pressure is used to move the gate progressively downstream along the sewer.

More recently, the frame supporting the HYDRASS gate has been built separately from the gate itself. This allows the sewer operator to move an individual HYDRASS gate around a series of sites where the frames are permanently installed.

Figure 4.2.11a. HYDRASS gate tipping in a sewer.

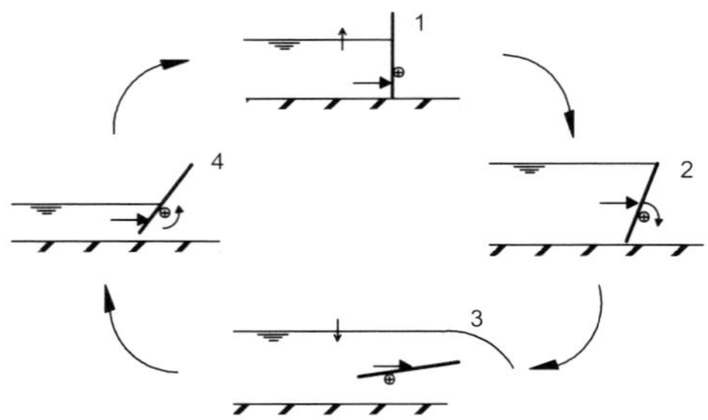

Figure 4.2.11b. HYDRASS gate functioning.

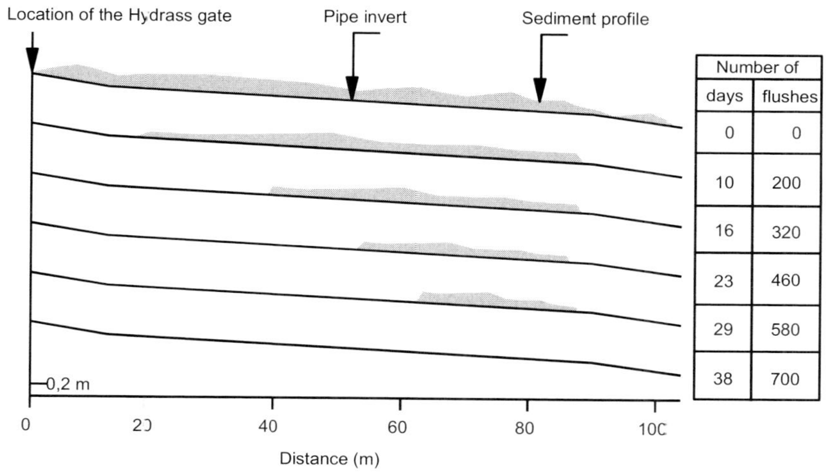

Figure 4.2.11c. Experiments with the Hydrass gate in Marseille (adapted from Chebbo *et al.*, 1996).

4.2.2.4 Decantation basins

The most economic way of dealing with in-sewer solids is at an outfall treatment plant (Butler and Clark, 1995). However, the problem is to ensure that the solids are actually transported to the treatment plant. The most straightforward method of separating solid and liquid phases is by sedimentation, and a number of operators have proposed the use of detention basins for the finer suspended solids, which are amenable to 'decantation' (Chebbo *et al.*, 1996). These basins are usually large (100 to 200 m^3/impervious ha) (Ruscassier *et al.*, 1998; Bertrand-Krajewski *et al.*, 2000) end-of-pipe structures that are, in effect, a type of treatment plant. Their utilisation can be a very efficient way to protect either environmental systems or to buffer shock loads to WWTP. However, they do not assist in the efficient operation of the sewerage system itself. More detailed information about the settling of solids in tanks is given in Section 4.3.4.

4.2.2.5 Sediment traps

4.2.2.51. Introduction

Within sewer networks, there is potential for more cost-effective operation by localising sedimentation in designated zones where large volumes of solids can be stored and accessed more readily than when dispersed throughout the sewer network. One potential solution to alleviate the problems caused by the bulk sediments (grit) responsible for much of the loss in hydraulic capacity, is to install 'silt' or sediment traps within the contributing sewers. Experience shows (Dartus *et al.*, 1990; Laplace, 1991; Lin, 1993; Ashley *et al.*, 1994; Verbanck *et al.*, 1994) that most of these sediments arrive by means of bed-load type transport mechanisms (Section 2.3). An on-line trap, comprising a sump in the sewer invert, would thus be expected to interrupt the bed-load movement, and depending upon size, would also remove some of the suspended solids. With modern suction methods of emptying such traps, this approach can be very effective in preventing problems caused by in-sewer deposition (Ashley *et al.*, 2002).

4.2.2.5.2. Types of traps

Various types of traps exist, for which there are no standardised designations. Hence, in order to avoid confusion in this Section it is necessary first to give some definitions and to suggest a terminology based on examples from the UK and France.

In the UK, **silt traps** are sumps in the sewer invert, whose width matches the incoming sewer. Typical examples are the Dundee traps (see Figure 4.2.12), whose volumes range from 2.7 to 13.4 m^3 (Ashley *et al.*, 1995). Despite being termed 'silt traps' these sumps in the invert of the sewers collect all types of solids moving near the bed, and also allow settlement of some suspended particles. There are very few such traps in the UK, with only twelve remaining in Dundee and a few isolated installations elsewhere. The Dundee traps were built during the last and previous centuries and originally there were many more that have been progressively filled-in (Ashley *at al.*, 1995). This was because their operation was deemed ineffective due to the need to regularly clean them out (there was no planned maintenance programme) and the risk of developing odours in the trap deposits and the odour impacts on the busy city centre.

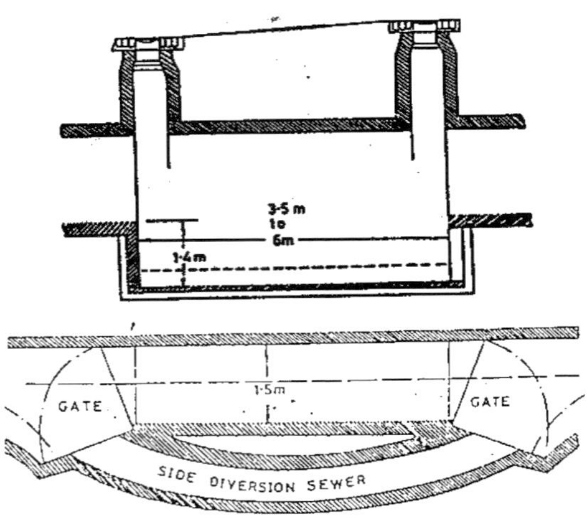

(a) Sketch of trap.

(b) Trap full of sediment (photograph Richard Ashley).

Figure 4.2.12. Example of silt trap in Dundee, UK.

In France, traditional **grit chambers** (*chambres de dessablement*) are usually parallelepipedic chambers located at the pipe invert, which also usually widen as the sewer enters the chamber (see Figure 4.2.13). Most of these are symmetrical, but other shapes have also been tested at both laboratory and field scales (STU, 1982; Charton, 1986). However, some of these chambers are not wider than the upstream sewer pipe and are similar to the silt traps in the UK. French grit chambers are usually larger than the silt traps in the UK. Their size ranges from some cubic metres to some hundreds of cubic metres. For example, in the Seine-Saint-Denis region near Paris, there are more than 100 grit chambers in the 650 km sewer system, with volumes varying between 10 and 400 m^3, with a mean volume of 100 m^3 (Paitry *et al.*, 1990). These chambers collect a wide range of types of particles, not only the grit fraction.

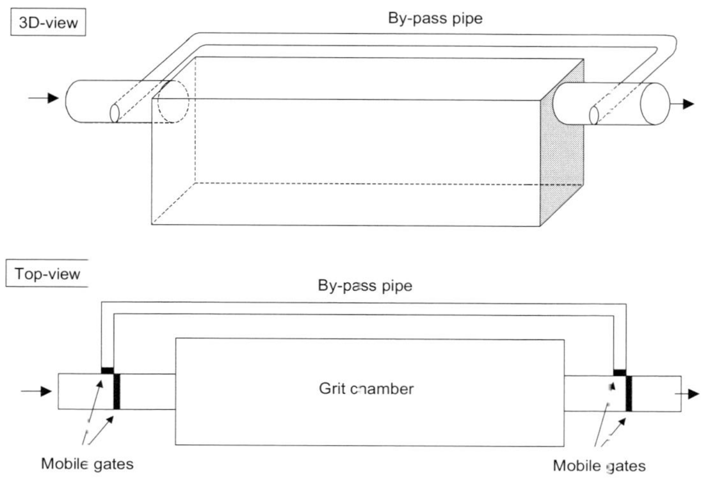

Figure 4.2.13. Example of traditional French grit chamber.

French grit chambers are more comparable with traditional fluvial settling basins as they have a high trap width to sewer width ratio. However, US studies and design guidelines for fluvial settling basins and recent re-assessment of the methods (Garde *et al.*, 1990) may not be applicable to combined sewers where the particle nature is highly variable (inhomogeneous, with temporal and spatial variations) and the temporal variability in hydraulic conditions is much greater across the basin. Basin width to flow depth aspect ratios are also much larger than the 1:1, typical of the Dundee silt traps.

Recently, a new concept of traps has been introduced in France, based on research into bed load transport in sewer systems (Laplace, 1991; Lin, 1993, Section 2.3). These new traps, named **bed load traps**, are completely different from the traditional grit chambers (see Figure 4.2.14). Bed load traps are sumps in the sewer invert, with a width equal to the upstream sewer pipe width. They are covered with plates and a transverse open slot of approximately 300 mm between the two plates allows the only the heaviest of the bed load solids to settle into the trap (Laplace *et al.*, 1993). In contrast with traditional grit chambers, bed load trap operation is not based on the settling of bed load and suspended solids, but on the selective interception of the heavier bed load solids moving along the sewer invert (Lin *et al.*, 1993; Bertrand-Krajewski *et al.*, 1996a, 1996b; Laplace and Felouzis, 1997). Their volume is relatively small, ranging from 1 to 5 m^3, so that a vacuum suction vehicle can cleanse one trap completely in a single load.

There are a small number of trap installations in Denmark, Germany and Malaysia (open sewers), but no published details or performance information.

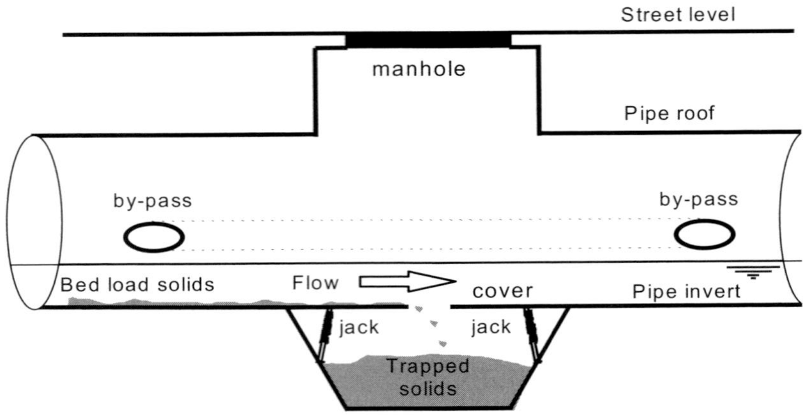

(a) Normal conditions

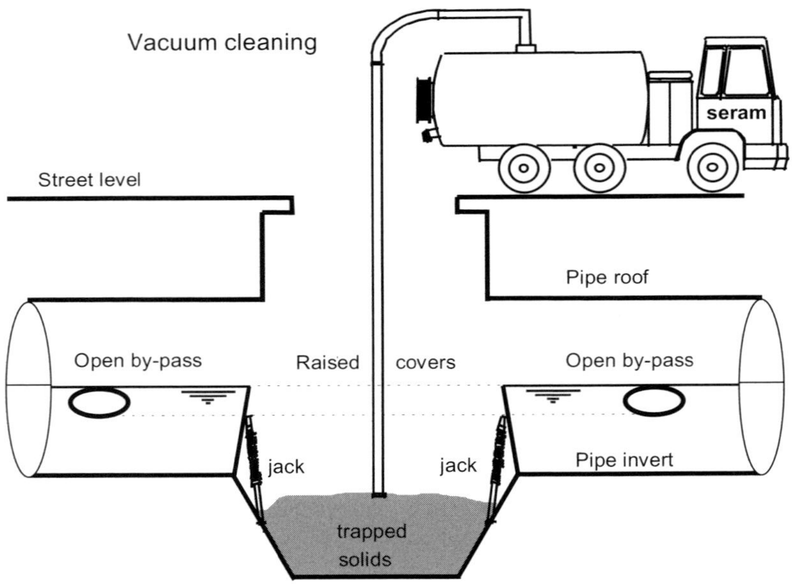

(b) Cleansing conditions

Figure 4.2.14. Example of French bed load traps (from Laplace *et al.*, 1998, reproduced by permission of GRAIE).

4.2.2.5.3 Use of traps

The use of invert traps as a method of sediment management is a traditional approach which had lost favour with many sewerage designers and operators due to the labour intensive and unpredictable nature of the maintenance required (Ashley *et al.*, 1995). In France, however, the utilisation of grit chambers is quite extensive (Paitry *et al.*, 1990) and research initiatives were begun in the 1980s to devise improved methods for their design and operation (STU, 1982; Dartus and Alquier, 1986; Charton, 1986; Dartus *et al.*, 1990).

Recent advances in knowledge about sewer sediments have resulted in the development of predictive models for sediment transport rates (Section 2.3). The opportunity thus exists to apply new approaches to the prediction of the effectiveness of trapping sewer sediments in sewers at strategic, accessible locations. Dynamic devices to 'flush' sewer sediments from slack to steeper, higher transport capacity sewers, or to traps as points of storage, are once again (having been used 100 years ago) being developed (see Lorenzen *et al.*, 1997, Lorenzen and Laplace, 1998; Chebbo *et al.*, 1996; also Section 4.2.2.3).

Nonetheless, little information is as yet available about the operation of in-sewer traps. The most extensive studies have been long-term studies of trap filling rates (Paitry, 1983; Paitry *et al.*, 1990) and trap efficiency (Dartus *et al.*, 1990; Dartus and Alquier, 1986) in France. In 1995, there were thirteen traps in operation in Dundee (UK). In at least one situation it is probable that earlier removal of a trap may be resulting in siltation at a downstream section, and within the last 5 years, one trap has been replaced with a more modern design incorporating appropriate operational facilities, thus facilitating easier cleaning and monitoring. Three new traps have also been installed within the former Tayside Region around Dundee, two of which are entirely ineffective due to inappropriate location as a consequence of a lack of clear design guidance (Fraser *et al.*, 2001).

Studies carried out over 8 years of the sedimentation in the main Dundee interceptor sewer have shown clearly the important effect of the silt trap at the head of the sewer (Figure 4.2.12). For periods of typically 3 months following emptying of the silt trap, the mass of sediments deposited downstream in the interceptor sewer are reduced. Measurements of the rate of bed-load transport into this trap during dry weather flow have shown that on average this is between 1–2 kg/hour and comprises some 12 % of the total solids in transport at any one time. Similar measurements in another silt trap at Dens Brae on a trunk sewer (1030 mm x 680 mm, slope up to 1/22) gave more variable results but generally mass transport rates as bed load in excess of 1 kg/hour during dry weather (Ashley *et al.*, 1994). Data collected for sediment transport during storms are more variable and whilst the suspended load transport rates increase at such times, it is possible that the bed-load mass transport rate decreases. Overall the material found in transport near the bed in the trunk sewer studied was much coarser than that in the interceptor sewer silt traps because the catchment and inlet sewer is much steeper (see Section 3).

Trap filling rate studies have been carried out in Dundee and a number of locations in France. These show that in open traps there is generally a preferential filling at the front of the trap, with a more gradual build-up in deposits toward the end of the trap. However, this preferential mode of deposition in silt traps/grit chambers depends on geometric and hydraulic conditions within traps/chambers, and preferential deposition at the downstream end and scouring effects have also been observed (Paitry and Maurin, 1989).

In Dundee, the deposits show evidence of consolidation with time, particularly near the silt trap inlet. It is probable that re-erosion occurs mainly from the deposits within the downstream part of the trap. Typical filling rates are shown by the 3-D plot (Figure 4.2.15), for a trap on a Dundee trunk sewer, monitored using scanning sonar techniques (Ashley *et al.*, 1993). Re-suspension in silt trap or grit chambers has also been confirmed by means of modelling (Dartus and Alquier, 1986; Boniface, 1996; Boniface *et al.*, 1998; Buxton *et al.*, 2001).

In Seine-Saint-Denis, qualitative measurements of deposits in grit chamber have shown that particle diameters decrease along the grit chamber. Particles ranging from 200 to 500 μm represent 35 to 45 % of the mineral fraction. The percentage of particles finer than 200 μm ranges from 12 to 40 %. There is also a variation of the trapped sediment water content, which increases along the chamber. A vertical gradient in the sediment density has also been observed, with relatively fluid and fine sediments at the top to very compact sediments at the bottom of the chamber. It appears that fine sediments progressively consolidate and hence lose some of the interstitial water (Paitry, 1983).

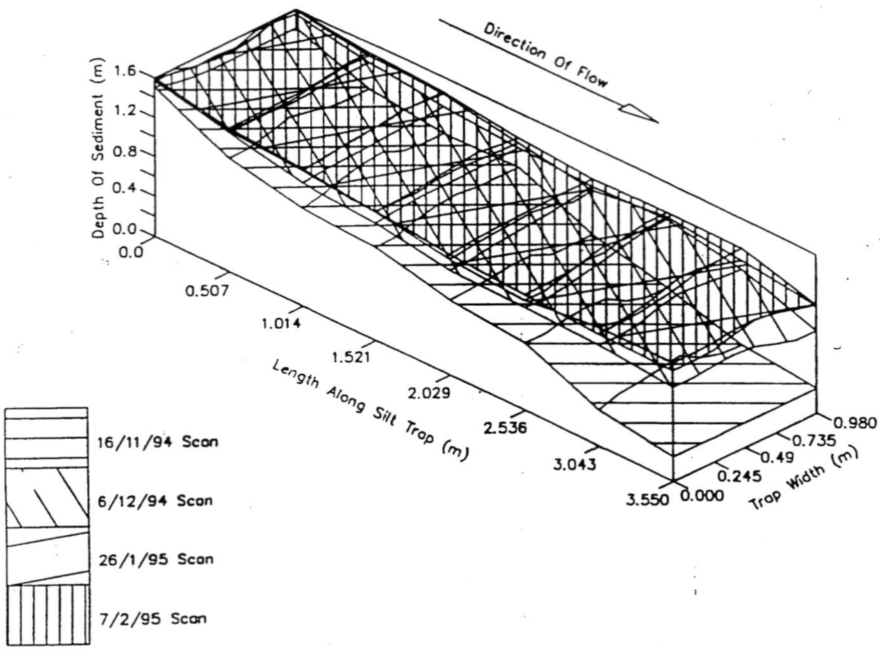

Figure 4.2.15. 3-D plot of trap filling rate in Dundee.

French grit chambers fill at varying rates, averaging some 0.56 % of chamber volume per day in combined sewers and 0.30 % in storm sewers (Dartus *et al.*, 1990), trapping some 65-85 % of the total mass in transport near the bed. Of the French grit chambers studied, it was found that they eventually filled on average up to 70 % of their capacity over periods of up to a year. The Dundee traps also fill on average to less than capacity. Traditionally grit chambers have been sized arbitrarily – possibly as grit tanks – removing suspended as well as bed loads. In Dundee, transport rates in the bed-load phase have been recorded as up to 34 % by mass of total solids in transport at any one time (see Section 3). It is to be expected that all of this material will be trapped, but the situation is complicated because of the temporal variations in solids transported and differences between dry and wet weather periods. For example, what is transported as 'bed-load' in dry weather (and may or may not be 'near-bed solids' (NBS) which tend to be primarily organic particles) may become 'suspended' load in wet weather (see Section 3). Consequently design criteria should be based on both bed and suspended loads, and account for temporal variations. Specifically, NBS should be excluded from depositing in traps, since disposal of sewer solids that comprise a high organic fraction are more problematic. Ideally, traps should collect only the grit particles, for subsequent processing and re-use once removed from the sewer (see Section 4.2.2.5.1).

Unfortunately, there are as yet no standard design methods available to size sediment traps or to estimate collection rates and nature of the solids, and hence emptying and disposal requirements (see Section 4.2.2.5.5).

A 1.6 m deep silt trap in Dundee on a 1350 x 920 mm combined sewer has been studied, with sediment data as given in Table 4.2.7. The Claverhouse samples relate to an effectively full trap, which had never been emptied (approximately 12 months) The Meadowside trap was emptied on 2 October 1994 and the samples are for a gradual build-up in deposits. On 6 December 1994, the deposits had increased from 800 mm (average) on 16 November to about 1200 mm (Figure 4.2.16), i.e. an increase of 800 mm in 42 days followed by 400 mm in 20 days; an average rate of filling between each survey of 0.07 m^3/day (1.25 %/day). This is equivalent to 130 kg/day in terms of the bulk density given in Table 4.2.6. Whilst the volume change is expected, this mass is unlikely and illustrates that the depth of deposit monitored by the sonar survey also comprised low density unconsolidated organic material.

Table 4.2.7 Dundee silt trap sediment samples

Trap	Date	Sediment depth (mm)	d50 (mm)	Organic (%)	Bulk density (kg/m^3)
Claverhouse					
Top	26/8/94	420/520	0.55	2.9	1678
mid-depth	26/8/94	(full)	0.35	11.7	1327
bottom	26/8/94		0.25	9.4	1229
Meadowside					
Top	16/11/94	800/1600	3.00	1.0	1878
Bottom	16/11/94		0.9	1.8	1748
Top	30/11/94	NR	2.00	1.5	1871
Bottom	0/11/94		1.5	1.4	1877

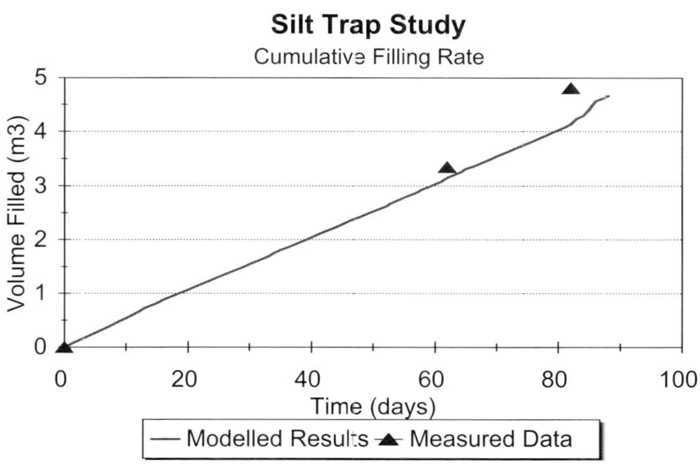

Figure 4.2.16. Filling rate of Dundee Meadowside trap.

4.2.2.5.4 Operation of traps

The main operational problems arising from the use of silt traps and grit chambers are wholly associated with emptying and include time taken, odour, disruption to traffic and inconvenience to the general public. Over the years the frequency of cleaning has decreased in Dundee as it was so labour intensive; while this was partially offset by the reduction in sediment ingress due to the provision of paved surfaces, the lack of maintenance resulted in an increase in deposits within the sewer network overall. At the start of the 1980s there was real concern as almost all the traps were in need of emptying and there was a build-up of some 600 mm of sediment throughout much of the main interceptor sewer. The large traps can usually be emptied in approximately 2 days at a cost of about £1400 per day. So far it has been possible to dispose of the material at the local landfill site as it is not designated hazardous, although under the UK's Environmental Protection Act (1990) a 'Duty of Care-controlled Waste Transfer Note' is required which details the description of the waste and the contractor collecting and disposing of the material. In France, solids extracted from grit chambers were also traditionally disposed of in landfills but, due to recent environmental considerations, they are now being more frequently re-used after screening and washing.

4.2.2.5.5 Trap design and filling rates

Trap design is a complex problem; despite advances from earlier understanding (e.g. Ashley *et al.*, 1992, 1995, 2002; Bertrand-Krajewski *et al.*, 1996b), knowledge is still not sufficiently developed to determine where sediments will be a problem in a system, or what the nature of those sediments is likely to be. It is clear that the predominant 'near-bed' solids will vary both temporally and spatially in a system. Until such time as 'universal' sewer sediment transport models are available, it is only possible to assume that during dry weather the solids will comprise a granular bed load in the steeper trunk sewers and to hypothesise that in flatter sewers, the organic near bed solids are replaced by granular solids during wet weather (Section 2.3).

Approaches so far are empirical and site-specific, and relate to the prediction of the solids in transport and the layout of the trap for most efficient collection of the solids. Ideally, as few organics as possible should be trapped, as this will affect the ultimate disposal route. Models to estimate trap fill rates have been developed based on field empirical observation and laboratory and computational modelling studies.

A prototype silt trap fill rate model using two 'near bed solids' (NBS) transport equations has been applied to a silt trap situated on a length of trunk sewer in the city centre area of Dundee (Fraser *et al.*, 1998). The model assumptions are:

- The silt trap collects/retains near bed material represented in 2 ways:
  - during dry weather: organic/inorganic mixed solids;
  - during storm flows: mainly inorganic, granular material.
- Flow characteristics for the site are accurately represented using a full-solution hydraulic simulation model for the Dundee City sewerage.
- Sediment characteristics, where not measured, can be accurately estimated using alternative data from Dundee.

Relationships to predict near bed solids transport rates were used to represent the masses moving in the lower part of the flow for both dry and wet weather respectively (Section 2.3.4.3.2). The main premise is that under dry weather, an approach based upon field data considering total solids best described the material moving near the bed. However, during storm flows, a large proportion of the organic content of the near bed material is transported in suspension, thus leaving predominantly larger inorganics at the near bed region. As no actual field measurements of the near bed solids during wet weather have been achieved by any of the on-going research programmes before 1999, other than speculatively (Ashley *et al.*, 1994), a modified laboratory-based relationship was used for this phase, which was derived from tests using granular material (Equation 2.3.12).

No attempt was made to account for the 'settlement' of suspended solids in the trap, sediment washout or consolidation. The model predictions for rate of trap filling have been compared with a limited set of measured data taken from a previous study (Figure 4.2.16). Actual rainfall data were available for this period and individual events were ranked in a hierarchy of significance to be used in conjunction with the sewer flow model.

The results, although clearly restricted by the use of estimated parameters, provide a level of accuracy which warrants further investigation and data collection. The effect of each of the parameters being used in the equations was evaluated via a series of sensitivity tests. The results were found to be sensitive only to the estimated near-bed velocity.

The French bed load traps have been studied at both reduced scale (Lin, 1993; Lin *et al.*, 1993; Bertrand-Krajewski *et al.*, 1996a) and later at full scale (Bertrand-Krajewski *et al.*, 1996b; Boniface, 1996; Laplace and Felouzis, 1997; Laplace *et al.*, 1998; de Prato, 1999; Laplace *et al.*, 2001). The main conclusions from these studies are:

- Bed load traps are efficient: approximately 70–80 % of the bed load solids are intercepted. This has been confirmed by measurements of deposits downstream of a bed load trap in Marseille, France. Figure 4.2.17a shows that deposits were accumulating in this trunk sewer with a rate of approximately 40 m^3 every 6 months before the building of the bed load trap. After the bed load trap had been installed, the deposits appeared as quasi steady compared with the previous situation, as shown in Figure 4.2.17b. Assuming that the amount of solids transported at the upstream end of this trunk sewer was constant during the two periods of measurements, the results clearly indicate that the bed load trap intercepts almost 100 % of the solids which previously settled in the trunk sewer downstream of the bed load trap.
- Compared with traditional grit chambers, the solids trapped are coarser and more mineral. For example, in the Boulogne 84.7 m^3 grit chamber, 24 to 31 % of the deposits were larger than 100 μm, and 5 to 13 % were larger than 2000 μm. After replacement of the grit chamber by bed load traps, 14 to 52.5 % of the deposits were larger than 2 mm, with mineral fraction between 66 to 83 % depending on dry or wet weather conditions (Boniface, 1996).
- All solids trapped remain trapped: neither erosion nor scouring effects have been observed. Both empirical pilot-scale experiments (Hamparian *et al.*, 1993; Hydraulic Models, 1995) and later 2D hydrodynamic modelling (Schmitt *et al.*, 1999) have shown that the best bed load trap design involves a central slot perpendicular to the sewer flow with two plates covering the trap located at the same height as one another. In such conditions, the interception efficiency is maximised, internal turbulence is minimised and hydraulic recirculation next to the downstream cover stays under control (see Figure 4.2.18).

Long-term monitoring of the filling of bed load traps shows that solids accumulate without scouring. Contrary to traditional grit chambers, the long-term accumulation of deposits within the bed load trap does not decrease with time and filling level, but remains reasonably constant (Laplace *et al.*, 1998). The filling increases according to the upstream bed load flux, until it is completely filled. Figure 4.2.19 illustrates this process in three successive filling cycles. During wet weather, the mass of solids intercepted by the bed load trap increases significantly, and pilot-scale experiments have shown that this increase was proportional to the rainfall depth as shown on Figure 4.2.20.

In the 1990s, the design of bed load traps was purely empirical because the estimation of the bed load flux to be intercepted was difficult. However, a recent attempt by Laplace *et al.* (2001), using both the bed load transport model developed by Laplace (1991) and additional assumptions (different accumulation rates in dry and wet weather, no scouring in the bed load trap, gave promising results. After calibration, the model satisfactorily reproduced the long-term (240 days of observation) filling of the bed load trap, as shown in Figure 4.2.21. The filling dynamics were closely linked to dry and wet weather periods. The bed load trap filled slowly during dry weather, and more rapidly during wet weather according to daily rainfall depth. Such a model could be used to size bed load traps and to evaluate their required cleansing frequency based on long time series of rainfall after verification and transposition to other bed load traps for further validation.

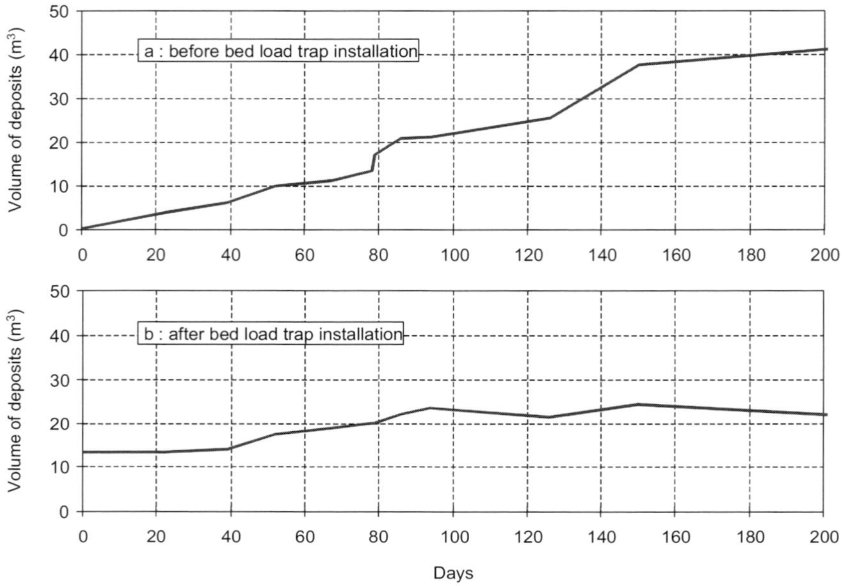

Figure 4.2.17. Accumulation of deposits in a trunk sewer downstream of a bed load trap in Marseille (from Laplace *et al.*, 1998, reproduced by permission of GRAIE).

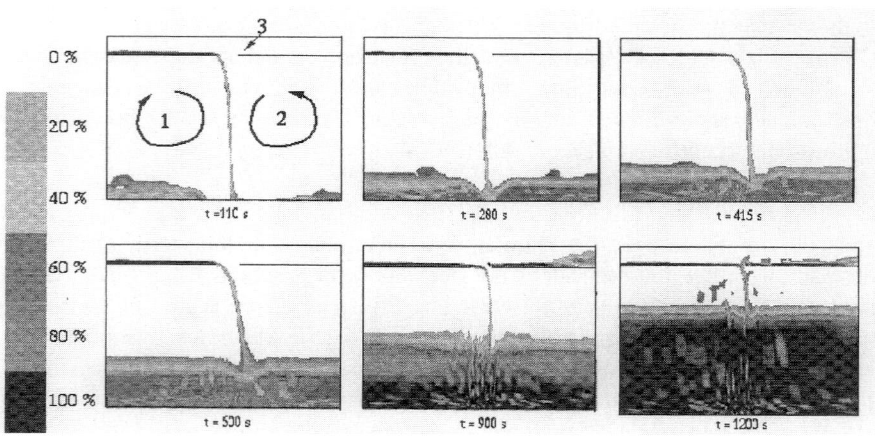

Figure 4.2.18. Modelling of bed load trap filling, with relative sediment concentration scale from 0 to 100 % (from Schmitt *et al.*, 1999).

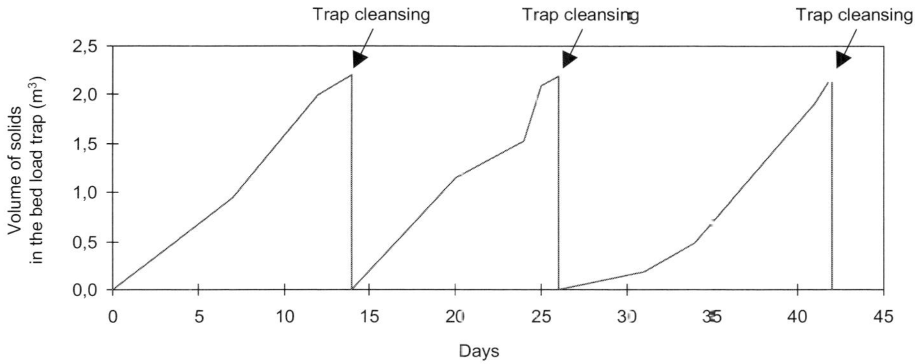

Figure 4.2.19. Accumulation of solids in a bed load trap in Marseille (from Laplace *et al.*, 1998, reproduced by permission of GRAIE).

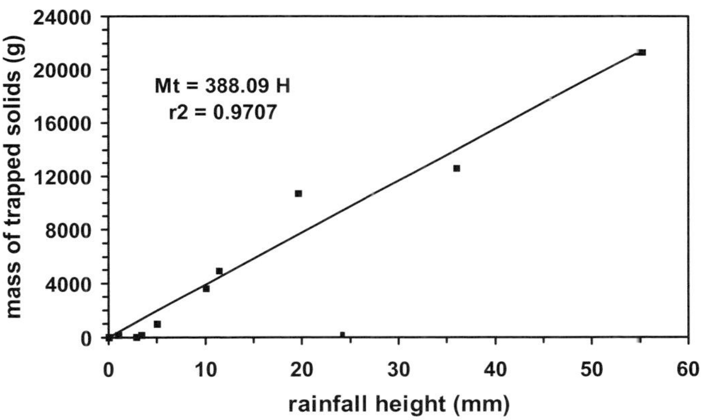

Figure 4.2.20. Relationship between mass of solids intercepted in pilot-scale bed load trap in Bordeaux, France and rainfall depth (from Bertrand-Krajewsk *et al.*, 1996a).

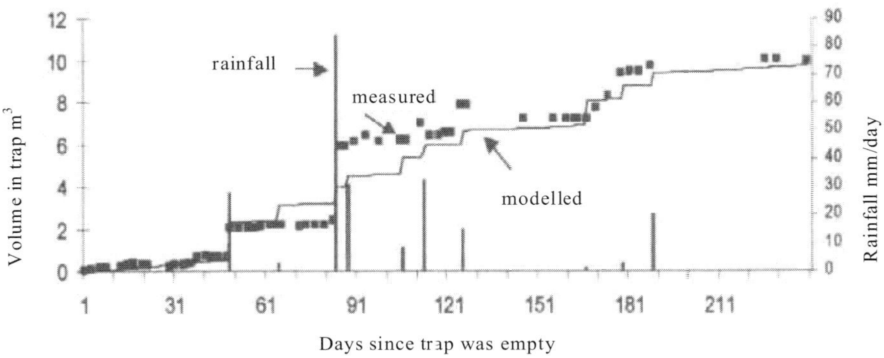

Figure 4.2.21. Modelling of filling dynamics of bed load trap in Marseille (from Laplace *et al.*, 2001, reproduced by permission of GRAIE).

4.2.2.6 Hydrogen sulphide control

Sediments provide a surface on which bacteria can form biofilms and reduce sulphates to sulphide; where temperatures are elevated there is a danger of sulphides and subsequent hydrogen sulphide gas being produced (Section 2.2). A requirement is that a gravity sewer should be designed to avoid sulphide formation. Water velocities which prevent solids deposition and promote high re-aeration are thus beneficial.

In pressure mains, sulphide formation may take place (Section 3). In such systems, hydrogen sulphide control may be needed. The following methods may be used:

1. Air, pure oxygen or nitrate may be added to prevent sulphate reducing conditions occurring.
2. The negative effects of sulphide can be avoided by adding metal salts which produce an insoluble sulphide. The most common salts used are iron sulphate or iron chloride.
3. Sulphide production may be reduced or prevented by removal or inhibition of the sulphate reducing bacteria. Chemicals having such effects are, e.g. alkaline substances, chlorine and hydrogen peroxide.

These control methods are described in detail elsewhere, e.g. Melbourne and Metropolitan Board of Works (1989) and USEPA (1974). It is important also to consider whether using these in-sewer methods may have other detrimental effects, on for example, downstream wastewater treatment. Pisano *et al.* (1998) and Fan *et al.* (2001) review the problem for gravity sewers and the potential to utilise flushing gates to control it. Flushing is shown to be very cost-effective.

The retention time of the effluents retained in the tank is a particularly important parameter in respect of the deterioration of effluent quality. Long retention times, in excess of 12 hours, may lead to problems of septicity and odour. Microbiological activity in both the effluent and in any sediment deposits on the bed of the tank results in an uptake of oxygen with a corresponding reduction in the concentration of dissolved oxygen in the effluent. The rate of oxygen uptake is a function of many parameters which include the initial concentration of DO, the constituents and pollutant strength of the effluent and the bed sediment deposits, temperature and the re-entrainment of oxygen. Equations have been developed (see Section 2.2) to estimate the uptake of DO and hence it is possible to predict, with limited confidence, the potential for the onset of septic conditions and the generation of hydrogen sulphide and methane in tanks.

4.2.3 Sewer solids treatment and disposal

4.2.3.1 Introduction

There is the potential to re-use solids removed from sewer systems and the European 1975 Directive, modified in 1991, exhorts member states to take measures to promote the re-use of such waste. In France, implementation of this Directive requires that from July 2002, landfills will admit only ultimate waste, i.e. waste that cannot be treated using contemporary technical and economic options. Other European countries have similar legislation. This section of the report examines the nature, sources and characteristics of the collected solids, and of the options for treatment and disposal.

4.2.3.2 Sources of solids

Solids extracted from sewer systems originate mainly from street surfaces, road gullies and sanitary sewer inputs (Delattre *et al.*, 1998). However, changes in the design of surface drainage systems utilising, for example, porous pavements, infiltration systems and storage and settling tanks etc., will result in changes in the origins and nature of solids to be 'recovered' or disposed. Because of these different and changing origins, the solids are very heterogeneous with characteristics depending on a range of factors: combined or separate systems, the nature of the catchment, traffic intensity, land use, urbanisation type and level, etc. (refer to Section 2.1). In a wide ranging study of solids removed from sewers in different French cities (Gousailles, 1994), the volumes of solids removed from sewer pipes were found to vary widely from site to site: from 2 litres/capita/year in Bordeaux to 17 litres/capita/year in the Paris area. The mean value was found to be 10 litres/capita/year or about 18 kg/capita/year. In France this leads to a total mass of removed solids of about 0.9 million tons/year.

4.2.3.3 Characteristics of silt trap solids

The heterogeneity of the solids removed from sewer pipes is typically broad. For example, an analysis of solids extracted from on-line Parisian grit chambers (see Section 4.2.2.5) showed that about 40 % of the mass was not fine granular and comprised a range of materials as detailed in Table 4.2.8.

Table 4.2.8. Composition of some Parisian grit chamber solids (Gousailles, 1994)

Type of solids	Mass percentage
Fruit pits and pips	2 %
Plastic	4 %
Aluminium	4 %
Nodules not identified	4 %
Stones	4 %
Cigarette filters	6 %
Greasy balls	6 %
Vegetable litter	11 %
Other, including sand	59 %
Total	100 %

The solids are typically made of a mineral central core covered by an organic layer which contained micro-pollutants such as heavy metals, hydrocarbons and organic compounds. The mineral fraction originated mainly from street degradation, urban construction sites, road sand blasting in winter and soil erosion. More detailed information about these solids and their attached pollutants is given in Sections 2.1 and 2.2. Table 4.2.9 shows that the solids are mainly mineral and have little potential for agricultural use, unlike sludge from wastewater treatment plants. Solids from combined sewers appear usually more polluted than solids removed from separate sewers.

4.2.3.4 Solids treatment processes

The typical treatment processes which may be used for solids removal from sewer systems may be divided into four main categories (Figure 4.2.22): screening, washing, classification and dewatering (Delattre *et al.*, 1998). These should be adapted according to the solids characteristics and to the required treatment objectives at any given site. Solids treatment facilities are usually sited at wastewater treatment plants and the way in which the solids are introduced into the treatment plant has a significant effect on treatment efficiency. For example, a floor band shuttle is better adapted to feeding the typical heterogeneous materials into the treatment unit when compared to a tractive shovel.

Table 4.2.9. Characteristics of solids removed from sewers in France (from Delattre and Bertrand-Krajewski, 1996)

	Source:	1	2	3	4	5
	Site:	Entzheim	Orléans	Amsterdam	Mean values	Bordeaux
	Sampling location:	Street	Gully pot	Combined sewer	Grit chamber	Settling tank
Parameters	Units					
Dry matter content	%		35		44	40
Organic fraction	%	5	16	30	16	17
Cd	mg/kg	3	1	1.3	9	2.2
Cr	mg/kg	29	44	27	63	
Cu	mg/kg	21	94	454	300	222
Hg	mg/kg		0.86	1.9	2	
Ni	mg/kg	20	28	33	45	
Pb	mg/kg		164	303	2250	337
Zn	mg/kg	230	520	562	1027	725
Hydrocarbons	mg/kg		128		886	
TKN	g N/kg	1				
phosphorus	g P/kg	0.74				

Sources: 1: Artières, 1987; 2: IRH Environnement, 1995; 3: Lassus and Salomé, 1993; 4: Gousailles, 1994; 5: Baladès and Bertrand-Krajewski, 1995.

(a) Screening (or sieving)

The objective of screening is to remove the larger materials, usually household refuse, which entered the wastewater system. The separation threshold is usually about 10 mm for coarse screening and about 5 mm for fine screening. The screenings (screened material) may be mixed with household refuse for incineration or landfill storage. The volume of process water which is required for effective screen operation is substantial: some 6 m^3 of process water for 1 ton of solids. The mass percentage of the screenings collected is typically about 15 to 35 % of the initial total mass M of solids entering the treatment facility. Rotative screens appear more effective than vibrating ones which clog more rapidly and need more frequent maintenance.

(b) Washing

The objective of washing the solids is to separate the mineral and organic phases. The process water from the screening stage may be re-used for this purpose. The typical technique consists of using one or two successive hydrocyclones, depending on solids characteristics and potential reuse of solids. The fine and organic particles are removed as supernatant in the process water, while coarse and mineral solids are removed at the bottom of the device. The organics in the process water are treated at the wastewater treatment plant.

(c) Classification

The classification step is only used if the separation between organic and mineral particles after the washing step is not satisfactory, according to the future reuse of the solids. It is normally not necessary when two successive hydrocyclones are used for washing, but was required in some old plants where the washing process was not efficient, especially for tomato seeds.

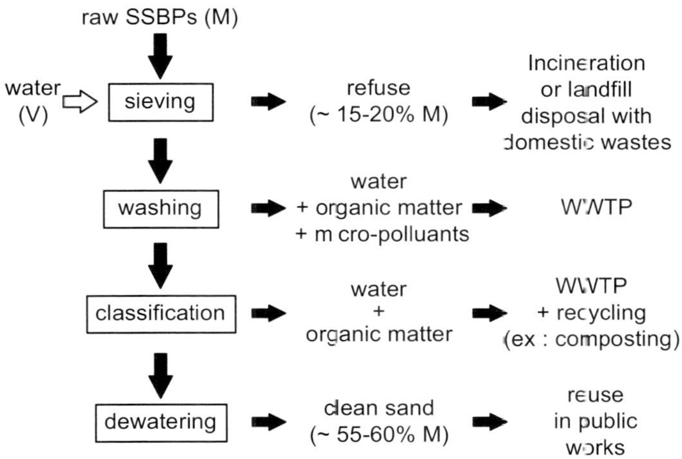

Figure 4.2.22. Typical treatment process and fluxes for solids (SSPB = sewerage solid by products) removed from sewers and inlets (from Delattre *et al.*, 1998, reproduced by permission of GRAIE).

(d) Dewatering

This step is required for all treatment plants, in order to remove as much water as possible from any separated sand. The efficiency depends on the performance of the preceding steps. Dewatering is usually achieved using a system of presses or vibrating bands. The mass of mineral solids collected at the outlet of the treatment system is typically 55 to 60 % of the initial mass M.

(e) Some examples of treatment facilities

In Blainville-sur-Orne, France, a SERFA treatment unit has been built at the 15000 EH municipal wastewater treatment plant. Its treatment capacity is of 10 000 tons of solids per year (CISE, 1997). The facility is a basic one, with one trommel for screening, two hydrocyclones for washing and a final dewatering step. Analyses have shown that 82 % of the mass of dry solids is made up of sand, the other 18 % being screening refuse. The unit requires 6 m^3 of process water per m^3 of sand. The clean sand has a dry matter content ranging from 82 to 88 %, and an organic content from 3 to 5 %. The sand is reused in public works. Two types of effluent are treated in the WWTP:

- the raw water issued directly from the solids to be treated (they are delivered to the treatment unit by vacuum lorries with water);
- the polluted process water.

The raw water, despite a limited volume of some m^3 per day, is very concentrated (1000 to 15 000 mg/L of BOD_5, with peak values at 40 000 mg/L). The polluted process water, with a volume of some tens of m^3 per day, is less concentrated (COD concentration ranging from 400 to 2800 mg/L, mean value approx. 1000 mg/L). Both types of water have a significant impact on the WWTP: the aeration had to be increased and the sludge volume index decreased. Micropollutants, and especially heavy metals, have no observable impact on the WWTP, but are present in relatively high quantities which may affect the sludge characteristics.

In Lyon, France, a similar treatment unit is located at the 475 000 EH Pierre-Bénite municipal wastewater treatment plant. In 1999, during the experimental phase, 11 000 tons of raw sand have been treated (Shoshany, 2000):

- 3300 tons were screening refuse, simply dewatered before disposal in controlled landfills;
- 7700 tons have been washed and treated.

These 7700 tons include:

- 2135 tons (28 %) of clean sand used in public works after a maturation period of 3 months, mainly to refill trenches for sewer pipes;
- 540 tons (7 %) of additional screening residues;
- 500 tons (65 %) of large and fine organic particles transported by the process water and treated in the WWTP.

Examples of the sand recovered are illustrated in Figure 4.2.23.

In Bordeaux, France, some preliminary studies have been carried out to evaluate the mass of solids to be treated (Delattre and Bertrand-Krajewski, 1996; Picard, 1997; Delattre *et al.*, 1998; Lucas *et al.*, 1999). On a 15 year term, it has been evaluated at a minimum of 27 500 tons of solids per year, including all sources of solids as given in Table 4.2.10. Depending on their source, these solids have variable pollutant loads (Delattre *et al.*, 1998) and, according to these loads and to the potential reuse, different treatment trains may be used (Lucas and Bourgogne, 2000). For comparison, previous data for Zürich, Switzerland are also given in Table 4.2.10 (Conradin, 1989). Both examples show that very large amounts of solids are accounted for in the frame of an integrated and sustainable management of pollutants in urban areas.

Table 4.2.10. Mass balance of solids removed from streets and sewer systems in Bordeaux, France (Picard, 1997) and Zürich, Switzerland (Conradin, 1989)

Origin of solids	City of Bordeaux	City of Zürich
From street sweeping and cleaning	18 000	10 105
Removed from sewer pipes and gully pots	5 000	1 325
Removed from grit chambers located in WWTPs or along sewer pipes	15 000	2 000
Removed from storage and settling tanks	3 000	-
Total (tons/year)	27 500	13 430

(f) In-sewer treatment

A moving plant, the ‘Sielwolf System’, has been developed in the city of Hamburg for the cleaning of thick layers of sediments from large sewers with diameters in the range 1.4 to 3.2 m. (Dinkelacker, 1992). Combined with a small shredding robot inserted into the sewer and extracting the material, the plant uses the sewage directly from the sewer. The system comprises a filter which separates the coarse materials (> 6 mm), and two hydrocyclones (see Figure 4.2.24). The Sielwolf and the slurry pump are moved along the sewer and can clean up to 9 m^3/h of solids. This plant avoids any need to remove the solids from the sewer for treatment elsewhere at a specific treatment plant.

a. View of the outlet of the vibrating dewatering belt downstream of the hydrocyclones. The washed sand coming out may be re-used.

b. View of the washed sand. It will be re-used in public works after a maturation period of 3 months.

Figure 4.2.23. Sand recovery plant in France.

4.2.3.5 Reuse of materials recovered from sewer solids treatment

As yet, the treatment of solids extracted from sewer systems has usually been limited to the separation of the liquid and solid phases. The evolution of environmental regulations, particularly restriction of dumping of sewer solids in landfills, requires new ideas to develop treatment plants which will account for the heterogeneity of sewer solids and their diversity in terms of quantity and quality.

4.2.3.5.1 Screening refuse

No re-use is made of screenings and the screening refuse is typically mixed with household refuse for landfill disposal or incineration.

4.2.3.5.2 Process water

Process water is usually discharged into the nearest WWTP for conventional biological treatment. However, this water can be highly polluted by metals and micropollutants and may have in some cases (e.g. solids removed from separate stormwater storage and settling tanks) very low biodegradability. It may have similar characteristics to trade effluents, and as a consequence may cause disruption to wastewater treatment plant processes (both sewage and sludge treatment) (CISE, 1997).

The two examples of sewer sediment cleaning process water shown in Table 4.2.11 demonstrate that some preliminary physical-chemical treatment would be required before transfer into the municipal WWTP. Simple settling would not be adequate due to the fact that the settling velocities of the finer suspended solids in the process water were very low.

The most efficient pretreatment is coagulation-flocculation, as used in Orléans, France, where alum (concentration 100 mg/l) with polymer flocculent (concentration 4 mg/l) is used to produce a sludge containing concentrated heavy metals and other micropollutants. This sludge has subsequently to be treated by incineration or disposal to landfill.

Table 4.2.11. Characteristics of process water in two French sites (IRH Environnement, 1995)

Site	Montgeron (mean values from 5 sets of samples)	Orléans (mean values from 3 sets of samples)
pH	7.7	7.7
TSS (g/l)	32,1	10
COD (mg O_2/l)	15 115	3668
BOD (mg O_2/l)	3684	550
COD / BOD	4.1	6.7
TKN (mg/l)	283	138
phosphorus (mg/l)	123.6	68.9
hydrocarbon (mg/l)	1261.8	73.2
Cd (mg/l)	0,65	0.04
Cr (mg/l)	7.6	2.5
Cu (mg/l)	13.5	4
Hg (mg/l)	0.003	
Ni (mg/l)	1.6	3.8
Pb (mg/l)	87	9.4
Zn (mg/l)	89	11.8

4.2.3.5.3 Cleaned sand – usage

There are several possibilities for the re-use of recovered and cleaned sewer sediment sand in construction and public works, such as:

- quarry rehabilitation;
- bottom layers in trenches for cables or sewer pipes (as e.g. in Lyon, France);
- alternating layers with household waste in landfills;
- landfill cover;
- sublayers for sport fields.

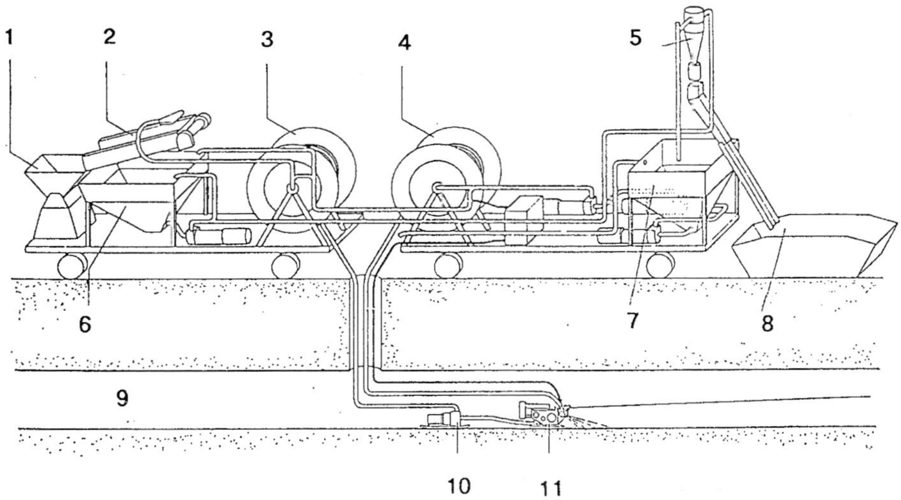

1 – coarse material; 2 – screening plant; 3 – hose reel sludge; 4 – hose reel flushing water; 5 – centrifugal separator; 6 – sludge water mixture; 7 – overflow water; 8 – separated solids (mainly sand); 9 – sewer pipe; 10 – slurry pump; 11 – Sielwolf

(a) Schematic view of the 'Sielwolf System'

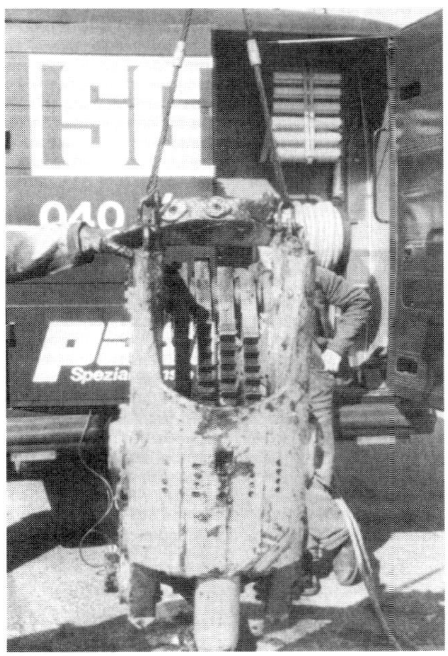

(b) The head

Figure 4.2.24. The Sielwolf moving solid in-sewer treatment plant (Blumberg and Bauer, 1984).

The French standard NF P 11-300 (1992) gives a classification of materials that can be used in embankments and highway structures. As cleaned sands resulting from sewer solids treatment are not specifically included in this standard, they are classified as 'other industrial by-products'.

Similarly in the Netherlands, the sand recovered from the Amsterdam treatment unit (see Table 4.2.12) is classified as Category B within the following system of classification.

- category A: good quality ground, that can be used for urbanisation;
- category B: ground with a small level of contamination, reserved for industrial usage;
- category C: ground which needs decontamination before re-use.

Table 4.2.12. Results of sewer solids treatment in Amsterdam, Netherlands (from Lassus and Salomé, 1993)

	Raw solids	After washing	After dewatering
Dry matter content	45 %	76 %	84 %
Organic content	19 %	1.5 %	0.8 %

The sand recovered in Amsterdam is re-used in the construction of embankments.

4.2.4 References

Ackers P. (1991) Sediment aspects of drainage and outfall design. *Proceedings of International Symposium on Environmental Hydraulics*, edited by Lee and Cheung, Balkema, Hong-Kong, Rotterdam, The Netherlands, December 1991, 29–29.

Ackers, J.C., Butler, D. and May, R.W.P. (1996) *Design of Sewers to Control Sediment Problems.* CIRIA Report 134.

AFNOR (1992). Norme NF P 11-300: *Exécution des terrassements – Classification des matériaux utilisables dans la construction des remblais et des couches de forme d'infrastructures routières.* AFNOR, Paris, France, September 1992, 24 pp.

ASCE/WPCF (1970) *Design and Construction of Sanitary and Storm Sewers.* ASCE Manuals and reports on engineering practice, No. 37.

Arthur, S. and Ashley, R.M. (1997) Near bed solids transport rate prediction in a combined sewer network. *Wat. Sci. Tech.*, 36(8–9), 129–134.

Arthur, S., Ashley, R.M., Tait, S. and Nalluri, C. (1999) Sediment transport in sewers: a step towards towards the design of sewers to control sediment problems. *Proc. Instn. Civ. Engrs. Wat., Marit. and Energy*, 136, 9-19 March, Paper 11606.

Artières, O. (1987) *Les dépôts en réseau d'assainissement: origine, caractéristiques, pollution, transport.* Thèse de doctorat, Université Louis Pasteur, Strasbourg, 214pp.

Ashley, R.M., Wotherspoon, D.J.J., Goodison, M.J., McGregor, I. and Coghlan, B.P. (1992) The deposition and erosion of sediments in sewers. *Wat. Sci. Tech.*, 26(5–6), 1283–1293.

Ashley, R.M., Longair, I.M., Wotherspoon, D.J.J. and Kirby, K. (1993) Flow and sediment movement monitoring in large sewers. *Wat. Sci. Tech.*, 28(11–12), 55–62.

Ashley, R.M., Arthur, S., Coghlan, B.P. and McGregor, I. (1994) Fluid sediment in combined sewers. *Wat. Sci. Tech.*, 29(1–2), 113–123.

Ashley, R.M., Watson, N. and McGregor, I. (1995) Resurrecting silt traps in combined sewers for solids control. *Proceedings of Novatech 95*, Lyon, France, 30 May–1 June, 139–146.

Ashley, R.M., Tait, S.J., Stovin, V.R., Burrows, R., Fraser, A., Buxton, A.P., Blackwood, D., Saul, A.J. and Blanksby, J.R. (2003) The utilisation of engineered invert traps in the management of near bed solids in sewer networks. *Wat. Sci. Tech.*, 47(4), 137–148.

ATV (Abwassertechnische Vereinigung) (1998) *A110 Richtlinien für die hydraulische Dimensionierung und den Leistungsnachweis von Abwässerkanälen und -leitungen.*

ATV-DVWK-Regelwerk (2001) *Hydraulische Dimensionierung und Leistungsnachweis von Abwässerkanälen und -leitungen, Arbeitsblatt*, ATV-DVWK-A 110. ISBN 3-935669-22-4.

Bachoc, A. (1992) *Le transfert des solides dans les réseaux d'assainissement unitaires.* PhD thesis, Institut National Polytechnique de Toulouse, France, Novembre, 292 pp.

Bagnold, R.A. (1954) *The physics of wind blown sand and desert dunes.* Methuen.

Baladès, A. and Bertrand-Krajewski, J.-L. (1995) Réflexions préliminaires pour le choix et l'instrumentation de deux bassins versants expérimentaux. CTIA Lyonnaise des Eaux, rapport de recherche GARIH, Bordeaux, France, August 1995, 49 pp.

Bechmann, G.-E. (1900) *Notice sur le service des eaux et de l'assainissement de Paris.*: Ch. Béranger, Paris, France, 521 pp.

Berlamont, J. and Torfs, H. (1996) Modelling (partly) cohesive sediment transport in sewer systems. *Wat. Sci. Tech.*, 33(9), 171-178.

Bertrand-Krajewski, J.-L. (2003) Sewer sediment management: some historical aspects of egg-shape sewers and flushing tanks. *Wat. Sci. Tech.*, 47(4), 109–122.

Bertrand-Krajewski, J.L., Briat, P. and Scrivener, O. (1993) Sewer sediment production and transport modelling: a literature review. *J. Hydr. Research*, 31(4), 435–460.

Bertrand-Krajewski, J.-L., Madiec, H. and Moine, O. (1996a) Two experimental sediment traps: operation and solids characteristics. *Wat. Sci. Tech.*, 33(9), 155–162.

Bertrand-Krajewski, J.-L., Madiec, H., Moine, O., Heneau, T., Tougne, P. and Schaal, C. (1996b) Assessment of experimental bed load sediment traps to replace usual grit chambers in sewer systems. *Proceedings of 7th International Conference on Urban Storm Drainage*, Hannover, Germany, 9–13 Sept. 1993, Vol 2, 737–742.

Bertrand-Krajewski, J.-L., El-Jawhari, A. and Chebbo, G. (2000) *Dimensionnement des ouvrages de traitement des rejets urbains de temps de pluie. Villeurbanne (France).* INSA de Lyon – Laboratoire URGC Hydrologie Urbaine et Marne-la-Vallée (France): Cereve, rapport de recherche pour l'Agence de l'Eau Seine-Normandie (commande no. 99-97044), Septembre, 108 pp.

Blumberg, D. and Bauer, W. (1984) Beseitigung von Ablagerungen in Abwasserkanälen grosser Durchmesser. *Korrespondenz Abwasser*, 31. Jahrgang, 12/84, 1063–1072.

Boniface, S. (1996) *Pièges à charriage en réseau d'assainissement.*: ENGEES, Rapport de Fin d'Etudes, Strasbourg, France, Novembre, 134 pp. + annexes.

Boniface, S., Daras, S., Martaud, M. and Santal, O. (1998) Les insuffisances des chambres de dessablement dans la lutte contre la pollution d'origine pluviale. *TSM*, 93(3), 50–56.

Brombach, H., Michelbach, S. and Wöhrle, C. (1992) Sedimentations- und Remobilisierungsvorgange im Abwasserkanal, Schlubericht des Teilprojektes 3 im BMFT-Verbundprojekt NIEDERSCHLAG, Eigenverlag Umwelt- und Fluid-Technik GmbH, Bad Mergentheim.

Butler, D. and Clark, P. (1995) *Sediment Management in Urban Drainage Catchments.* Construction Industry Research and Information Association (CIRIA), Report No. 134, London, UK, 93 pp. ISBN 0869174093.

Butler, D., May, R. and Ackers, J. (2003) Self-cleansing sewer design based on sediment transport principles. *Journal of Hydraulic Engineering*, ASCE, April, 276–282.

Buxton, A., Tait, S., Stovin, V. and Saul, A.J. (2001) The performance of engineered invert traps in the management of sediments in combined sewer systems. *Proceedings of Novatech 2001*, Lyon, France, 25–27 June, 989–996.

Charton, S. (1986) Etude sur modèle réduit d'un ouvrage de dessablement. *TSM*, 81(7–8), 367–376.

Chebbo, G., Laplace, D., Bachoc, A, Sanchez, Y. and LeGuennec, B. (1996) Technical solutions envisaged in managing solids in combined sewer networks. *Wat. Sci. Tech.*, 33(9), 237–244.

CIRIA (1986) *Sediment movement in combined sewerage and storm-water drainage systems. Phase 1 – Project report.* CIRIA research project No. 366.

CISE (1997) *Etude du lavage des sables issus des réseaux d'assainissement.* CISE report for the Seine Normandie Water Agency, Paris, France.

Conradin, F. (1989) Study on catch basins in Switzerland. *Proceedings of URRC ASCE Conference 'Urban stormwater quality enhancement – Source control, retrofitting and combined sewer technology'*, Davos, Switzerland, 22–27 October 1989, 199–214.

Dartus, D. and Alquier, M. (1986) A conceptual model optimizing the maintenance of grit chambers in combined sewers. *Urban Runoff Pollution*, edited by H.C. Torno, J. Marsalek and M. Desbordes, Springer Verlag, Berlin, Germany, NATO ASI Series, G10, 725–737.

Dartus, D., Paitry, A., Leblic, J. and Bachoc, A. (1990) Filling-up of grit chambers. *Proceedings of 5th International Conference on Urban Storm Drainage*, Osaka, Japan, 23–27 July, 903–908.

Delattre, C. and Bertrand-Krajewski, J.-L. (1996) *Caractérisation et traitabilité des sous-produits solides de l'assainissement.*: GARIH – CTIA Lyonnaise des Eaux, research report, Bordeaux, France, November, 130 pp.

Delattre, C., Bertrand-Krajewski, J.-L., Picard, V., Berga, P. and Baladès, J.-D. (1998) Characteristics and treatability of deposits from urban catchments, sewer systems and storage tanks. *Proceedings of Novatech 98*, Lyon, France, 4–6 May, 1, 559–566.

de Prato, K. (1999) *Etude du piège à charriage.* Marseille (France): SERAM, rapport de stage, June 1999, 49 pp.

Dinkelacker, A. (1992) Cleaning of sewers. *Wat. Sci. Tech.*, 25(8), 37–46.

Dyer, K.R. (1986) *Coastal and Estuarine Dynamics.* Wiley, Chichester.

European Standard EN 752-4 (1997) *Drain and sewer systems outside buildings: Part 4. Hydraulic design and environmental considerations.* CEN (European Committee for Standardisation), Brussels, Belgium, September.

Fan, C.-Y., Field, R., Pisano, W.C., Barsanti, J., Joyce, J.J. and Sorensen, H. (2001) Sewer and tank flushing for sediment, corrosion and pollution control. *ASCE J. Water Resources Planning and Management*, May/June, 194–201.

Fan, C.-Y., Field, R. and Lai, F.-H. (2003) Sewer sediment control: overview of an environmental protection agency wet weather flow research programme. *J. Hydraulic Engineering*, ASCE, 129(4), 253–259.

FMC Corporation Central Engineering Laboratory (1966) *Feasibility of a Period Flushing System for Combined Sewer Cleansing*, EPA/11020DN08/67, USEPA, Washington D.C.

FMC Corporation (1972) *A Flushing System for Combined Sewer Cleansing*, EPA/11020DN03/72, USEPA, Washington D.C.

Fraser, A.G., Ashley, R.M., Sutherland, M. and Vollertsen, J. (1998) Sewer solids management using invert traps. *Wat. Sci. Tech.*, 37(1), 139–146.

Fraser, A.G., Ashley, R.M., Sakrabani, R. and Johnstone, F. (2001) Managing sewer solids for the reduction of foul flush effects. *Wat. Sci. Tech.*, 45(3), 265–272.

Garde, R.J., Ranga Raju, K.G. and Sujudi, A.W.R. (1990) Design of settling basins. *J. Hydr. Res.*, 28(1), 81–91.

Gendreau, N., Le Guennec, B., Poinot-Chazal, J.-P. and Sechet, P. (1993) Sediment motion under flood wave: use of a partial release of stored water by a weir. *Proceedings of 6th International Conference on Urban Storm Drainage*, Niagara Falls, Canada, 12–17 September, 760–765.

Gérard, C. (1999) *Aide au diagnostic des réseaux d'assainissement: analyse et modélisation des relations entre la structure du réseau et les risques d'envasement.* PhD thesis, INSA de Lyon, Villeurbanne, France, Mars, 379 pp.

Gousailles, M. (1994) Devenir des produits extraits des réseaux d'assainissement et des ouvrages de prétraitement. *TSM*, 12, 704–710.

Grande, N. and Novak, G. (1995) Cost analysis of different methods of cleaning CSO and waste water equalisation tanks. *WEFTEC'95, 68th Annual Conf. and Exposition*, Miami Beach, Florida, USA, October 1995.

Grande, N. *et al.* (1995) New innovation for cleaning sediments in sewers and in storage tanks. *WEF Speciality Conference – Sewers of the Future*, Houston, Texas.

Hamparian, S., Presle, C. and Sikora, B. (1993) *Etude du vidage des bassins de dessablement durant les orages.* Communauté Urbaine de Lyon, Direction de l'Eau et de l'Assainissement, Rapport de stage, Lyon, France, 24 pp.

Hydraulic Models (1995) *Etude hydraulique de la station des Quatre Ecluses à Coudekerque.* Lyonnaise des Eaux, Rapport d'étude Hydraulic Models, Bordeaux, France, 36 pp.

IRH Environment (1995) *Ville d'Orléans: étude de faisabilité du traitement des sables d'assainissement et de voirie.* IRH Environnement, rapport, Beaucouze, France, December 1995, 86 pp.

Kaufman, H.L. and Lai, D. (1978) *Conventional and Advanced Sewerage Design Concepts for Dual Flood and Pollution Control.* EPA/600/2-78-090, USEPA, Washington D.C.

Lainé, S., Phan, L., Malabat, D. and Duffros, B. (1998) Flush cleaning of sewers using the Hydrass valve. *Proceedings of UDM '98 Developments in urban drainage modelling*, London, UK, 21-24 September, 1, 391–398.

Laplace, D. (1991) *Dynamique du dépôt en collecteur d'assainissement.* PhD thesis, Institut National Polytechnique/ENSHEEIT, Toulouse, France, 202 pp + annexes.

Laplace, D., Bachoc, A. and Sanchez, Y. (1993) Solutions techniques pour gérer les dépôts en collecteurs visitables. *TSM*, 88(10), 519–523.

Laplace, D. and Felouzis, L. (1997) Le piège à charriage: une technique innovante pour l'entretien des réseaux d'assainissement. *TSM*, 93(9), 83–88.

Laplace, D., Bertrand-Krajewski, J.-L., Chebbo, G. and Felouzis, L. (1998) Les pièges à charriage: de la théorie à la pratique. *Proceedings of Novatech 95*, Lyon, France, 4–6 May 1998, Vol. 2, 329-336.

Laplace, D., de Prato, K. and Felouzis, L. (2001) Modélisation globale du remplissage d'un piège à charriage. *Proceedings of Novatech 2001*, Lyon, France, 25–27 June, 981–987.

Lassus, C. and Salomé, C. (1993) *Centre de traitement des boues de curage de la ville d'Amsterdam: Compte rendu de visite du 8 mars 1993.* Agence de l'Eau Seine-Normandie, Nanterre, France, March 1993, 22 pp.

Lin, H.S. (1993) *Le transport solide en collecteur unitaire d'assainissement et sa modélisation.* PhD thesis, Ecole Nationale des Ponts et Chaussées, Paris, France, 281 pp + annexes.

Lin, H.S., Le Guennec, B., Dartus, D. and Bachoc, A. (1993) Development of a specific method for the bed-load measurement in sewer trunks. *Proceedings of 6th International Conference on Urban Storm Drainage*, Niagara Falls, Canada, 12–17 September, 766–777.

Lindholm, O.G. (1984) General discussion on papers in Session B2:1 'Pollutant loads from combined sewer systems'. *Proc. 3rd Int. Conf. Urban Storm Drainage*, Gothenburg, Sweden, June, Vol. 4, 1602.

Lorenzen, A., Willms, M. and Dinkelacker, A. (1992) The Göttingen Boat. *Wat. Sci. Tech.*, 25(8), 57-62.

Lorenzen, A., Ristenpart, E. and Pfuhl, W. (1997) Reinigung von Abwasserkanälen durch Schwallspülung: Teil 1. Feldversuche in einem Mischwasserkanal. *Korrespondenz Abwasser*, 44(11), 1994–2002.

Lorenzen, A., Ristenpart, E. and Pfuhl, W. (1996) Flush cleaning of sewers. *Wat. Sci. Tech.*, 33(9), 221–228.

Lorenzen, A. and Laplace, D. (1998) Reinigung von Abwasserkanälen durch Schwallspülung: Teil 2. Werkzeuge für die Schwallreinigung. *Korrespondenz Abwasser*, 45(1), 39–46.

Lucas, E., Sermanson, A., Baladès, J.-D., Berga, P., Bertrand-Krajewski, J.-L. and Bourgogne, P. (1999) Les rejets solides et les réseaux d'assainissement. *TSM*, 94(5), 18–27.

Lucas, E. and Bourgogne, P. (2000) *Présentation de la démarche de la Communauté Urbaine de Bordeaux.* GRAIE, actes des rendez-vous du GRAIE, La gestion des produits de curage de réseaux et de bassins, Lyon, France, 12 May, 23 pp.

Lysne, D.K. (1969) Hydraulic design of self-cleaning sewage tunnels. *Journal of Sanitary Engineering Division of ASCE*, Vol. 95.

Macke, E. (1982) *About Sedimentation at Low Concentrations in Partly Filled Pipes.* Mitteilungen, Leichtweiss – Institut für Wasserbau der Technischen Universität Braunschweig, Heft 76.

Ma, Tianlin (2003) A steampower approach for preventing accumulation of solids and associated pollutants for newly designed sewers. PhD thesis, Université Libre de Bruxelles.

May, R.W.P. (1993) *Sediment Transport in Pipes and Sewers with Deposited Beds.* Hydraulics Research Report SR 320.

May, R.W.P. (1996) Development of design methodology for self-cleansing sewers. *Proc. Int. Conf. on Sewer Solids - Characteristics, Movement, Effects and Control*, 5–8 September 1995.

Melbourne and Metropolitan Board of Works (1989) *Hydrogen sulphide control manual – septicity, corrosion and odour control in sewerage systems.* Techological Standing Committee on Hydrogen Sulphide Corrosion in Sewerage Works, Vols 1 and 2.

Minister of Interior (1977) *Instruction technique relative aux réseaux d'assainissement des agglomérations.* Ministère de l'Intérieur, circulaire interministérielle, IT 77284 INT, Paris, France, 62 pp + annexes.

Nalluri, C. and Ab Ghani, A. (1993) Sediment transport over deposited beds in sewers. *Proc. 6th Int. Conf. on Urban Storm Drainage*, Niagara Falls, Ontario, Canada, September, 625–630.

Nalluri, C. and Alverez, E.M. (1992) The influence of cohesion on sediment behaviour. *Wat. Sci. Tech.*, 25(8), 151–164.

Newitt, D.M., Richardson, J.F., Abbott, M. and Turtle, R.B. (1955) Hydraulic conveying of solids in horizontal pipes. *Trans. Inst. Chem. Engrs.*, 33, 93–113.

Ogden, N.H. (1898) Flushing in sewer pipes. *Transactions of the American Society of Civil Engineers*, 40, 1-30.

Paitry, A. (1983) *Etude du fonctionnement des chambres de dessablement sur le réseau unitaire.* Direction Départementale de l'Equipement de Seine-Saint-Denis, Arrondissement Opérationnel No. 3, rapport d'étude, Rosny-sous-Bois, France, 52 pp + annexes.

Paitry, A. and Maurin, J.-M. (1989) Les chambres de dessablement: un patrimoine, une gestion. *TSM*, 84(11), 585–590.

Paitry, A., Bertilotti, R. and Kovacs, Y. (1990) Computer database for the management of grit chambers. *Proceedings of 5th International Conference on Urban Storm Drainage*, Osaka, Japan, 23–27 July, 915–920.

Parente, M., Stevens, K. and Eiker, C. (1995) Evaluation of new technology in the flushing of detention facilities. *WEF 68th Annual Conference*, Miami, Florida.

Picard, V. (1997) *Traitement des sous-produits solides de l'assainissement: optimisation sous contraintes technico-économiques appliquée à la Communauté Urbaine de Bordeaux.* Université Montesquieu Bordeaux IV, mémoire de DESS Aménagement du territoire et économie du développement local, September, Bordeaux, France, 52 pp + annexes.

Pisano, W.C. *et al.* (1979) *Dry Weather Deposition and Flushing for CSO Pollution Control.* EPA/600/2-79-133, USEPA, Washington D.C.

Pisano, W.C., Barsanti, J., Joyce, J. and Sorensen, H. (1998) *Sewer and tank sediment flushing: Case studies.* EPA/600/R-98/157; NTIS PB 99-127839INZ. USEPA, Cincinnati.

Pisano, W.C., White, S.L., O'Riordian, O., Ayotte, F.J., Barsanti, J.R. and Carr D.L. (2003) Automated sewer and drainage flushing systems in Cambridge Massachusetts. *Journal of Hydraulic Engineering*, ASCE, 129(4), 260–266.

Ristenpart, E. (1996) Modelling of sediment erosion in a combined sewer. *Proc. 7th Int. Conf. on Urban Storm Drainage*, Hannover, Germany, September.

Ruscassier, G., Chebbo, G., Saget, A. and Bertrand-Krajewski, J.-L. (1998) New design method of treatment devices: storage tanks and on-stream treatments. *Proceedings of Novatech 98*, Lyon, France, 4–6 May, 1, 451–459.

Sanchez, Y. (1987) Curage des ouvrages visitables des réseaux d'assainissement: synthèse des principales méthodes appliquées en France. *TSM*, 82(2), 91–97.

Scandiaconsult (1974) *Synopsis of research programme for self-cleansing sewers*. Orrje, Stockholm.

Schilling, W. and Kollatsch, D.T. (1990) Reduction of combined sewer overflow pollution loads by detention of sanitary sewage. *Wat. Sci. Tech.*, 10/11, 205–212.

Schmitt, F., Milisic, V., Bertrand-Krajewski, J.-L., Laplace, D. and Chebbo, G. (1999) Numerical modelling of bed load sediment traps in sewer systems by density currents. *Wat. Sci. Tech.*, 39(9), 153–160.

Shoshany, M. (2000) *ELSA ensemble de lavage des sables*. GRAIE, actes des rendez-vous du GRAIE La gestion des produits de curage de réseaux et de bassins, Lyon, France, 12 May, 9 pp.

Skipworth, P.J., Tait, S.J. and Saul, A.J. (1999) Erosion of sediment beds in sewers: model development. *J. Environ. Eng.*, 125-6, 566–573.

STU (1982) *Quelques ouvrages annexes aux réseaux d'assainissement: déversoirs d'orage, chambres de dessablement, siphons.* Ministère de l'Urbanisme et du Logement, Service Technique de l'Urbanisme, Paris, France, October, 77 pp.

Sykora, B. (1989) *Vanne cyclique autocurante à décantation.* INPI, demande de brevet d'invention no 2643971, Paris, France.

UK Environmental Protection Act (1990) see http://www.hmso.gov.uk/acts/acts1990/.

USEPA (1974) *Process design manual for sulphide control in sanitary sewerage systems.* USEPA, Technology transfer, Washington, D.C.

USEPA (1999) *Collection Systems O & M Fact Sheet. Sewer Cleaning and Inspection*. EPA 832-F-99-031, 11 pp.

Verbanck, M., Ashley, R.M. and Bachoc, A. (1994) International workshop on origin, occurrence, and behaviour of sediments in sewer systems: summary of conclusions. *Water Research*, 28(1), 187-194.

Wotherspoon, D.J.J. (1994) *The movement of cohesive sediment in a large combined sewer*. PhD thesis, University of Abertay, Dundee.

Yao, K.M., Sewer line design based on critical shear stress. *Journal Environmental Engineering Division of ASCE*, Vol. 100, 1974.

4.3 CONTROL OF SEDIMENTS IN COMBINED SEWER OVERFLOWS AND TANKS

4.3.1 The function of CSOs

In Europe, the Urban Wastewater Treatment Directive (1991) acknowledges the need for combined sewer overflows; the primary functions of a CSO may be defined as follows:

1. to act as a hydraulic control on the system and to restrict the continuation flow to the downstream sewer to a prescribed value (setting).
2. to maintain a continuation flow of a constant magnitude (equal to that of the setting) irrespective of the inflow magnitude;
3. to provide a relief overflow to allow the spill (usually to the nearest watercourse) of any inflow having a magnitude greater than that of the setting and to prevent flooding in the upstream catchment;
4. to operate only at the prescribed flow conditions;
5. to be efficient in the retention of solids and associated pollutants for treatment and to create a flow pattern within the chamber that is conducive to the separation and retention of gross and finely suspended solids;
6. to be self-cleansing and maintenance free by avoiding blockage and complication in design.

The European Urban Wastewater Treatment Directive (1991) placed the responsibility on these member States to decide measures to limit the pollution discharged from CSOs to receiving waters. Typically the measures to be considered for CSO chamber design are based on three criteria:

- a dilution ratio between the quality of the sewer flow and that in the receiving stream;
- the capacity of the downstream sewer in relation to the dry weather flow;
- a specified number of overflows per year.

Design practice to meet these criteria vary widely but the present trend is one of integrated pollution control on a catchment-wide scale. For example, in Germany design guidance is given in ATV 128 (1992) whilst in the UK the recommended design procedures are outlined in the Urban Pollution Management Manual (FWR, 1994a, 1998). The US EPA specifies nine minimum controls for the effective operation of CSOs:

1. Proper operation and regular maintenance programmes for the sewer system and CSO outfalls.
2. Maximum use of the collection system for storage.
3. Review and modification of pretreatment to ensure that CSO impacts are minimised.
4. Maximisation of flow to the treatment works for treatment.
5. Elimination of CSOs during dry weather.
6. Control of solid and floatable material in CSOs.
7. Pollution prevention programmes to reduce contaminants in CSOs.
8. Public notification to ensure that the public receives adequate notification of CSO occurrences and CSO impacts.
9. Monitoring to effectively characterise CSO impacts and the efficacy of CSO controls.

Engineers in the USA are guided by a series of EPA reports:

EPA 832-B-95-002, Combined Sewer Overflows – Guidance for Long – Term Control Plan
EPA 832-B-95-003, Combined Sewer Overflows – Guidance for Nine Minimum Controls
EPA 832-B-95-004 Combined Sewer Overflows – Guidance for Screening and Ranking Guidance.

It remains common worldwide practice, however, to design the system and the CSO structure based on a multiple of the dry weather flow passed forward to the treatment works.

4.3.2 Sediment controls in CSOs

With respect to the control of sewer solids and floatables, the retention efficiency performance of the CSO is a critical design parameter to achieve effective solids separation through good engineering design. Several types of chamber are recommended for use in practice; these include stilling ponds with end weirs, high-side weir chambers and vortex, swirl and dynamic separators (FWR 1994b). The principal operation of the stilling pond and side weir chambers is to create tranquil flow conditions which provide zones where the solids may separate with the floating solids rising to the water surface (to be retained within the chamber by a scumboard or baffle), whilst the solids with a fall velocity sink towards the bed and dry weather flow channel of the chamber and onto subsequent treatment. Vortex chambers and dynamic separators rely on the action of centrifugal forces to separate the solids.

Several major systems incorporating large scale hydrodynamic separators have been constructed in Germany, the United States and the UK and performance evaluations of these systems have been widely reported, for example, Andoh (1994), Pisano and Brombach (1993, 1995), Muhs (1995), Hedges *et al.* (1992), Field and O'Connor (1996), Konicek *et al.* (1995), Villeneuve *et al.* (1994), Pisano (1997a) and Saul (1999). These performance evaluations have identified that the hydraulic and solids retention performance of each of these types of devices was similar but absolute comparisons of chamber performance have not been possible due to the wide variation of chamber location and differences in the operational regime and quality of the sewer flow at each location. To address this issue a series of full-scale trials are ongoing in New York into the performance of the latest design of the US EPA Swirl Concentrator, the Fluidsep device and a Storm King hydrodynamic separator (Field and Sullivan, 1999; Moutal and Gaffoglio, 2000).

At a smaller scale, many researchers have carried out laboratory-based studies using synthetic particles to evaluate chamber performance and to develop improved designs of chamber, for example, Sharpe and Kirkbride (1959), Smisson (1967), Field (1974), Halliwell and Saul (1980), Saul and Delo (1982), Balmforth *et al.* (1984), Brombach (1987, 1992) and Saul *et al.* (1993). From these studies performance comparisons between chamber types have been reported but significant differences between model scale and the types of particles used in the tests of each researcher do not allow an accurate and fair comparison of absolute chamber performance. However, in respect of the removal of such gross solids as pant liners, release tapes, condoms and cotton bud sticks, Saul *et al.* (1993) reported on a laboratory performance evaluation of the four commonly found types of CSO structure in the UK, namely the extended stilling pond, the high-side weir chamber, the vortex overflow with peripheral spill and a Storm King hydrodynamic separator. At the design inflow rate for the chambers the retention efficiency for these particles was little better than the ratio of the continuation flow to inflow, i.e. each type of chamber divided the particles in the same ratio as the flow split and effectively no selective separation of particles was observed to occur.

Important work has been completed to assess the performance of full-scale structures under controlled conditions using live sewage, e.g. Balmforth (1990) and Saul (1999). In this latter respect the operation of full-scale chambers has been tested as part of the UKWIR CSO Research Group programme of work, completed at the National CSO Test Facility, Wigan Wastewater Treatment works. This has highlighted that the performance of the updated designs of chamber outlined in FR 0488 (Balmforth *et al.*, 1994), in respect of the retention of total solids, i.e. solids collected in 6 mm mesh sacks, is little, if any better than that of the flow split. A large proportion of the pollutants found in combined sewer flow will therefore be discharged to the receiving watercourse.

To assist in the design process and the understanding of the pollution retention performance in CSO chambers, advances in Computational Fluid Dynamics (CFD) have allowed the accurate prediction of such performance to be better understood, e.g., Harwood and Saul (1999) and Richardson (2000). This work, in association with laboratory and fieldwork studies used for calibration and verification, has shown that it is feasible to predict the comparative performance of different geometry chambers. It has also been possible to study the impact of changes to the geometry of CSO structures. Other recent work that has concentrated on the effects of scale (Hübner and Geiger, 1995; Michelbach and Weiss, 1996; Fenner and Tyack, 1997; Luyckx *et al.*, 1998; Saul, 2000a) has shown that it is not generally possible to predict scale effects in CSO chambers using a single scale law, but that it is feasible to develop hybrid relationships linking two or more of the scale laws to define the performance of each type of chamber.

The spill of particles to the receiving watercourse or coastal water creates a visual aesthetic nuisance often leading to public complaint. Frequently the regulatory requirements and consents to discharge make reference to required standards for aesthetic pollution. For example, in the UK, the Environment Agency AMP2 – Asset Management Plan (National Rivers Authority, 1993) guidelines required that solids separation should be related to the retention of solids greater than either 6 mm in two dimensions or that equivalent to the retention performance of a 10 mm bar screen. Where 6 mm solids separation is required, the spilled effluents should not contain a significant quantity of solid matter greater than 6 mm in two dimensions. It is recognised, however, that site specific criteria should apply and that not all of the spilled flow should be subject to 6 mm separation. The following criteria were laid down to define the flow rate to be used for the hydraulic design of the 6 mm solids separation:

- For time series (long duration sequences) storms:
  At a maximum flow equivalent to 80 % of the flow volume that would be discharged in an annual time series;
- For design storms:
  At a flow rate equal to 50 % of the volume that would be discharged in a 1 in 1 year design event.

In addition, the remaining flow, up to that resulting from a 1 in 5 year storm should also be subject to the 10 mm solids separation standard.

In the USA, the presumptive EPA approach is to adopt primary treatment equivalency as acceptable wet weather treatment, with anything less considered as untreated overflow. This equivalency criterion is the subject of much interpretation and frequently a percent TSS treatment for the complete volume of a design storm event (specified return period and duration) is used to establish performance targets for CSOs. Similarly in Germany the guideline ATV 128 (1992) recommends 85 % settleable solids capture at a maximum load of 10 m/hr.

The retention of the solids within the sewer system is therefore an extremely important aspect of system design. Such retention performance is known to be a function of the hydraulic and the solids separation performance of individual CSO structures and previous research has shown that the retention efficiency of conventional designs of CSO chambers may fall short of the regulatory guidelines. Hence, to reduce the environmental impact of these solids three alternative strategies may be adopted: source control (solids and flow), increases to the size of the CSO chamber or the inclusion of a storage tank or the incorporation of screens within the CSO chamber.

4.3.3 Screening technologies for the capture and retention of CSO solids

Screens are employed at the spill exit from CSOs to reduce the quantity of gross solids released to the receiving watercourse and the term 'screen' in this context is used to describe any type of device that is designed to separate solids in a spilled flow of combined sewage. Figure 4.3.1 shows a screen at an overflow outlet.

Figure 4.3.1. Screen fitted to a CSO overflow outlet (photograph John Blanksby).

The adoption of screening technology in CSO chambers is not a new concept. Krejci and Baer (1990), Brombach and Pisano (1996) and Pisano (1997b) highlighted that screens were effective in the retention of solids. However, the performance of screens and their impact on other aspects of solids and sediment retention are not fully understood. For example, do the solids and sediments change their characteristics as they meet or are passed through the screen and, if so, what is the subsequent change in their impact on the receiving water?

The solids separation performance of different screen installations is influenced by the screen configuration, the local installation arrangements, the flow regime and the solids content of the flow. There are currently several types of screen that are marketed specifically for CSO installations.

4.3.3.1 Sacks

The simplest type of screening is provided by mesh sacks, which may be installed as individual sacks or as a mesh that extends at a rate appropriate to the amount at which the solids are collected. The efficiency of solids retention is high but the sacks require manual replacement, and, in the case of individual sacks, this is likely to be following each individual event which may not be practicable at many CSO locations.

4.3.3.2 Bar and mesh screens

Conventional parallel bar screens are frequently used in CSO chambers. Typically the bars are 10 to 20 mm thick and have apertures of 6 mm, 12 mm and 25 mm between the bars. The bars may be straight or curved and the screens are usually fixed to the wall of the spill weir. Such screens, and in particular the screens with 6 mm bar spacing, are prone to blinding and hence they are used in combination with a mechanism for cleaning, usually in the form of a mechanically driven raking system. The operation of these types of screens was shown by Riding *et al.* (1994) to exhibit a wide range in performance efficiency for the retention of large gross solids and a relatively inefficient performance in the retention of 6 mm solids.

Significant developments have taken place in bar screen technology as reported by Pisano and Brombach (1995), Saul (2000b) and Thompson and Saul (2002). These screens are constructed from post stressed stainless steel plate with continuous 4 mm slots through which the spill flow is discharged. They may be positioned horizontally, for both upward and downward flow directions, or vertically, which is the most usual application for CSO chambers. A series of mechanically driven combs move along the 4 mm slots in the front face of the screen to transfer the retained solids laterally to the downstream end of the device where a scraper device removes the collected solids to a region from where they are transported to the continuation flow for subsequent treatment.

Similarly, there have been significant developments in mesh screen technology. The performance of a mesh screen with 6 mm aperture opening was reported by Krejci and Baer (1990). Large volumes of CSO solids were effectively collected but the screen was prone to rapid blinding and difficulties were experienced in keeping the mesh clean. These findings were confirmed by Saul (2000b) and Thompson and Saul (2002) who tested a number of different types of static screen at the UK National CSO Test Facility. As a consequence cleaning mechanisms have been introduced to maintain the screen clean during operation. These screens often comprise a stainless steel semi-circular mesh screen of aperture size in the range 4–6 mm that is fixed to the spill flow weir of the CSO chamber. The screen is positioned such that the spilled flow falls vertically through the screen thus creating no additional head loss. In general these screens, for example the Romag screen, are continuously cleaned by a motor driven helical screw which sits inside the screen and, as the screw rotates, the collected screenings are transported along the length of the screen to a collection chamber from where they may be periodically retrieved or continuously returned into the continuation flow downstream of the chamber (see Thompson and Saul, 2002).

Rotating mesh drum screens have been developed (Pisano and Brombach, 1995) where the solids are retained on the outer face of the screen whilst the spilled flow is able to pass through the screen and be discharged to the receiving water. As the drum rotates the gross solids are removed from the face of the drum by a scraper, brush or high-pressure water jet for subsequent discharge into the continuation flow. The drum's rotation may be either self powered or, more usually, motorised.

A continuous deflective system (Wong *et al.*, 1996) has been developed to remove gross solids from stormwater discharges. The mechanism by which the unit separates and retains gross solids is by deflecting the inflow and associated pollutants away from the main flow stream into a pollutant separation and containment chamber. The chamber consists of a sump in the lower section and an upper separation section. Gross solids are separated within the chamber by a perforated screen allowing the filtered water to pass through to a volute return section. The fluid and associated solids contained within the separation chamber are kept in continuous motion by the circular flow action generated by the incoming flow. This has the effect of keeping the solids in the containment chamber from blocking the perforated screen. The heavier solids ultimately settle into the containment sump. The filtration system consists of a large pressed perforated screen that acts as a filter screen with an outer volute outlet passage. The perforations in the separation screen are elongated in shape, being 3.8 mm wide by 10 mm long. Smaller mesh screens have been successfully used for handling solids within industrial wastes. Units of 3–4 m in diameter have been constructed handling about 1.1 m^3/s. Current research is investigating continuous underflow withdrawal from the lower section to induce vortex action for settleable solids removal and hence the technology may be useful for CSO applications, although this is not yet proven.

Rotating disc screens (Saul 2000b) have a set of circular discs that are mounted on a vertically held spindle and adjacent sets of discs and spindles are positioned to form a screen along the full length of the spill weir. The spacing between the discs is adjusted to give the required size of aperture through which the spilled flow is discharged. Each spindle is rotated by an electrically driven motor such that the front face of the disc rotates towards the downstream end of the chamber. Collected solids are subsequently transferred across the face of each disc to the downstream end of the chamber from where they are safely transferred into the continuation flow.

Other new types of CSO screens include band screens, for example the Robbins and Myers screen that is mounted vertically on the spill weir and the Longwood upward flow band screen. These screens have a mechanically rotated continuous band of mesh through which the spill flow has to pass. These screens were shown to effectively retain solids with dimension greater than 6 mm in two directions (Saul, 2000b; Thompson and Saul, 2002).

4.3.3.3 Self-powered screens

Screens with self-powered cleaning mechanisms include the Steinhardt brush screen and the Hydro International Jet screens (rotary and linear versions). The Steinhardt screen has a plastic brush with long and flexible bristles which is used to strain the solids from the spilled flow. The bristles pass over a stainless steel bar that is so positioned that the collected solids are flicked into a trash screen holding area from where they are transferred back into the sewer flow downstream of the device when the storm subsides.

The Hydro Jet screen (Andoh *et al.*, 1999; Faram and Andoh, 1999) may be incorporated into the Storm King Hydrodynamic separator (rotary version) or into side weir or end weir chambers (linear version). In this device a siphon is used to periodically backflush a 6 mm mesh screen that is used to retain the solids in the spilled flow. The device has been shown to maintain self-cleansing operation and to effectively remove all solids greater than 6 mm (in two dimensions) from the spilled flow.

The technologies described above and other types of screen are specifically designed to effectively separate gross solids, but with respect to the retention of the finer fractions of suspended solids and associated pollutants, attention has focused on the introduction of storage facilities to retain these solids.

4.3.4 Storage tanks

One methodology to ensure an increased retention of the finely suspended solids and pollutants in solution is to incorporate additional storage into the system. Such storage may be in the form of an enhanced CSO structure or a storage volume in the form of a tank. Tanks may be on-line or off-line, with the latter designs of tank usually incorporating a pumped return feed to the downstream sewer system. Reviews of current practice have been presented by Cant (1991), Roebuck (1989) and Brombach (1989).

Three common philosophies for tank design have emerged:

- tanks that are designed to be self-cleansing;
- tanks that are designed specifically to retain sediments as a settlement tank;
- tanks for the specific retention of the load first flush (see Section 3.4).

In the case of self-cleansing tanks, design criteria for their effective operation have been developed; Saul and Ellis (1990), Ellis (1992), Stovin (1996) and a recent study, Stovin and Saul (2000), reported that, in general, the performance of such tanks in UK practice, in respect of their self-cleansing behaviour, was satisfactory. Figure 4.3.2 shows deposits in a tank following a storm.

Figure 4.3.2. Sediment deposits in a tank (photograph Hans Brombach).

Ruscassier *et al.* (1998), De Cock *et al.* (1998), Kluck (1997), Vaes *et al.* (1999), Lafond *et al.* (1996) and others have examined the design and performance of storage settlement tanks and these have been shown to be effective for the retention of sediments and associated pollution. Current research is attempting to refine understanding on the pollutant retention performance of both on-line and off-line tanks and of the interactions between sediments and water quality pollution, particularly in large tanks with long retention times and/or when the tanks are emptied in downstream wastewater treatment plants for further treatment. In the latter respect better understanding of the interactions associated with the previously described 'in sewer processes' is needed; here the residence time distribution of the sediments and solutes within the system and of the interaction between these processes and any re-aeration are key parameters (see Section 2.5).

In respect of load first flush tanks, Brombach *et al.* (1999) reported on the operation of a large number of devices that have been installed in Germany. These have been shown to perform well in that they reduced the frequency of spill operation, thereby reducing the spill of pollution and solids.

The initial problem is associated with the definition of the volume of the load first flush to be stored and the calculation of tank volume is the topic of much debate and argument (see Section 3.4). However, once the volume of tank has been decided the selection of the geometry of the tank to be adopted is a function of the required mode of operation.

Two general arrangements of storage tank are commonly found in sewer systems; these may be classified as *on-line* and *off-line* storage tanks. On-line tanks form an integral part of the sewer system, consisting of an enlarged flow section that fills when the inflow exceeds the required continuation flow to the downstream sewer. The flow into the downstream sewer is usually controlled using an orifice plate, penstock, throttle pipe or vortex flow regulator. Pumps are rarely used in on-line tanks. Off-line tanks are physically separate from the sewer system, with the flow entering the tank via an overflow device incorporated into a diversion structure. The flow to the downstream sewer is controlled in a similar way to that of the on-line tank. The stored flows are returned to the system either by gravity, where the site allows, or by pumping. Design decisions include the selection of the appropriate volume and geometry of tank and of the appropriate maintenance strategy. Each has an influence on the build-up of sediments in the tank.

4.3.4.1 Volume of tank

Approaches to the sizing of storage facilities have progressed from a simple consideration of the hydraulic capacity of the downstream sewer system, to methods which either account for the temporal variation in storm water quality (in which an attempt is made to retain the worst polluted storm volume), or consider an integrated catchment wide pollution control approach, backed up by sophisticated modelling tools. These diverse approaches have developed over the past four decades and the most recent ones are linked with various models and local specific conditions. As such it is not possible to present standard values but some orders of magnitude may be given:

- tanks based on load first flush concepts and/or on retention of easily settelable solids usually have a specific volume of approx. 20 to 40 m^3/ha (ATV 128, 1992; FNDAE, 1986; Michelbach, 1995);
- tanks based on total suspended solids loads and accounting for stormwater quality variations have greater volumes of approx. 80 to 100 m^3/ha (Ruscassier *et al.*, 1998; Bertrand-Krajewski *et al.* 2000; Bertrand-Krajewski and Chebbo, 2002).

Pollutant removal objectives are different in each case, and should be the key criteria to size locally adapted storage tanks.

The need for catchment-wide integrated pollution control is now widely recognised and current design practice recommends the use of dynamic models to estimate the appropriate size of storage volume. Due regard is taken of the location of the tank and tank volumes are calculated not only on sewer and treatment works performance criteria, but also with consideration for the assimilative capacity and quality objectives of the receiving watercourse.

4.3.4.2 Sediments in tanks

The inclusion of a storage volume as part of the sewer system results in low flow velocities that are conducive to the deposition of sediment within the tank. As a consequence major research initiatives have been directed to the collection and assimilation of data describing the nature and behaviour of sediments found in combined sewers and storage tanks. For example, Crabtree (1988) presented a five category classification of the types of combined sewer sediments (Section 2.2.3.3), for which Type E sediments, fine grained organic and mineral deposits, were specific to storage tanks. These sediments are found to largely comprise sand-sized particles, with a mean wet bulk density of 1460 kg/m^3, and an average organic content of 22 %. Sample variability is broad, and this is attributable to differences in tank design and operation. Deposits of sediments in storage chambers may lead to blockages, surcharging, flooding and premature spill operation. In addition, the ageing of deposited sediments, and their subsequent re-entrainment in later storms, can lead to poor quality continuation flow and overflow spilled discharges, and may also lead to odour and septicity problems (Section 3). It is therefore important that sediment deposition and the build up of sediments in storage tanks are controlled. Their further disposal and reuse is also a significant aspect in an integrated and sustainable approach (Section 4.2.2.5).

4.3.4.3 Design of tanks to control sediments

Two alternative philosophies and design approaches may be considered for the control of sediments in storage tanks. The first is to design the tank to maintain self-cleansing operation, with the creation of a flow pattern in which the flow velocities adjacent to the bed have a sufficient magnitude to erode and transport any deposited sediment, whilst the second approach is to operate the tank as a settlement tank and manage subsequently the settled deposits of sediment. The former designs are biased towards major capital investment with low revenue expenditure, whereas tanks designed to the latter concept have a specific maintenance element and incur an ongoing revenue cost, although the initial capital investment may be comparatively lower.

Three types of storage tank are commonly used in present day practice:

- oversized sewers and large diameter tunnels;
- rectangular plan shaped tanks;
- circular shaft tanks.

Knott and Taylor (1985) reviewed British and European design practice for the geometric configuration of tanks. It was concluded that a large number of chamber types with a range of layout and bed configuration had been constructed and that site constraints, the allowable level of surcharge, the layout of the existing sewer system and the setting were major design factors. In many instances the flair of the individual design engineer had been incorporated into the design of chambers. Almost all UK tanks have no odour control.

In Japan, Yoshimura and Furuyama (1990) observed that large underground tunnels are commonly used in many Japanese cities.

Since the middle of the 1970s, some 13 000 CSO tanks of median size ranging from about 1300 to 25 000 m^3 have been constructed in Germany; a review of the design of these tanks was reported by Brombach and Michelbach (1996). It was shown that the majority of tanks were rectangular or circular in plan. Most are flow-through tanks built according to the ATV 128 guidelines. Of these, some 3 % are covered and there is little odour control. Typically these tanks drain out in less than a day and are well maintained. Many tanks incorporate tipping flushers and flushing gates to rapidly handle the sediments at the end of each storm event. In Switzerland, about 650 tanks have been constructed; approximately 25 % are covered and few have odour control. These tanks utilise cleaning equipment in the form of tipping flushers, flushing gates and mechanically raked screens.

In the USA and Canada, less than 200 tanks have been constructed, but most of these tanks are larger than those in Europe, with an average volume ranging from approximately 6500 to 180 000 m^3. Approximately one-third of tanks are located in industrial areas or open space and are not covered. These open tanks are generally cleaned within one day and few odour problems have been reported. Most of the new designs use tipping flushers or flushing gates. Covered tanks are used in sensitive urban areas.

Design concepts vary widely but there are two variations. If covered, tanks with interior catwalks are designed so that workers can get inside after an event and wash floors with cannons or hoses. This type of tank is termed a covered 'open' tank.

An alternative concept is the 'closed' tank, with the tank sealed (one metre headroom) with underground ventilated personnel galleries with air sealed connections into the tank. In each case the design philosophy is to incorporate the correct type of flushing equipment, with redundancies, so that it is not necessary to enter the tank for the purpose of cleaning; i.e. the tanks are self-cleansing but with a minimum maintenance requirement for the cleaning systems.

Many tanks are located immediately downstream of CSO chambers or they incorporate a relief overflow as part of the tank structure. In tanks which are designed to be self-cleansing (without mechanical cleaning systems) it is preferable that the weir is located at the upstream end of the tank (with the advantage that any heavily polluted concentration first flush that may occur at the start of the storm event may be retained within the tank downstream of the weir), whereas for tanks designed to settle out the sediments the overflow weir should be located at the downstream end of the tank.

The deposition of sediment within a storage tank has been shown to be a function of the chamber layout and internal geometrical configuration, the flow conditions when the tank is full, the flow regime as the tank fills and empties, and the nature of the bed load and suspended sediments in the flow which enters the tank. For example, Michelbach (1995) showed that if sediments with the characteristics shown on Figure 2.2.6 were to enter a tank with a design surface loading of 10 m^3/m^2h, equivalent to a sediment settling velocity of 2.8 mm/s, approximately 80 % of the pollutant load which is connected to the settleable solids will be retained. It should be noticed here that settleable solids correspond to the fraction of solids which settle in an Imhoff cone during 2 hours, and that this fraction may represent between 40 to 85 % of the total suspended solids in stormwater, depending on rainfall, catchment, type of sewer and so on.

The geometric features that influence the volume and location of sediment deposits include the longitudinal gradient, the length to breadth ratio, the gradient of transverse benching and the presence of dry weather flow channel(s) (Saul and Ellis, 1992). The internal bed benching configuration must control the flow within the tank in such a way that either any sediment deposited on the benching during a storm is washed into the DWF channel as flow in the water level in the tank subsides, or any deposited material is scoured from the benching as the DWF channel is overtopped when the tank fills at the time of the subsequent storm. Knott and Taylor (1985) stated that there were no established design methods for achieving the self-cleansing operation of tanks; however they recommended that the longitudinal gradient of tanks should be between 1 in 25 and 1 in 100, and that benching gradients of 1 in 3 should be used for off-line tanks, and between 1 in 4 and 1 in 10 be used for on-line tanks. They also stated the importance of minimising obstructions at invert level as experience shows that deposits may accumulate downstream of access ladders, safety protection, etc.

Roebuck (1989) suggested that the minimum criteria for self-cleansing conditions were a longitudinal gradient of 1 in 100 and a central DWF channel with flow velocities of 1.5 m/s. The recommendations for benching slope were dependent upon tank geometry: 1 in 2 for an off-line tank sewer, 1 in 5 for an on-line tank sewer and 1 in 10 for a storage tank.

Similarly, for on-line tanks, Saul and Ellis (1992) recommended that:

1. The self-cleansing operation requires high bed velocity.
2. The filling process is critical.
3. A DWF channel is essential.
4. A single DWF channel is preferable.
5. The slope of the channel should be as steep as practically possible.
6. Transverse benching towards the DWF channel should be included but the gradient of this benching is not critical.
7. A drop in invert to create supercritical flow in the DWF channel is an advantage.
8. Long narrow tanks are preferable.
9. The tank width should not exceed 4 m.
10. Maintenance criteria are major design considerations.

It follows, therefore, that in situations where the required storage volume is large, it is preferable to use a tank which incorporates a number of dry weather flow channels or consists of a series of compartments, with the flow diverted into successive compartments one at time. In this latter respect, the first on-line tank will operate most frequently, hence this design of system has the advantage that any required maintenance operations may be carried out as duty standby.

The hydraulic performance of a tank is critical to self-cleansing performance and it has been observed that highly complex flow patterns occur in tanks, particularly during time varying flow conditions. The magnitude and volume of the flow rates (continuation flow and spilled flow corresponding to a given inflow) in a tank of known geometry, may be simply estimated by using the continuity equation, particularly in on-line tanks. However, the hydraulic performance of off-line tanks is more complex due to the need for a diversion structure and the requirement for a gravity or pumped return feed from the tank to the sewer system. In pumped systems, the selection of the type and number of pumps and their mode of operation is critical to the hydraulic performance of the tank and it is only feasible to pump the return feed from the storage tank to the downstream sewer at a time when there is available capacity in that part of the system. Sediment deposits are therefore likely to occur in off-line tanks. These sediments need to be managed in an efficient and cost effective way.

4.3.4.4 Management of deposited sediments

The alternative strategy is to design tanks as settlement tanks and to manage the deposited sediments as part of a planned maintenance schedule. This is now the preferred design option in many countries. Cleansing procedures are required to remove the deposited sediments. Cleaning should ideally occur soon after the event which fills the chamber. In the UK, where the design for self-cleansing operation is implemented, it appears common practice for cleaning to occur relatively infrequently or only in response to an identified problem. The inherently unpredictable timing of storm events implies that flexible maintenance programmes and/or automatic cleaning systems are preferable. It would also seem to imply that the development of rigid rules for the frequency of cleaning would be both difficult and highly inappropriate.

In many cases **manual cleaning** is employed either in the form of physical shovelling and brushing or with the aid of small **mechanical scrapers or dozers**. More recently there has been an increase in the use of **high pressure water jets** and sprays to facilitate mechanical cleaning. In tanks, these washdown systems may be triggered automatically to respond at a time immediately prior to the final stages of drawdown or when the tank has recently emptied. Systems include **sparge pipe** nozzle systems connected either to a moveable bridge or to the wall of the structure. Frequently these systems are connected to a clean or rainwater filled collection tank, fitted externally to the structure, with a pumped outlet to create a series of high pressured water jets which are targeted to remove the deposited sediment.

4.3.4.4.1 Tipping buckets/flushers

The tipping bucket or flusher is a cylindrical stainless steel bucket which is positioned adjacent to the upstream wall or roof of the tank, and, if necessary, at suitable locations along the length of the tank. A typical installation is shown in Figure 4.3.3. The buckets are positioned at an elevation above the top water level in the tank and following an event, river water, groundwater, potable water, or sewage is pumped at low rates into the tipping flusher.

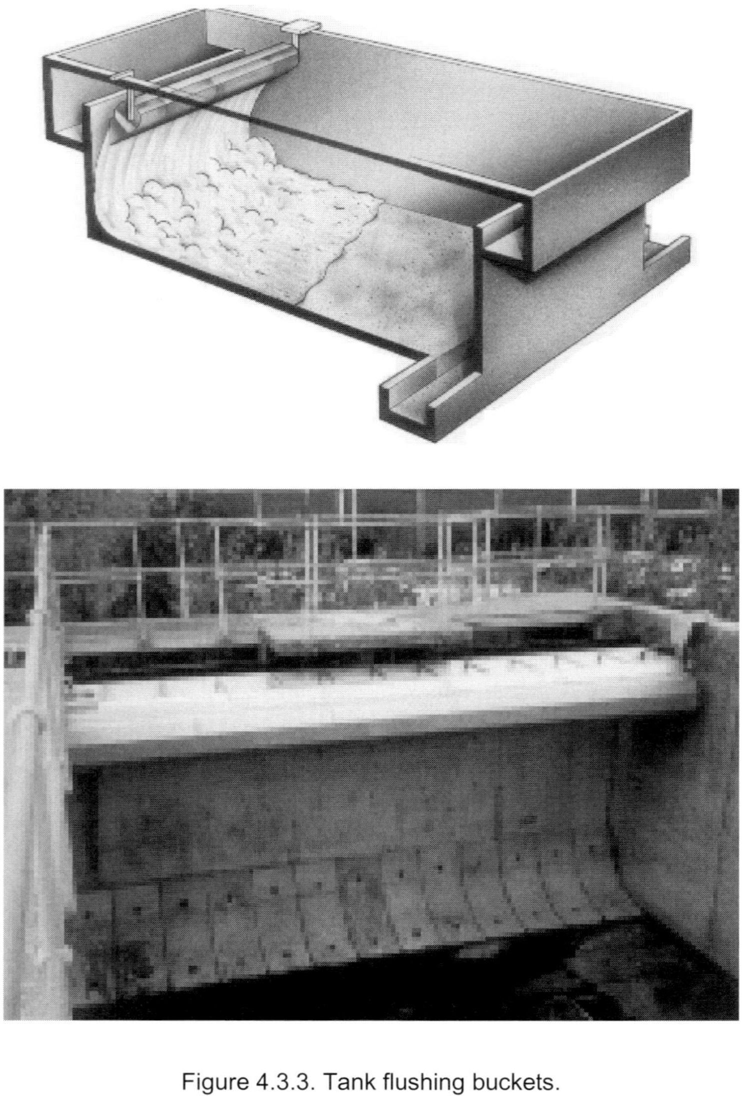

Figure 4.3.3. Tank flushing buckets.

Just prior to overtopping of the tipping flusher by the filling water, the centre of gravity shifts and causes the tipping flusher to rotate about its central axis, thereby spilling its contents along the back wall of the tank. Alternatively, the flusher may be filled following the previous flush and a holding mechanism is released in response to a trigger from a water level or other sensor following the draindown of the subsequent storm. The bucket tips under its own weight and the large volume of water is released to flush the sediment from the bed of the tank. Immediately below the bucket, the tank wall at the intersection of the wall and the invert to the tank is optimally shaped to maximise the velocity and hence the impact of the flush wave to re-suspend and remove the sediments from the benching of the tank. The bed of the tank is sloped to assist in the movement of these sediments which are subsequently transported, by the force of the flush water, to a sump or to the downstream sewer. The amount of water required to clean a tank depends on the distance to be flushed, the height of the tipping flusher above the tank floor, and slope of tank floor in the direction of flushing. There is little advantage to be gained by increasing the effective flushing height above 5.5 m and tipping flusher capacity varies from 150 to 1500 l/m of flusher length.

The maximum effective cleaning span of a low capacity tipping flusher is 10 m and the maximum flushing length is about 50 m. Beyond this distance the flush wave cannot be sustained and tends to break up into rivulets at greater flushing lengths. Experience with flushing tanks in the length range 50–80 m has highlighted that two flushes are often necessary to satisfactorily remove the deposited sediments from the bed of the tank.

Tipping flusher systems require flushing lanes of equal width to that of the flusher, allowing sufficient space for flusher support. In wide tanks in which multiple flushers are used, each flush lane requires a training wall about 0.4 m high along the length of the tank, stopping at the end discharge channel. The tank side walls should be square with the floor to ensure that the wall edges are cleaned. Such systems have been used extensively in Germany, and in North America there are approximately 25 tanks with flushers, most of which are in the State of Michigan.

4.3.4.4.2 Hydroself flushing system

Another recent innovation for cleaning storage tanks is the HYDROSELF system which has been in use in Europe since the 1990s. It is a simple method that uses a washwater storage area and hydraulically operated flap gates to create a cleaning wave to re-suspend tank and sewer sediment. Details of the system are shown in Figure 4.3.4.

This system consists of a hydraulically operated flap gate, a flush water storage area created by the erection of a concrete wall section, a float or pump to supply hydraulic pressure, and valves controlled by either a float system or an electronic control panel. The water level in the sewer or tank is used to activate the release and/or closure of the gate via a permanently sealed, float controlled hydraulic system. The flushing system is designed to operate automatically whenever the in-system water level reaches a predetermined level, thereby releasing the gate and causing a ‘dambreak’ flushing wave to occur. Activation by remote control is also possible. Depending on the physical constraints, the flushing system may include from one to many gates. The units may operate either in a fully automated mode where the receding water level triggers the start of the pre-programmed flushing sequence, or in manual mode where the tanks are flushed at the discretion of an operator. This system has been used extensively in Europe with 132 installations for cleaning sediments from CSO tanks (Grande and Novak, 1995). The first flushing gate system installed in North America was reported by Parente *et al.* (1995) whilst Pisano *et al.* (1998) reported that in an evaluation of eight CSO storage tank installations the tipping flusher and flushing gate technology appeared to be the most cost effective means of flushing solids and debris from tanks.

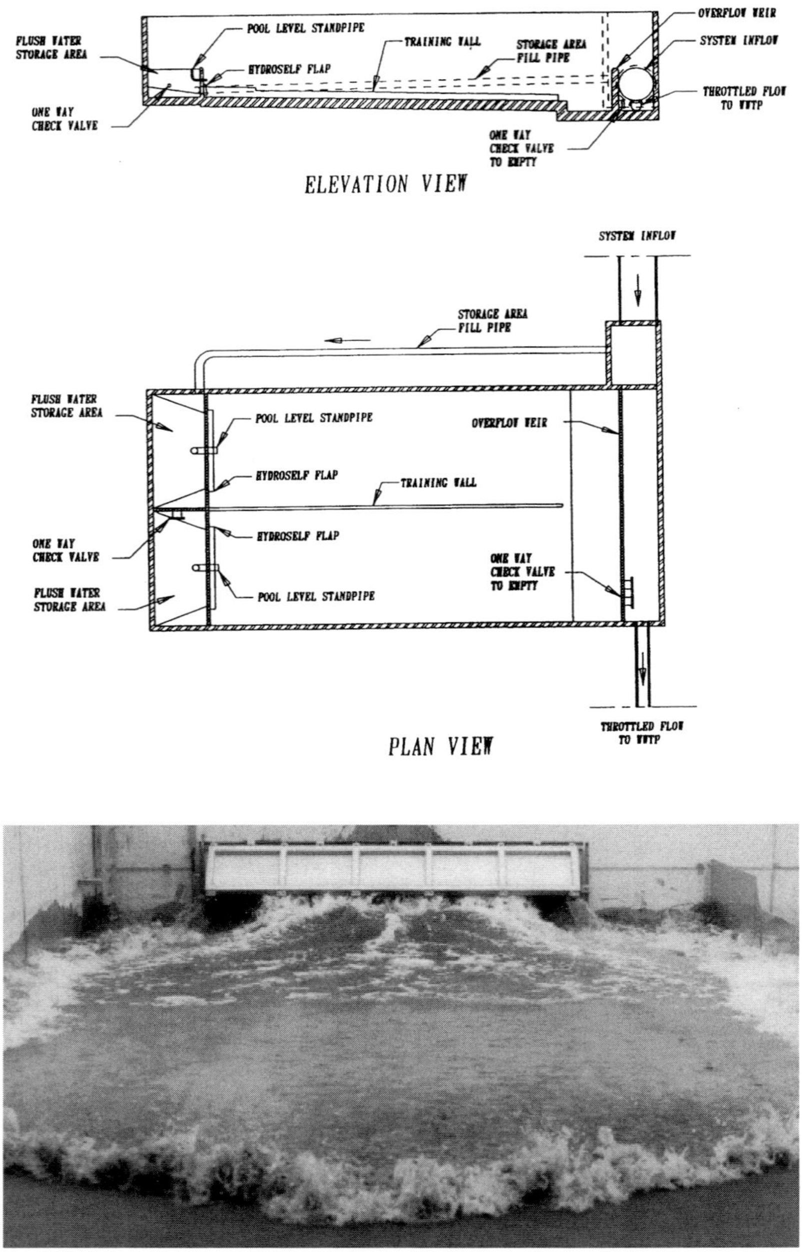

Figure 4.3.4. Hydroself flusher in a tank.

4.3.4.4.3 Prevention of deposition – agitators, etc.

An alternative approach is not to allow the sediments to settle onto the bed of the chamber and to utilise a system that maintains the solids and the sediments in suspension. Systems which may be used for this purpose include mechanical agitators in the form of paddles, propellers, brushes, jets and pumps, or the use of air injection systems. These devices maintain the wastewater stored in the tank in a state of high turbulence and the resultant continuous motion prevents deposition. Little is known about design criteria for these types of system and further research on their application is required.

4.3.4.4.4 Case study

The cleaning of circular tanks at the Lincoln Park facility in Decatur, Illinois was reported by Drake and Hunsinger (1990) and Pisano (1997a). The system consists of two mechanical screening facilities, a 7500 m^3 open circular load first flush storage tank, an 11 m diameter vortex flow dividing chamber (two asymmetric flow inputs of 11 000 l/s and 7200 l/s are divided into four 4550 l/s waste streams), four 13.6 m diameter vortex solids separators (FluidsepR), and a treated effluent to the Sagamon River. Diverted wet-weather flows (WWF) (two inputs) are first passed through mechanically cleaned, automatically controlled, catenary screens. A manually cleaned bar screen is provided for emergency bypass. Downstream of each screen chamber are two liquid-level actuated, motorised sluice gates directing flow into the load first flush retention tank. The tank diameter is 36 m with a side water depth of 8.5 m. The tank is equipped with mixers and aerators and the tank floor slopes to a circular gutter draining to a pumping station. When the tank water level rises to a pre-set level, control gates direct any additional inflows to the vortex flow divider with outputs into the four vortex solids separators. Foul underflows from the bottom of the four separators are pumped into the load first flush tank. The pumping station also dewaters the tank and the separators after an event.

Clean out of the circular load first flush tank includes several novel design and operational concepts. First, the two main gravity feeds to the tank are tangentially fed during an event. Secondary flow currents are established moving most of the solids to the centre zone of the tank. After an event the cleaning operation involves two steps. The contents of the tank are kept in motion while being slowly pumped to the interceptor. This circulation feature is accomplished by a feed from the centre of the tank with a separate recirculation loop with a tangential return. When the tank is fully drained, no sediment is typically observed within 5 m of the perimeter wall after which the sediments grade to a depth of approximately 150 mm in the centre region of the tank. Two high pressure water jets on opposite sides of the tank are then used to cleanse the remaining sediments to the centre well in about 6 minutes. The operation of this system has been successful for several years.

4.3.4.4.5 Impact on sewer system downstream of tank

It is conceivable that a system with a number of sediment control devices which operate following a storm event could result in significant flush of pollution, either in conjunction with normal DWF or, where storage tanks in some parts of the system are widely dispersed, in combination with dilute stored and released stormwater. The impact on the sewer system downstream of the storage tank needs to be assessed both in respect of the solids carrying capacity of the downstream pipe (this must be adequate to take the large concentrations of sediments discharged with the flush wave), and in respect of the time of travel of the pollutant flush to the downstream WWTP. Sewage flush quality can be determined from a knowledge of the amounts and nature of sediment re-entrained. Where tank flushing is designed to occur at the start of successive storms, the re-entrained material will add to the load first flush.

4.3.5 Lamella settling with chemicals

As an alternative or a complement to storage and settling tanks, whose construction in densely urbanised areas is not always easy due to space limitations, more compact lamella settlers, originally used as primary settlement tanks at WWTPs, have been specifically adapted since the beginning of the 1990s to remove suspended solids and associated pollutants from urban wet weather discharges.

In order to reach high removal efficiencies in compact devices, the basic idea consists of using a lamella settler where settling velocities of suspended solids are increased by chemicals (floculant and coagulant), sludge recycling and/or addition of sand.

A coagulating agent (usually ferric chloride, at concentrations typically ranging from 15 to 100 g $FeCl_3/m^3$) is added to the effluent to be treated. The mixing of the effluent and the coagulating agent occurs within a first fully stirred compartment where rapid mixers facilitate the contact of the ferric chloride with the solid particles. In a second plug flow type compartment, a flocculating agent (usually an anionic polymer, at concentrations ranging typically from 1 to 2 g/m^3) is added to increase the size of the flocs. Low-speed mixers are used in this second compartment. The effluent then flows through modules of lamellas where the solids and the flocs settle over the bottom of the tank.

As the settling efficiency increases when solid concentration increases and/or when settling velocities increase, two technical solutions have been developed by manufacturers. In the first, the settled sludge is thickened and recycled in the flocculation compartment (see Figure 4.3.5). This solution is used in both Densadeg® and Delreb® plants, where the concentration of the recycled sludge is approx. 50 to 60 g/l with the Densadeg® (Delporte *et al.*, 1995) and 30 to 40 g/l in the Delreb® (Bridoux *et al.*, 1998). If the sludge concentration increases too much, the excess sludge is extracted for further treatment in a wastewater treatment plant.

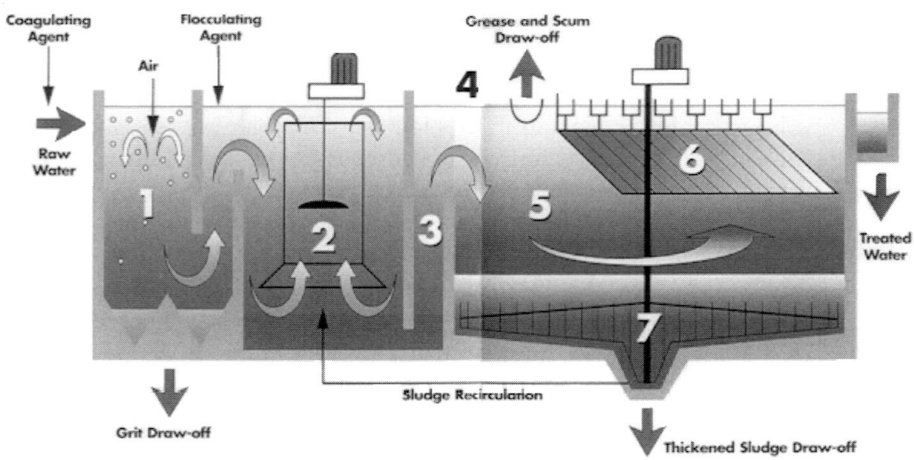

Figure 4.3.5. Operation of the Densadeg 4D (from http://www.enquip.com/CSO-SSO.htm/).

The second solution, used in the Actiflo® plants, is to add sand in the flocculation compartment (the sand mean diameter is 100–150 µm, and its concentration in the plant is 2-4 g/L). The sand is then included in the flocs. The settled solids are extracted without thickening, then enter a hydrocyclone where the sand and the sludge are separated (see Figure 4.3.6). The sand is re-used in the flocculation compartment and the sludge is further treated in a wastewater treatment plant (OTV, 1994; Le Poder and Binot, 1995).

For all of these plants, the concentration of coagulant and the type and concentration of polymer are adjusted after site specific jar-tests. Their injection may be regulated by the flow and/or the turbidity of effluent. The upwards flow velocities through the lamella compartment are very high, and range from 70 to 120 m/h, or even more in some pilot tests.

Various experiments in France have shown that these devices are very efficient in both separate and combined sewer systems, and that mean TSS removal efficiency is ranging from 85 to 91 % (Le Poder and Binot, 1995; Zobrist *et al.*, 1995; Briat and Delporte, 1995; Delporte *et al.*, 1995; Bridoux *et al.*, 1998). As the sludge is to be treated rapidly after the settling, such devices are frequently built at the same place as wastewater treatment plants where the sludge treatment train already exists.

Since storm events occur for typically less than 15 % of the time in any one year, such lamella settlers may be used under two completely different conditions, in order to increase their economic performance:

- during dry weather periods, the lamella settler is used for tertiary phosphorus removal and to reduce the remaining TSS concentration in the effluent;
- during wet weather periods, the flow exceeding the WWTP capacity is by-passed through an overflow structure and directly treated in the lamella settler. The tertiary phosphorus removal is then temporarily stopped.

As an example, this mode of functioning has been adopted in the new Paris, France wastewater treatment plant of Colombes. However, in some locations where the receiving water protection is very important and where no wastewater treatment plant exists, such lamella settlers may be built specifically. This increases the costs and the plant operation complexity, however, especially the sludge removal.

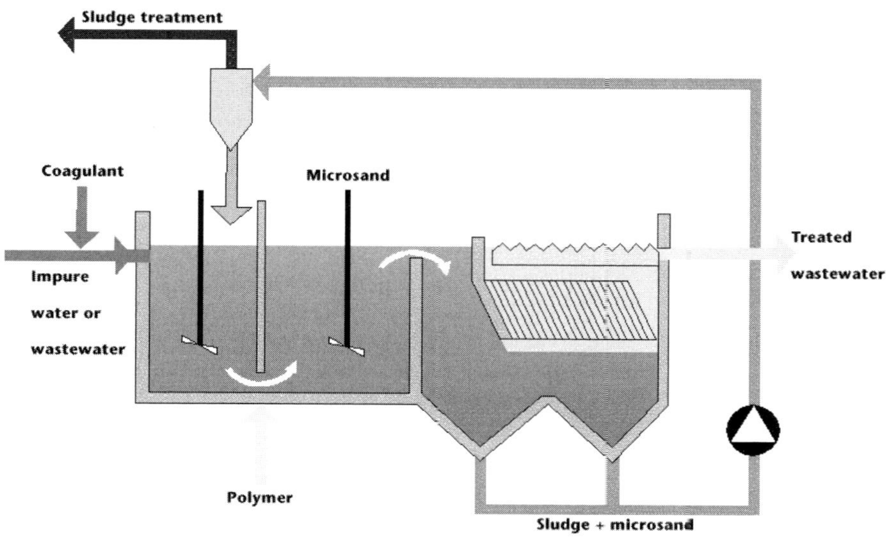

Figure 4.3.6. Layout of the Actiflo (from http://www.kruger-as.com/doc/Products/Actiflo.htm/).

4.3.6 References

Andoh, R.Y.G. (1994) The Stormking™ Overflow Hydrodynamic Separator. *Hydro Research and Development Ltd Seminar: Alleviating the Problems of Combined Sewer Overflows within the Piped System*, Warrington, April 1994.

Andoh, R.Y.G., Smith, B.P. and Saul, A.J. (1999) The Screen Efficiency of a Novel Self Cleansing CSO. *8th Int. Conf. on Urban Storm Drainage*, Sydney.

ATV-Arbeitsblatt A 128 neu (1992) *Richtlinien für die Bemessung und Gestaltung von Regenentlastungsanlagen in Mischwasserkanälen*. Gesellschaft zur Förderung der Abwassertechnik, St. Augustin.

Balmforth, D.J., Lea, S.J. and Sarginson, E.J. (1984) Development of a Vortex Storm Sewage Overflow with Peripheral Spill. *3rd Int. Conf. on Urban Storm Drainage*, Gothenburg, Sweden.

Balmforth, D.J. (1990) The pollution aspects of storm sewage overflows. *Journ. IWEM,* 4(3), 219–226.

Balmforth, D.J., Saul, A.J. and Clifforde, I.T. (1994) *Guide to the Design of Combined Sewer Overflow Structures*. Foundation for Water Research, Report No. FR 0488, Bucks, UK, November 1994, 75 pp.

Bertrand-Krajewski, J.-L., El-Jawhari, A. and Chebbo, G. (2000) *Dimensionnement des ouvrages de traitement des rejets urbains de temps de pluie*. INSA de Lyon – Laboratoire URGC Hydrologie Urbaine et Marne-la-Vallée (France): Cereve, rapport de recherche pour l'Agence de l'Eau Seine-Normandie (commande n° 99-97044), Villeurbanne, France, September 2000, 108 pp.

Bertrand-Krajewski J.-L. and Chebbo, G. (2002) Sizing ratios for stormwater treatment facilities. *Proceedings of 9th International Conference on Urban Drainage*, 8–13 September, Portland, USA, 17 pp.

Briat, P. and Delporte, C. (1995) Le traitement des rejets urbains urbains par temps de pluie au moyen de la décantation lamellaire. *L'Eau, l'Industrie, les Nuisances*, 182, 59–61.

Bridoux, G., Villeroux, A., Riotte, M. and Huau, M.-C. (1998). Optimisation d'un décanteur lamellaire au traitement des eaux pluviales. *Proceedings of Novatech 98*, Lyon, France, 4–6 May, 1, 429-436.

Brombach, H. (1987) Liquid-solid separation at vortex storm overflows. *4th Int. Conf. on Urban Storm Drainage*, Lausanne.

Brombach, H. (1989) Equipment and instrumentation for CSO control. *Engineering Foundation Conference: Urban Stormwater Quality Enhancement-Source Control, Retrofitting and Combined Sewer Technology*, Davos, Switzerland.

Brombach, H. (1992) Solids removal from CSO's with vortex separators. *Novatech 92, International Conference on Innovative Technologies in the Domain of Urban Water Drainage*, Lyon, France, November, 447-459.

Brombach, H. and Pisano, W.C. (1996) Operational experience with CSO sieving treatment. *Proc. 7th Int. Conf. on Urban Storm Drainage*, Hannover, September, 1007–1012.

Brombach, H. and Michelbach, S. (1996) Sewer solids and efficiency of clarifier type CSO tanks. *Proc. 7th Int. Conf. on Urban Storm Drainage*, Hannover, September, 1013–1018.

Brombach, H., Steinriede, D., Weib, G. and Wöhrle, C. (1999) Measured overflow activity of combined sewer overflow tanks. *Proc. 8th Int. Conf. on Urban Storm Drainage*, Sydney, September, 2059–2066.

Cant, J. (1991) *Detention tank design and maintenance*. Water Research centre Report UM 1233, March.

Crabtree, R.W. (1988) *A classification of combined sewer sediment types and characteristics*. WRc Report No. ER324E.

De Cock, W., Blom, P. and Berlamont, J. (1998) The feasibility of flocculation in a storm water sedimentation tank. *Proc. Novatech*, Lyon, France, 379–386.

Delporte, C., Pujol, R. and Vion, P. (1995) Eaux pluviales et décantation lamellaire optimisée. *Proceedings of Novatech 95*, Lyon, France, 30 May–1 June, 233–242.

Drake, J. and Hunsinger, C. (1990) CSO treatment utilizing revolutionary swirl concentrator. *Sanitary District of Decatur, Ill., Central States WPCA Annual Conference*, Wisconsin, USA.

Ellis, D.R. (1992) *Performance of storage tank chambers*. PhD thesis, University of Manchester, UK.

EPA 832-B-95-002 (1995) *Combined Sewer Overflows – Guidance for Long - Term Control Plan*, September, USEPA, Office of Wastewater Management, Washington.

EPA 832-B-95-003 (1995) *Combined Sewer Overflows – Guidance for Nine Minimum Controls*, May, USEPA, Office of Wastewater Management, Washington.

EPA 832-B-95-004 (1995) *Combined Sewer Overflows – Guidance for Screening and Ranking Guidance*, August, USEPA, Office of Wastewater Management, Washington.

Faram, M.G. and Andoh, R.Y.G. (1999) Evaluation and optimisation of a novel self cleaning combined sewer overflow screening system using computational fluid dynamics. *Proc 8th Int. Conf. on Urban Storm Drainage*, Sydney, September.

Fenner, R.A. and Tyack, J.N. (1997) Scaling laws for hydrodynamic separators. *ASCE Journal of Environmental Engineering*, 123(10), October. 1019–1026.

Field, R. (1974) Design of a combined sewer overflow regulator/concentrator. *Journal WPCF*, 46(7), 1722–1741.

Field, R. and O'Connor, T.P. (1996) Swirl technology:enhancement of design, evaluation and application. *ASCE Journal of Environmental Engineering*, 122(8), August, 741–748.

Field, R. and Sullivan, D. (1999) Overview of EPA's wet weather flow research programme. *Proc. 8th Int. Conf. on Urban Storm Drainage*, Sydney, September, 1159?–1605.

FWR (1994a) *Urban Pollution Management Manual*. Foundation for Water Research, Report FR/CL0002, November, 1994.

FWR (1994b) *Guide to the Design of Combined Sewer Overflow Structures*, Foundation for Water Research, Report FR 0488, November.

FWR (1998) *Update of Urban Pollution Management Manual*. Foundation for Water Research.

FNDAE (1986) *Stormwater Tanks in Sewer Systems*. FNDAE, Technical Report 6, Paris, France, 61 pp (in French).

Grande, N. and Novak, G. (1995) Cost analysis of different methods of cleaning CSO and waste water equalisation tanks. *WEFTEC'95, 68th Annual Conf. and Exposition*, Miami Beach, Florida, USA, October.

Halliwell, A.R.H. and Saul, A.J. (1980) The use of hydraulic models to examine the performance of storm sewage overflows. *Proc. ICE*, 69(2), 245–259, June.

Harwood, R. and Saul, A.J. (1999) The influence of particle size on particle retention efficiency performance. *Proc 8th Int. Conf. on Urban Storm Drainage*, Sydney, 1–10.

Hedges, P.D., Lockley, P.E. and Martin, J.R. (1992) A Field Study of an Hydrodynamic Separator CSO. *Novatech 92, Int. Conf. on Innovative Technologies in the Domain of Urban Water Drainage*, Lyon, France, November, 349–358.

Hübner, M. and Geiger, W.F. (1995) Review of hydrodynamic separator efficiencies for practical applications. *Proceedings of Novatech 95*, Lyon, France, 30 May–1 June 1995, 155–162.

Kluck, J. (1997) *Modelling of flow and settling in storm water sedimentation tanks.* PhD thesis, Technical University of Delft, Germany.

Knott, G.E. and Taylor, A.J. (1985) *The use of detention tanks for flow attenuation – a review of current practice*. WRc Report No. 171E, WRc Ltd.

Konicek, Z., Pryl, K. and Suchanek, M. (1995) Practical application of vortex flow separators in the Czech Republic. *Int. Conf. on Sewer Solids – Characteristics, Movement, Effects and Control*, University of Abertay, Dundee, September, 280–289.

Krejci, V. and Baer, E. (1990) Screening structures for CSO treatment in Switzerland. *Proc. 5th Int. Conf. on Urban Storm Drainage*, Osaka, Japan, July.

Lafond, J.M., Herouin, E. and Chebbo, G. (1996) Comparison of suspended solid transport models application to storage-treatment structures. *Proc. 7th Int. Conf. on Urban Storm Drainage*, Hannover, Germany, September.

Le Poder, N. and Binot, P. (1995) Traitement des eaux pluviales par décantation de flocs physico-chimiques lestés par du microsable. *Proceedings of Novatech 95*, Lyon, France, 30 May–1 June 1995, 243–250.

Luyckx, G., Vaes, G. and Berlamont, J (1998) Experimental investigation on the efficiency of a side weir overflow. *Proc Novatech*, Lyon, May

Michelbach, S. (1995) Origin, resuspension and settling characteristics of solids transported in combined sewage. *Wat. Sci. Tech.,* 31(7), 69–76.

Michelbach, S. and Weiss, G. (1996) Vortex separator: dimensionless properties and calculation of annual retention efficiencies. *Wat. Sci. Tech.*, 33(9), 261–267.

Moutal, H. and Gaffoglio, R. (2000) Alternative vortex technology demonstration project. *Proc. IWA 1st World Water Congress*, July.

Muhs, G.P. (1995) Evaluation of two CSO treatment facilities with comparisons between swirl concentrator and vortex separator performance. *WEFTEC'95, 68th Annual Conf. and Exposition*, Miami Beach, Florida, USA, October.

National Rivers Authority (1993) *Guidelines for AMP2*. Periodic Review. Version 2, UK.

OTV (1994) *Dépolluer les eaux pluviales*. Tec et Doc, Paris, France, 349 pp. ISBN 2-87777-417-1.

Parente, M., Stevens, K. and Eiker, C. (1995) Evaluation of new technology in the flushing of detention facilities. *WEF 68th Annual Conference*, Miami, Florida.

Pisano, W.C. (1997a) Evaluation of 416 mgd Lincoln Park Fluidsep vortex solids separator facility. *Decatur, Illinois, New England Water Environment Federation Annual Conference*, Boston, Ma, January.

Pisano, W.C. (1997b) Comparative assessment: vortex separators, rotary sieves and 'combing' screens for CSO floatables control. *WEFTECH '97*.

Pisano, W.C. and Brombach, H. (1993) Operational experience with vortex solids separators for CSO control. *6th Int. Conf. on Urban Storm Drainage*, Niagara Falls, Canada, 2, September, 1651-1656.

Pisano, W.C. and Brombach, H. (1995) New European innovations for CSO control. *WEF New and Emerging Technologies for Collection and Treatment of Wastewater: A global perspective*, Toronto, Canada.

Pisano, W.C., Barsanti, J., Joyce, J. and Sorensen, H. (1998) *Sewer and tank sediment flushing: case studies*. Report EPA/600/R-98/157, December.

Richardson, J. (2000) Virtual hydraulics – 3D modelling for better benefits: CFD for sewer systems and CSOs. *Proc. IMUG*, Prague, April.

Riding, E., Balmforth, D.J. and Green, M. (1994) *Monitoring the performance of screens at CSOs*. Report No. UM1433, Wrc Ltd, April.

Roebuck, I.H. (1989) Practical considerations of the design of detention tanks. *Municipal Engineer*, 6, 231–239.

Ruscassier, G., Chebbo, G., Saget, A. and Bertrand-Krajewski, J.-L. (1998) New design method of treatment devices: storage tanks and on-stream treatments. *Proceedings of Novatech 98*, Lyon, France, 4–6 May, 1, 451–459.

Saul, A.J. (1999) Comparative performance of CSO chambers. *UKWIR Report WW08/99/01*.

Saul, A.J. (2000a) Scale effects in CSO. *UKWIR Report WW08/97/008*.

Saul, A.J. (2000b) Screen performance: proprietary designs. *UKWIR Report WW08/99/08*.

Saul, A.J. and Ellis, D.R. (1990) Storage tank design in sewerage systems. *Proc. 5th Int. Conf. on Urban Storm Drainage*, Osaka, Japan.

Saul, A.J. and Delo, E.A. (1982) Laboratory tests on a storm sewer overflow chamber with unsteady flow. *Proc. 1st Int. Conf. on Urban Storm Drainage*, Southampton, England.

Saul, A.J. and Ellis, D.R. (1992) Sediment deposition in storage tanks. *Wat. Sci. Tech.*, 25(8), 189–198.

Saul, A.J., Ruff, S.J., Walsh, A.M. and Green, M.J. (1993) *Laboratory studies of CSO performance*, WRc Report No. UM 1421, October.

Sharpe, D.E. and Kirkbride, T.W. (1959) Storm water overflows: the operation and design of a stilling pond. *Proc. ICE*, 13.

Smisson, B. (1967) Design, construction and performance assessment of vortex overflows. Paper 8, *Symposium on Storm Sewage Overflows*, Inst. Civ. Engrs.

Stovin, V.R. (1996) *The Prediction of Sediment Deposition in Storage Tanks Based on Laboratory Observation and Numerical Simulations.* PhD thesis, University of Sheffield.

Stovin, V.R. and Saul, A.J. (2000) Computational fluid dynamics and the design of sewage storage chambers. *Journal of CIWEM*, 14, April.

Thompson, B. and Saul, A.J. (2002) *CSO screen efficiency (proprietary designs) 1997–2001*, UKWIR Report 02/WW/08/13.

Urban Waste Water Treatment Directive (1991) *Official Journal of the Council of the European Communities*, 91/271/EEC, Brussels, Belgium, L135, 30 May.

Vaes, G., Luyckx, G., De Cock, W. and Berlamont, J. (1999) A conceptual model for the impact of storage sedimentation basins. *Proc. ICUSD*, Sydney, September, 1415–1422.

Villeneuve, J.P., Gaume, E. and Michaud, F. (1994) Efficiency evaluation of an installed swirl separator. *Canadian Journal of Civ. Eng.*, 21(6) , December, 924–930.

Wong, T.H.F., Wooton, R.M. and Fabian, D. (1996) A solid seperator using a continuous deflective system. *Proc. 7th Int. Conf. on Urban Storm Drainage*, Hannover, Germany, September.

Yoshimura, M. and Furuyama, S. (1990) Improvement of the combined sewer system with stormwater tanks in Yokohama City. *Proc. 5th Int. Conf. on Urban Storm Drainage*, Osaka, Japan, July.

Zobrist, C., Herviou, P., Poujol, T., Herremans, L. and Bernard, C. (1995) Etude sur le traitement des eaux de temps de pluie par décantation lamellaire: le pilote industriel de Massy. *Proceedings of Novatech 95*, Lyon, France, 30 May–1 June, 217–224.

5

Recent developments and future perspectives related to sewer solids

It is clear from this Scientific and Technical report that since the end of the 1980s, scientific knowledge about the nature, behaviour, effects and management of sewer solids has advanced significantly. From early studies that concentrated largely on the determination of the nature of the particles, together with limited characteristic chemical properties, research has now advanced to a more comprehensive and integrated perspective that includes a range of processes involving reactions between the solids and the aqueous and dissolved phases in the wastewater as well as movement related phenomena. For example, separate consideration of the sewer states in which aerobic and anaerobic processes prevail has not been adequate to properly understand how systems perform when oxygen may become limited, as for storage provision. Hence new models based on the better conceptual understanding of physical and biological processes have recently been developed to account more effectively for the wider range of process states that may occur in a sewer. These models have been shown to be effective at designing and operating major sewers, such as the 50 km Emscher intercepting sewer system in Germany (Bjerre *et al.*, 1998) and the 25 km Estoril sewer in Portugal (Hvitved-Jacobsen *et al.*, 2002). Nonetheless, the majority of reported in-sewer studies produce results that relate only to the site or immediate vicinity within which the work has been carried out. Figure 5.1 illustrates sediment in the main Dundee Interceptor sewer. Much of the early work on characterising sediments was undertaken in a 200 m length of this sewer. It is still not clear how much of this work may be used to define what happens elsewhere.

Figure 5.1. Sediments on the inside of a bend in the Dundee Interceptor sewer (photograph Richard Ashley).

Figure 5.2. Gross solids collected on screen at outlet from combined sewer overflow (photograph Adrian Saul).

In many studies there is widespread concern about the implication of sewer solids as a major contributor to the pollution that may be caused by discharges from combined sewer overflow spills. Early studies in Germany (Dauber and Novak, 1983), the UK (Rees and

White, 1993) and more recently at different catchment scales in France (Gromaire *et al.*, 2001), have shown conclusively that sewer solids are major contributors to the pollutants transported in suspension and thence available to be spilled from CSOs during storm flows. Moreover, the major contributor is the erosion of previously deposited in-sewer solids. The fact that CSOs and associated chambers, where installed, have negligible effect on separation of these solids into the flow passed to treatment has only been acknowledged in the past decade (Chapter 3). Figure 5.2 illustrates some of the more obvious solids that can collect on outlets from CSOs as a result of ignoring the warnings illustrated in Figure 5.3.

The increasing utilisation of in-sewer storage to reduce combined sewer overflow spills has consequences for the retention of more solids within systems, and also increases retention times. The first effect will result in more in-sewer sedimentation and increased effects due to the larger gross solids. There will be an increased potential for downstream sewer blockage or reduction in flow capacity and also more possibility for biodegradation and the onset of anaerobic conditions. Subsequent storms may washout the elevated stored solids overloading the screen and grit processes at WwTW inlets. Alternatively, where the deposition in the sewer downstream of the overflow is significant, the reduction in flow capacity may cause the upstream overflow to spill sooner than envisaged. Figure 5.4 illustrates how a screen at a downstream treatment plant can become damaged due to just one type of solid.

Nonetheless, many operators still choose to ignore semi-permanent in-sewer deposits, believing they do not cause operational problems. To be fair, researchers tend not to publish in the journals (or other outlets) most read by sewer operators. The gap between the knowledge generators and the users cannot be easily narrowed because of the needs and actions of the separate groups. Researchers should make more effort to disseminate their knowledge in a more practical form that can be more directly used by operators. This Scientific and Technical Report is a first step towards this goal, which aims to provide a complete picture of the field of sewer solids.

Despite the lack of concern of many operators about the effect of semi-permanent in-sewer deposits, these deposits do reduce flow capacity, by both a reduction in cross-sectional area and by increasing the boundary drag (Chapter 3). Historical knowledge about the build-up of deposits in particular locations in many older sewer systems shows that these often recur within a very short timescale following cleaning, and are usually due to poor system design or construction. Techniques and knowledge in the past may not have been adequate to prevent this. Hence the acceptance by operators that these deposits have to be tolerated, and do not necessarily lead to problems, is understandable. However, there are already changes in stakeholder awareness of sewer flooding in European cities and a decrease in its acceptability. Recent major floods in Europe and likely future global climate changes have also raised stakeholder concerns about increases in precipitation. It is likely that rainfall patterns may alter within decades and hence historically based empirical predictions of system behaviour, such as recurrent depositional areas in sewers, may no longer be reliable. These factors will require operators to take a more proactive approach to the management of in-sewer sedimentation in the near future.

Ironically, much of the new knowledge about the best way of managing sewer solids has been a rediscovery of approaches that have been in use for millennia (e.g. Bertrand-Krajewski, 2003), albeit with more advanced technological opportunities. Unfortunately, application of this new knowledge by sewer managers and operators remains relatively limited worldwide, due to a lack of awareness or a mistrust of new ideas. In addition, system designers, relying often on computational model predictions, tend not to be questioning enough of the outputs that are visually impelling from GIS or GUI platforms, not realising that these outputs are often based on imperfect process knowledge.

Figure 5.3. Vacuum toilet with exhortation not to use as a rubbish bin (photograph Richard Ashley).

Figure 5.4. Cotton bud sticks clogging a screen (photograph former North of Scotland Water Authority)

.In terms of current practice, there are four types of problem with sewer flow quality models and their usage:

1 Deficiencies in model performance due to lack of fundamental knowledge. This should be apparent from this Scientific and Technical Report (Sections 2.2 and 2.3).
2 Errors in model coding, limiting parameters and assumptions, usually arising from the need to deal with phenomena for which knowledge is lacking. This is due to the deterministic approach that requires a lot of data and effective process model description (Sections 2.3 and 4.2).

3 Spurious model calibration and 'verification' using inadequate, misleading or inappropriate data. Here, the verification of models using infrequent data and samples extracted from a single depth within the flow field can be understood to be of little value (Section 2.3).
4 Unrealistic use of models by extrapolation outside the range of the coded relationships or outside the calibration achieved by data collection. Many of the relationships used in models have been developed in other situations, often for rivers (Section 2.3).

Of concern is the use of these tools without a proper appreciation of the complexity of the problems. In some countries, however, there is a wariness of attempting quality process modelling in sewer systems (Gromaire *et al.*, 2002). A recent study has identified the limitations of the current generation of sewer flow quality models (Bouteligier *et al.*, 2002) and highlighted the need to collect a lot of local data for these to be at all effective.

Data collection is a major problem in this field. Actual fieldwork in sewers is fraught with difficulties and that is why many studies have concentrated on using laboratory facility surrogates. Whilst this is instructive, scaling up results to full-scale is problematic and field studies will continue to be needed. Current techniques requiring sampling, storage and laboratory analysis are expensive and not likely to lead to reliable results. Of major importance is the development of new on-line quality sensors that may be used reliably in sewers in both wet and dry weather conditions. Recent developments in sensor technologies for WTP monitoring (Vanrolleghem and Lee, 2003) could be extended to sewers and there are signs that new equipment is beginning to emerge that can be robust enough for such purposes (Automation in Water Quality Monitoring, 2003).

From the above it can be seen that there are two major challenges for the experts in this field. It will become increasingly important to convince system managers and operators to utilise better the new knowledge if increasingly stringent emission standards and performance efficiencies continue to be expected by stakeholders, as is demanded by, for example, the European Water Framework Directive. For example, the future utilisation of sewers as linear reactors may be an effective way of improving overall system performance without large capital investments.

It is also important to ensure that those who design and model systems where the processes related to sewer solids are important, are fully aware of the limitations of the knowledge base that they are using for their designs, and by extension, are properly aware of the core models being used to represent the processes in the software packages being supplied. It is likely, however, that the need for designers and modellers to become more expert will only become a necessity once environmental and other regulators take a more sceptical approach to the schemes they are asked to approve. In turn this requires more knowledge about sewer solids and processes within these and wider stakeholder groups.

A more holistic approach is now advocated for the management of wastewater systems as a whole, and recent initiatives are looking at wastewater systems in an integrated way, considering the sewer, treatment plant and receiving water as a continuum (e.g. Schütze *et al.*, 2002). The INTERURBA conferences, the first held in 1992 and the second in 2001, have tracked the development of the integrated approach to wastewater systems management, from a concept to what is now much more of a presumed standard approach (INTERURBA II, 2002).

Nonetheless, tradition still tends to keep the sewer system and the (downstream) treatment plant in separate operational compartments. Around the world there are many instances of separate operators of the sewer network and the receiving treatment plant, often as a consequence of the way in which these services are financed. In many cases this results in there being no incentive to operate them conjunctively to be as efficient as possible.

The management of sewer solids is just one part of the proper and sustainable provision of water and wastewater services. It is a key component in the control of human and environmental health and physical risk due to flooding. The future, with the major uncertainty of climate change, with at-source separation of wastewaters and reductions in developed world per capita water use, may force a re-appraisal of our current paradigms for what we do with sewer solids. Perhaps, as was said in Chapter 1, these solids are destined for extinction, with ultimate local or on-site systems for sanitary wastes. At the very least the careless use of urban drainage systems (out-of-sight-out-of-mind) has to be addressed if services are to continue to become more sustainable and at the same time affordable.

Perhaps the greatest challenge remaining is the effective extension of these services to the majority of the world's population who do not have secure sanitation and flood protection (Figure 5.5), but this is going to require an integrated approach that covers social, political, economic and environmental factors (e.g. Seppälä, 2002) as well as the scientific and technical knowledge presented in this report.

Figure 5.5. Open drain in Addis Ababa with large debris (photograph Michel Verbanck).

References

Automation in Water Quality Monitoring (2003) *Proceedings of 1st IWA International Conference on Automation in Water Quality Monitoring*, Vienna, Austria, 21–22 May 2002. See *Wat. Sci. Tech.*, 47(2).

Bertrand-Krajewski, J.-L. (2003) Sewer sediment management: some historical aspects of egg-shape sewers and flushing tanks. *Wat. Sci. Tech.*, 47(4), 109–122.

Bjerre, H.L., Hvitved-Jacobsen, T., Schlegel, S. and Teichgraber, B. (1998). Biological activity of biofilm and sediment in the Emscher river, Germany. *Wat. Sci. Tech.*, 37(1), 9–16.

Bouteligier, R., Vaes, G. and Berlamont, J. (2002) Urban drainage quality modelling – case study. Proceedings of 3rd International Conference on Sewer Processes and Networks, Paris, France, 15-17 April 2002, 263–270.

Dauber, L. and Novak, B. (1983) *Quellen und Mengen der Schmutszstoffe in Regenabflussen einer stadtischen Mischkanalisation.* Eidgenossische Technische Hochschulen (ETH), Zurich, Separatum Nr.927.

Gromaire, M.-C., Garnaud, S., Saad, M. and Chebbo, G. (2001) Contribution of different sources to the pollution of wet weather flows in combined sewers. *Wat. Res.*, 35(2), 521–533.

Gromaire, M.-C., Cabane, P., Bertrand-Krajewski, J.-L. and Cebbo, G. (2002) Operational use of urban drainage pollutant load models: results from a French survey. *Proc. Conf. on Sewer Operation and Maintenance*, 24-26 November, University of Bradford. CD Rom.

Hvitved-Jacobsen, T., Vollertsen, J. and Matos, J. S. (2002) The sewer as a bioreactor – a dry weather approach. *Wat. Sci. Tech.*, 45(3), 11–24.

INTERURBA II (2002) Interactions between Sewers, Treatment Plants and Receiving Waters in Urban Areas. Conference held 19–22 February 2002, Lisbon, Portugal. Selected Proceedings. *Wat. Sci. Tech.*, 45(3).

Rees, A. and White, K.N. (1993) Impact of combined sewer overflows on the quality of an urban watercourse. *Regulated Rivers: Research and Management*, 8, 83–94.

Schütze, M., Butler, D. and Beck, M.B. (2002) *Modelling, Simulation and Control of Urban Wastewater Systems*. Springer Verlag, London.

Seppälä, O.T. (2002) Effective water and sanitation policy reform implementation: need for systemic approach and stakeholder participation. *Water Policy*, 4, 367–388.

Vanrolleghem, P.A. and Lee, D.S. (2003) On-line monitoring equipment for wastewater treatment processes: state of the art. *Wat. Sci. Tech.*, 47(2), 1–34.

Appendix

Sewage settling velocity – standardisation of terminology and data presentation

A.1 INTRODUCTION

Many of the settlement and transport processes associated with sewage solids are still poorly understood. No single representative value of specific gravity can be assigned to sewage solids, as is the case for fluvial sediments, and their transport and settlement processes are complicated by the flocculation and aggregation of particles. For these reasons, although there are a variety of techniques for its measurement, particle size cannot be employed in defining the settlement characteristics of the majority of sewage solids. As a consequence, settling velocity has steadily grown in importance as the parameter best representing the influence of gravity on these solids. Applications utilising the parameter include: computer modelling both for quality management purposes and for simulating transport in sewers, and theoretical, laboratory and field studies investigating the solids removal potential of sedimentation devices.

Over the past decades a number of research groups have independently developed equipment for measuring the settling velocity of sewage solids. However, it is clear from a study of the available literature that different terms are employed by researchers to describe various aspects of their work, and their results are presented in a variety of ways. If the outcome of research on the settling velocity of sewer solids is to be of value to other researchers and practitioners working both in the field and in modelling, there is a need for a common 'language' so that confusion is avoided and the origins and relevance of settling velocity data are fully understood.

Since many of the foremost researchers in the field were present in September 1994 at the 'International Conference on Sewer Solids – Characteristics, movement, effects and control' held at the University of Abertay Dundee in Scotland, a workshop was convened to discuss these issues. This Appendix is based on the outcome of that workshop.

A.2 STANDARDISATION OF TERMINOLOGY AND UNITS OF MEASUREMENT

Although it is recognised that there will always be cases where the use of standard terms is not practical, it is recommended that the following definitions are adhered to wherever possible.

1 **Settling velocity**: the average velocity with which a particle settles through a liquid under the influence of gravity, determined by timing the fall over a known distance.

Whilst the term *settling velocity* can be applied to both discrete and flocculating particles, the expression *terminal velocity* should be reserved for discrete particles and the steady state case where acceleration has ceased and the gravitational force = the frictional drag force.

In order both to comply with the SI system and to provide a sensible order of magnitude, the settling velocity should be expressed in units of **mm/sec.** To conform with recognised sedimentation practice, a sign convention of **positive downwards** should be followed. This will therefore indicate a particle that sinks or is a *sinker*. A negative settling velocity thus implies a rising particle or *floater*.

2 Equipment developed for the determination of settling velocity characteristics can be divided into two distinct categories:

- those where the sewage sample is introduced to the device in a fully mixed state are referred to as **homogeneous methods**;
- those where the solids are concentrated prior to being loaded into the top of the device are referred to as **top-induced methods**.

A.3 PRESENTATION OF RESULTS

The settling velocity results should be presented in accordance with the standard illustrated in Figure A.1, which is referred to as a Sewage Settling Velocity Grading Curve. In Figure A.1, the vertical axis is the percentage of solids by weight with a settling velocity less than the value shown on the horizontal axis, and is plotted to a natural scale. The horizontal axis is the settling velocity in mm/sec, plotted to a log scale, and with values increasing from left to right. However, if the proportion of floaters (with a negative settling velocity) are to be included it will be necessary to employ a linear scale for the horizontal axis.

It may also be convenient to present data as a bar or frequency diagram, as illustrated in Figure A.2, or an alternative format that best illustrates a particular aspect of the results. Regardless of the format adopted the standard sign convention and units should be adhered to.

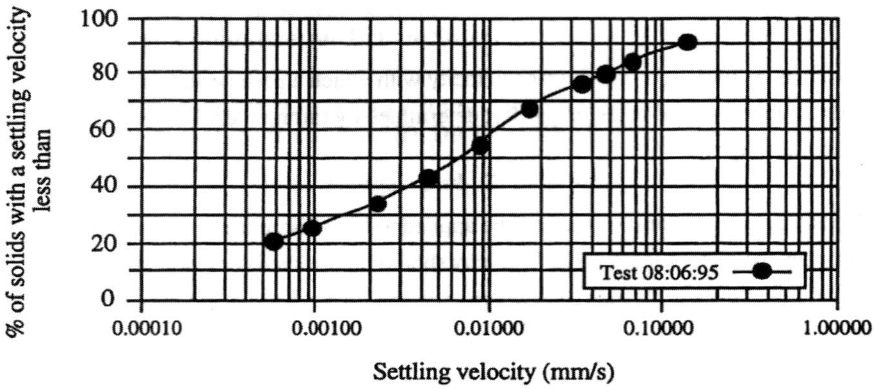

Figure A.1. Example illustrating the standard convention for presentation of results – sewage settling velocity grading curve.

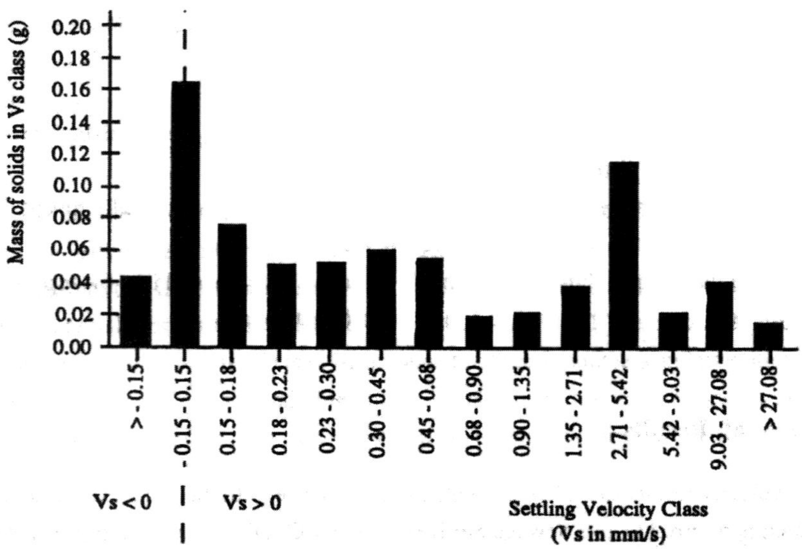

Figure A.2. Example of settling velocity results presented as bar diagram.

A.4 STATEMENT OF METHODOLOGY

To enable a potential user to assess the relevance of sewage settling velocity data, the presentation of results should always be accompanied by a brief statement detailing the method used and whether or not a standard procedure was adhered to. The Statement of Methodology should include the following:

1. name of apparatus;
2. principle of operation;
3. method of sample collection;
4. sample storage and handling;
5. sample preparation;
6. variation from standard procedure;
7. reference to give reader access to apparatus design and standard procedure.

An example is presented in Table A.1.

Table A.1. Example of a Statement of Methodology accompanying the presentation of sewage settling velocity data

Apparatus	Aston Column
Principle of operation	Top-induced
Method of sample collection	Single grab sample
Sample storage and handling	Sample conveyed to laboratory and stored for 24 h in refrigerator
Sample preparation	Solids concentrated for 3 h
Modification to procedure	None
Reference	Tyack, J.N. (1995), 'The Effect of Catchment Characteristics on Sewage Settling Velocity Grading', PhD Thesis, Aston University, Birmingham, UK.

A.5 APPARATUS FOR DETERMINING SEWAGE PARTICULATE SETTLING VELOCITY GRADING CURVES

Currently, there is no one universally accepted apparatus for determining sewage settling velocity. Indeed, even for a given apparatus it is not uncommon for different users to either modify the equipment or the experimental procedure (or do both) to meet their own needs. Therefore, to ensure that they are compatible with user requirements, it is essential that anyone wishing to utilise published settling velocity results ascertains the exact procedure used in their determination, prior to employing them.

The various types of settling velocity apparatus which have been used in recent years are listed in Table A.2, with the main characteristics of those in current use detailed in Table A.3 and illustrated in Figures A.3a to A.3e.

Table A.2. Apparatus used to determine the settling velocity of sewage

1	Bottom Withdrawal Method	Owen, 1976
2	Multi Port Method	Camp, 1946; Dalrymple *et al.*, 1975
3	Owen Tube	Owen, 1976
4	CERGRENE 1992 – Column	Chebbo, 1992
	– Andreasen's Pippette	Chebbo, 1992
5	UFT Device	Michelbach and Wöhrle, 1993
6	Aston Column	Tyack *et al.*, 1996
7	VICAS Protocol	Gromaire and Chebbo, 2003

Lucas-Aiguier *et al.* (1998) compared three selected methods (CERGRENE 1992, Aston method, Multi Port method) when tested with the same sample with the method developed by UFT in order to quantify the influence of the experimental procedure on the settling velocity grading curves. The results of the comparative tests indicate that the settling velocity curves using identical sewage samples are different. Hence there are differences between different samples and the test procedures. The influence of the experimental procedure for separating a sample by settling velocity fractions is demonstrated. It is possible to conclude that for a better comparison between tests for the characterisation of sewage, across various international studies, there is a need to define a protocol which could be used as a standard in order to eliminate the influence of test procedures.

Table A.3. Main characteristics of sewage settling velocity devices in current use

Settling velocity device	*Principle of operation*	*Original design objective*	*Sample preparation*	*Liquid in device and sample volume*	*Settlement distance and column diameter*	*Known modifications to device and/or procedure*	*Key references*
Multi-port method (Figure A.3a)	Homogeneous	Design of settling tanks	Wet sieving 6 mm	Mixed sewage approx. 30–200 litre sample	2000–3700 mm 150-300 mm dia	i. Procedure adapted for comparability studies by CEGEO, Canada ii. Dimensions and procedure modified by Pisano in USA	Camp, 1946; Dalrymple *et al*., 1975; Pisano *et al*., 1990; Gagné *et al*., 1995
CERGRENE 1992 Column (Figure A.3b)	Top-induced	Analysis of solids > 50 μm in sewer systems	Wet sieving to separate particles into > and < 50 μm	Water approx. 20 litre sample	1815 mm 50.8 mm dia	Procedure adapted for comparability studies by Aston Uni., UK, by CERGRENE and by CTIA, France	Chebbo, 1992
CERGRENE 1992 Andreasen Pipette (not shown)	Homogeneous	Analysis of solids < 50 μm in sewer systems	Wet sieving to separate particles into > and < 50 μm	Water 20 litre sample	200 mm 50.8 mm dia	Procedure adapted for comparability studies by CERGRENE and by CTIA, France	Chebbo, 1992; Lucas-Aiguier *et al*., 1998
UFT Device (*Umwelt- und Fluid Technik*) (Figure A.3c)	Top-induced	Analysis of urban runoff, sediment and slime from combined sewer	Settleable solids separated from sample by settlement for 2 hours in an Imhoff Cone	Water 1 litre sample	700 mm5 0 mm dia above cone	i. Procedure adapted for comparability studies by Aston Uni., UK, by CERGRENE and CTIA, France, and by CEGEO, Canada ii. Dimensions and procedure modified by Pisano, in USA	Michelbach and Wöhrle, 1993; Lucas-Aiguier *et al*., 1998; Gagné *et al*., 1995; Pisano, 1996
Aston Column (Figure A.3d)	Top-induced	Analysis of both sinker and floater suspended solids fractions in storm sewage	Sinkers and floaters separated and concentrated by standing in column for 3 hours	Sewage liquor 5 litre sample	1750 mm 50 mm dia	i. Used for foul sewage, and ii. larger dia for chemical analysis Aston Uni., UK	Tyack *et al*., 1996
CERGRENE 1995 Column (Figure A.3e)	Homogeneous	Rapid analysis of foul and storm sewage	None	Mixed sewage 3 litres	650 mm 70 mm dia	Under development	Unpublished

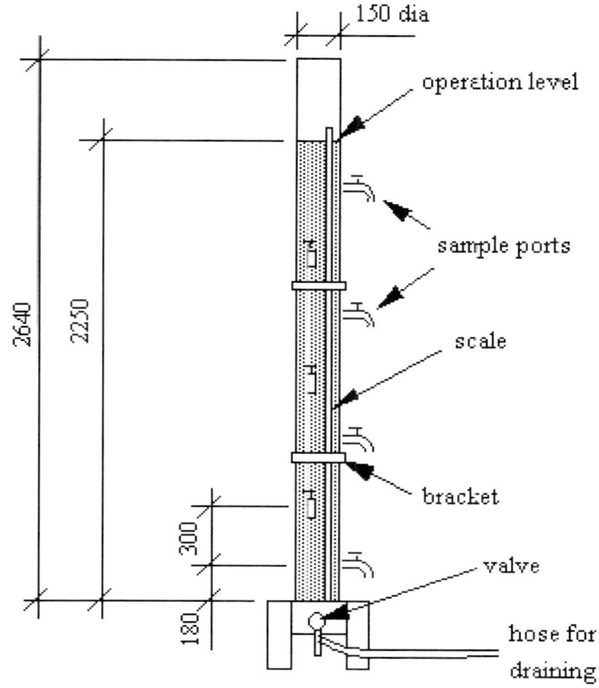

(a) Multi-port method (all dimensions in mm).

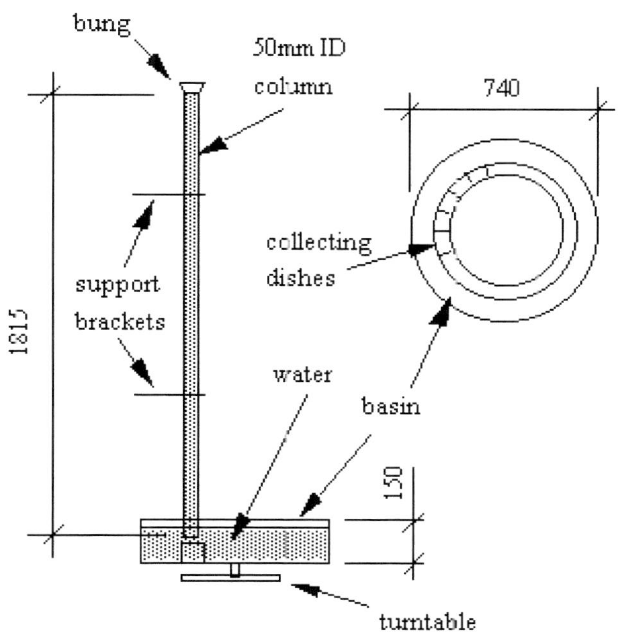

(b) CERGRENE 1992 Column (all dimensions in mm).

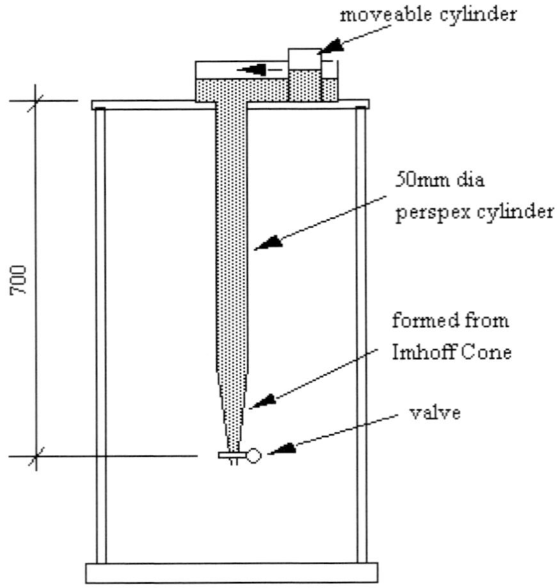

(c) UFT device (all dimensions in mm).

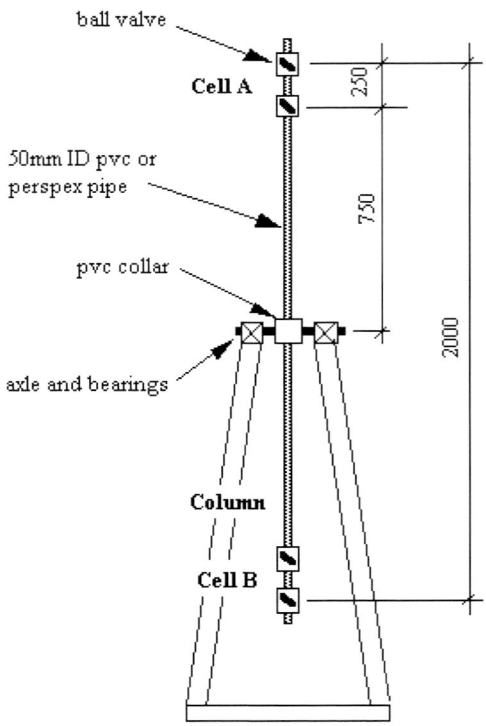

(d) Aston Column (all dimensions in mm).

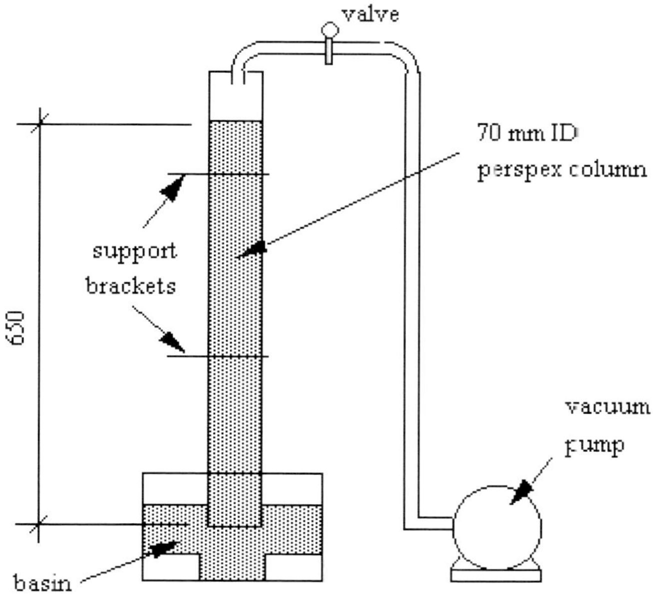

(e) CERGRENE 1995 column (all dimensions in mm).

Figure A.3. Sewage settling velocity devices in current use.

A.6 REFERENCES

Camp, T.R. (1946) Sedimentation and the design of settling tanks. *Trans. ASCE*, 111, Paper No. 2285, 895–958.

Chebbo, G. (1992) *Solides des rejets pluviaux urbains – Caractérisation et traitabilité*. PhD thesis, Thèse de l'Ecole Nationale des Ponts et Chaussées.

Dalrymple, R.J. *et al.* (1975) *Physical and Settling Characteristics of Particulates in Storm and Sanitary Wastewaters*, USEPA Report No. EPA-670/2-75-11, NTIS No PB242-01.

Gagné, B., Meunier, C., Bordeleau, F. and Comeau, Y. (1995) *Vitesses de chute des particules: essais comparatifs des méthodes Allemande et Américaine*. Unpublished report, CEGEO, Canada.

Gromaire, M.-C. and Chebbo, G. (2003). *Manuel d'utilisation du protocole VICAS*. Technical report (in French), CEREVE and AESN, 20 pp.

Lucas-Aiguier, E., Chebbo, G., Bertrand-Krajewski, J.-L., Gagné, B. and Hedges, P. (1998) Analysis of the methods for determining the settling characteristics of sewage and stormwater solids. *Wat. Sci. Tech.*, 37(1), 53–60.

Michelbach, S. and Wöhrle, C. (1993) Settleabile solids in a combined sewer system: measurement, quantity, characteristics. *Wat. Sci. Tech.*, 17(5/6), 407–427.

Owen, M.W. (1976) *Determination of the settling velocities of cohesive muds*. Report No. IT 161, Hydraulics Research, Wallingford, UK.

Pisano, W.C. (1996) Summary: United States 'sewer solids' settling characterization methods, results, uses and perspective. *Wat. Sci. Tech.*, 33(9), 109–115.

Pisano, W.C., Thibault, N. and Forbes, G. (1990) The Vortex Solids Separator. *Water Environment and Technology*, WPCF, 17(5-6), 64–71.

Tyack, J.N., Hedges, P.D. and Smisson, R.P.M. (1996) The relationship between settling velocity grading and the characteristics of the contributing catchment. *Wat. Sci. Tech.*, 33(9), 135–142.

Index